T0091830

Large-Scale Machine Learning in the Earth Sciences

Chapman & Hall/CRC
Data Mining and Knowledge Discovery Series

Series Editor: Vipin Kumar
University of Minnesota
Department of Computer Science and Engineering
Minneapolis, Minnesota, U.S.A.

AIMS AND SCOPE

This series aims to capture new developments and applications in data mining and knowledge discovery, while summarizing the computational tools and techniques useful in data analysis. This series encourages the integration of mathematical, statistical, and computational methods and techniques through the publication of a broad range of textbooks, reference works, and handbooks. The inclusion of concrete examples and applications is highly encouraged. The scope of the series includes, but is not limited to, titles in the areas of data mining and knowledge discovery methods and applications, modeling, algorithms, theory and foundations, data and knowledge visualization, data mining systems and tools, and privacy and security issues.

PUBLISHED TITLES

Accelerating Discovery: Mining Unstructured Information for Hypothesis Generation
Scott Spangler

Advances in Machine Learning and Data Mining for Astronomy
Michael J. Way, Jeffrey D. Scargle, Kamal M. Ali, and Ashok N. Srivastava

Biological Data Mining
Jake Y. Chen and Stefano Lonardi

Computational Business Analytics
Subrata Das

Computational Intelligent Data Analysis for Sustainable Development
Ting Yu, Nitesh V. Chawla, and Simeon Simoff

Computational Methods of Feature Selection
Huan Liu and Hiroshi Motoda

Constrained Clustering: Advances in Algorithms, Theory, and Applications
Sugato Basu, Ian Davidson, and Kiri L. Wagstaff

Contrast Data Mining: Concepts, Algorithms, and Applications
Guozhu Dong and James Bailey

Data Classification: Algorithms and Applications
Charu C. Aggarawal

Data Clustering: Algorithms and Applications
Charu C. Aggarawal and Chandan K. Reddy

Data Clustering in C++: An Object-Oriented Approach
Guojun Gan

Data Mining: A Tutorial-Based Primer, Second Edition
Richard J. Roiger

Data Mining for Design and Marketing
Yukio Ohsawa and Katsutoshi Yada

Data Mining with R: Learning with Case Studies, Second Edition
Luís Torgo

Data Science and Analytics with Python
Jesus Rogel-Salazar

Event Mining: Algorithms and Applications
Tao Li

Foundations of Predictive Analytics
James Wu and Stephen Coggeshall

Geographic Data Mining and Knowledge Discovery, Second Edition
Harvey J. Miller and Jiawei Han

Graph-Based Social Media Analysis
Ioannis Pitas

Handbook of Educational Data Mining
Cristóbal Romero, Sebastian Ventura, Mykola Pechenizkiy, and Ryan S.J.d. Baker

Healthcare Data Analytics
Chandan K. Reddy and Charu C. Aggarwal

Information Discovery on Electronic Health Records
Vagelis Hristidis

Intelligent Technologies for Web Applications
Priti Srinivas Sajja and Rajendra Akerkar

Introduction to Privacy-Preserving Data Publishing: Concepts and Techniques
Benjamin C. M. Fung, Ke Wang, Ada Wai-Chee Fu, and Philip S. Yu

Knowledge Discovery for Counterterrorism and Law Enforcement
David Skillicorn

Knowledge Discovery from Data Streams
João Gama

Large-Scale Machine Learning in The Earth Sciences
Ashok N. Srivastava, Ramakrishna Nemani, and Karsten Steinhaeuser

Machine Learning and Knowledge Discovery for Engineering Systems Health Management
Ashok N. Srivastava and Jiawei Han

Mining Software Specifications: Methodologies and Applications
David Lo, Siau-Cheng Khoo, Jiawei Han, and Chao Liu

Multimedia Data Mining: A Systematic Introduction to Concepts and Theory
Zhongfei Zhang and Ruofei Zhang

Music Data Mining
Tao Li, Mitsunori Ogihara, and George Tzanetakis

Next Generation of Data Mining
Hillol Kargupta, Jiawei Han, Philip S. Yu, Rajeev Motwani, and Vipin Kumar

Rapidminer: Data Mining Use Cases and Business Analytics Applications
Markus Hofmann and Ralf Klinkenberg

Relational Data Clustering: Models, Algorithms, and Applications
Bo Long, Zhongfei Zhang, and Philip S. Yu

Service-Oriented Distributed Knowledge Discovery
Domenico Talia and Paolo Trunfio

Spectral Feature Selection For Data Mining
Zheng Alan Zhao and Huan Liu

Statistical Data Mining Using SAS Applications, Second Edition
George Fernandez

Support Vector Machines: Optimization Based Theory, Algorithms, and Extensions
Naiyang Deng, Yingjie Tian, and Chunhua Zhang

Temporal Data Mining
Theophano Mitsa

Text Mining: Classification, Clustering, and Applications
Ashok N. Srivastava and Mehran Sahami

Text Mining and Visualization: Case Studies Using Open-Source Tools
Markus Hofmann and Andrew Chisholm

The Top Ten Algorithms in Data Mining
Xindong Wu and Vipin Kumar

Understanding Complex Datasets: Data Mining with Matrix Decompositions
David Skillicorn

Large-Scale Machine Learning in the Earth Sciences

Edited by
Ashok N. Srivastava
Ramakrishna Nemani
Karsten Steinhaeuser

CRC Press
Taylor & Francis Group
Boca Raton London New York

CRC Press is an imprint of the
Taylor & Francis Group, an **informa** business

CRC Press
Taylor & Francis Group
6000 Broken Sound Parkway NW, Suite 300
Boca Raton, FL 33487-2742

© 2017 by Taylor & Francis Group, LLC

CRC Press is an imprint of Taylor & Francis Group, an Informa business

No claim to original U.S. Government works

Printed on acid-free paper

International Standard Book Number-13: 978-1-4987-0387-1 (Hardback)

Library of Congress Cataloging-in-Publication Data

Names: Srivastava, Ashok N. (Ashok Narain), 1969- editor. | Nemani, Ramakrishna, editor. | Steinhaeuser, Karsten, editor.
Title: Large-scale machine learning in the earth sciences / [edited by] Ashok N. Srivastava, Dr. Ramakrishna Nemani, Karsten Steinhaeuser.
Description: Boca Raton : Taylor & Francis, 2017. | Series: Chapman & Hall/CRC data mining & knowledge discovery series ; 42 | "A CRC title, part of the Taylor & Francis imprint, a member of the Taylor & Francis Group, the academic division of T&F Informa plc."
Identifiers: LCCN 2017006160 | ISBN 9781498703871 (hardback : alk. paper)
Subjects: LCSH: Earth sciences–Computer network resources. | Earth sciences–Data processing.
Classification: LCC QE48.87 .L37 2017 | DDC 550.285/6312–dc23
LC record available at https://lccn.loc.gov/2017006160

Visit the Taylor & Francis Web site at
http://www.taylorandfrancis.com

and the CRC Press Web site at
http://www.crcpress.com

Contents

Foreword ix

Editors xi

Contributors xiii

Introduction xv

1   Network Science Perspectives on Engineering Adaptation to Climate Change and Weather Extremes 1
Udit Bhatia and Auroop R. Ganguly

2   Structured Estimation in High Dimensions: Applications in Climate 13
André R Goncalves, Arindam Banerjee, Vidyashankar Sivakumar, and Soumyadeep Chatterjee

3   Spatiotemporal Global Climate Model Tracking 33
Scott McQuade and Claire Monteleoni

4   Statistical Downscaling in Climate with State-of-the-Art Scalable Machine Learning 55
Thomas Vandal, Udit Bhatia, and Auroop R. Ganguly

5   Large-Scale Machine Learning for Species Distributions 73
Reid A. Johnson, Jason D. K. Dzurisin, and Nitesh V. Chawla

6   Using Large-Scale Machine Learning to Improve Our Understanding of the Formation of Tornadoes 95
Amy McGovern, Corey Potvin, and Rodger A. Brown

7   Deep Learning for Very High-Resolution Imagery Classification 113
Sangram Ganguly, Saikat Basu, Ramakrishna Nemani, Supratik Mukhopadhyay, Andrew Michaelis, Petr Votava, Cristina Milesi, and Uttam Kumar

8   Unmixing Algorithms: A Review of Techniques for Spectral Detection and
    Classification of Land Cover from Mixed Pixels on NASA Earth Exchange 131
    Uttam Kumar, Cristina Milesi, S. Kumar Raja, Ramakrishna Nemani, Sangram Ganguly,
    Weile Wang, Petr Votava, Andrew Michaelis, and Saikat Basu

9   Semantic Interoperability of Long-Tail Geoscience Resources over the Web 175
    Mostafa M. Elag, Praveen Kumar, Luigi Marini, Scott D. Peckham, and Rui Liu

Index 201

Foreword

The climate and Earth sciences have recently undergone a rapid transformation from a data-poor to a data-rich environment. In particular, massive amounts of climate and ecosystem data are now available from satellite and ground-based sensors, and physics-based climate model simulations. These information-rich data sets offer huge potential for monitoring, understanding, and predicting the behavior of the Earth's ecosystem and for advancing the science of global change.

While large-scale machine learning and data mining have greatly impacted a range of commercial applications, their use in the field of Earth sciences is still in the early stages. This book, edited by Ashok Srivastava, Ramakrishna Nemani, and Karsten Steinhaeuser, serves as an outstanding resource for anyone interested in the opportunities and challenges for the machine learning community in analyzing these data sets to answer questions of urgent societal interest.

This book is a compilation of recent research in the application of machine learning in the field of Earth sciences. It discusses a number of applications that exemplify some of the most important questions faced by the climate and ecosystem scientists today and the role the data mining community can play in answering them. Chapters are written by experts who are working at the intersection of the two fields. Topics covered include modeling of weather and climate extremes, evaluation of climate models, and the use of remote sensing data to quantify land-cover change dynamics. Collectively, they provide an excellent cross-section of research being done in this emerging field of great societal importance.

I hope that this book will inspire more computer scientists to focus on environmental applications, and Earth scientists to seek collaborations with researchers in machine learning and data mining to advance the frontiers in Earth sciences.

Vipin Kumar, PhD
Department of Computer Science and Engineering
University of Minnesota
Minneapolis, MN

Editors

Ashok N. Srivastava, PhD, is the vice president of Big Data and Artificial Intelligence Systems and the chief data scientist at Verizon. He leads a new research and development center in Palo Alto focusing on building products and technologies powered by big data, large-scale machine learning, and analytics. He is an adjunct professor at Stanford University in the Electrical Engineering Department and is the editor-in-chief of the AIAA Journal of Aerospace Information Systems. Dr. Srivastava is a fellow of the IEEE, the American Association for the Advancement of Science (AAAS), and the American Institute of Aeronautics and Astronautics (AIAA).

He is the author of over 100 research articles, has edited four books, has five patents awarded, and has over 30 patents under file. He has won numerous awards including the IEEE Computer Society Technical Achievement Award for "pioneering contributions to intelligent information systems," the NASA Exceptional Achievement Medal for contributions to state-of-the-art data mining and analysis, the NASA Honor Award for Outstanding Leadership, the NASA Distinguished Performance Award, several NASA Group Achievement Awards, the Distinguished Engineering Alumni Award from UC Boulder, the IBM Golden Circle Award, and the Department of Education Merit Fellowship.

Dr. Ramakrishna Nemani is a senior Earth scientist with the NASA Advanced Supercomputing Division at Ames Research Center, California. He leads NASA's efforts in ecological forecasting to understand the impacts of the impending climatic changes on Earth's ecosystems and in collaborative computing, bringing scientists together with big data and supercomputing to provide insights into how our planet is changing and the forces underlying such changes.

He has published over 190 papers on a variety of topics including remote sensing, global ecology, ecological forecasting, climatology, and scientific computing. He served on the science teams of several missions including Landsat-8, NPP, EOS/MODIS, ALOS-2, and GCOM-C. He has received numerous awards from NASA including the Exceptional Scientific Achievement Medal in 2008, Exceptional Achievement Medal in 2011, Outstanding Leadership Medal in 2012, and eight group achievement awards.

Karsten Steinhaeuser, PhD, is a research scientist affiliated with the Department of Computer Science and Engineering at the University of Minnesota and a data scientist with Progeny Systems Corporation. His research centers around data mining and machine learning, in particular construction and analysis of complex networks, with applications in diverse domains including climate, ecology, social networks, time series analysis, and computer vision. He is actively involved in shaping an emerging research area called *climate informatics*, which lies at the intersection of computer science and climate sciences, and his interests are more generally in interdisciplinary research and scientific problems relating to climate and sustainability.

Dr. Steinhaeuser has been awarded one patent and has authored several book chapters as well as numerous peer-reviewed articles and papers on these topics. His work has been recognized with multiple awards including two Oak Ridge National Laboratory Significant Event Awards for "Novel Analyses of the Simulation Results from the CCSM 3.0 Climate Model" and "Science Support for a Climate Change War Game and Follow-Up Support to the US Department of Defense."

Contributors

Arindam Banerjee
Department of Computer Science and
  Engineering
University of Minnesota, Twin Cities
Minneapolis, Minnesota

Saikat Basu
Department of Computer Science
Louisiana State University
Baton Rouge, Los Angeles

Udit Bhatia
Northeastern University
Boston, Massachusetts

Rodger A. Brown
NOAA/National Severe Storms Laboratory
Norman, Oklahoma

Soumyadeep Chatterjee
Data Science
Quora Inc.
Mountain View, California

Nitesh V. Chawla
University of Notre Dame
Notre Dame, Indiana

Jason D. K. Dzurisin
University of Notre Dame
Notre Dame, Indiana

Mostafa M. Elag
Ven Te Chow Hydrosystems Laboratory
Department of Civil and Environmental
  Engineering
University of Illinois
Urbana, Illinois

Auroop R. Ganguly
Northeastern University
Boston, Massachusetts

Sangram Ganguly
NASA Ames Research Center
Moffett Field, California
and
Bay Area Environmental Research Institute
(BAERI)
West Sonoma, California

André R. Goncalves
Center for Research and Development in
  Telecommunications
Campinas, Sao Paulo, Brazil

Reid A. Johnson
University of Notre Dame
Notre Dame, Indiana

Praveen Kumar
Ven Te Chow Hydrosystems Laboratory
Department of Civil and Environmental
  Engineering
University of Illinois
Urbana, Illinois

Uttam Kumar
NASA Ames Research Center/Oak Ridge
  Associated Universities
Moffett Field, California

Rui Liu
National Center for Supercomputing Applications
University of Illinois at Urbana-Champaign
Urbana, Illinois

Luigi Marini
National Center for Supercomputing Applications
University of Illinois at Urbana-Champaign
Urbana, Illinois

Amy McGovern
School of Computer Science
University of Oklahoma
Norman, Oklahoma

Scott McQuade
George Washington University
Washington, DC

Andrew Michaelis
NASA Ames Research Center
Moffett Field, California
and
University Corporation at Monterey Bay
California State University
Seaside, California

Cristina Milesi
EvalStat Research Institute
Palo Alto, California

Claire Monteleoni
George Washington University
Washington, DC

Supratik Mukhopadhyay
Department of Computer Science
Louisiana State University
Baton Rouge, Louisiana

Scott D. Peckham
Institute of Arctic and Alpine Research
University of Colorado
Boulder, Colorado

Corey Potvin
Cooperative Institute for Mesoscale
    Meteorological Studies and NOAA/National
    Severe Storms Laboratory
University of Oklahoma
Norman, Oklahoma

S. Kumar Raja
EADS Innovation Works
Airbus Engineering Centre India
Bangalore, India

Vidyashankar Sivakumar
Department of Computer Science and
    Engineering
University of Minnesota, Twin Cities
Minneapolis, Minnesota

Thomas Vandal
Northeastern University
Boston, Massachusetts

Petr Votava
NASA Ames Research Center
Moffett Field, California
and
University Corporation at Monterey Bay
California State University
Seaside, California

Weile Wang
California State University
Seaside, California
and
NASA Ames Research Center
Moffett Field, California

Introduction

Our society faces an unprecedented challenge in understanding, modeling, and forecasting Earth system dynamics that involve complex interactions among physical, biological, and socioeconomic systems. Addressing these challenges requires the ability to incorporate potentially massive data sets taken at multiple physical scales from sensors that are geospatially distributed, which can provide information at different sampling intervals. In addition, computer simulations generate data that need to be reconciled with observational data at different spatial and temporal scales.

In recent years, researchers in Earth science, computer science, statistics, and related fields have developed new techniques to address these challenges through interdisciplinary research. The purpose of this book is to provide researchers and practitioners in these fields, a broad overview of some of the key challenges in the intersection of these disciplines, and the approaches that have been taken to address them. This highly interdisciplinary area of research is in significant need for advancement on all fronts; our hope is that this compilation will provide a resource to help further develop these areas. This book is a collection of invited articles from leading researchers in this dynamic field.

The book begins with the chapter entitled, "Network Science Perspectives on Engineering Adaptation to Climate Change and Weather Extremes" by Bhatia and Ganguly. The authors observe that extreme climate and weather events can be correlated and can cause significant stress on human-built and natural systems, which in turn can lead to negative impacts on human populations and the economies in which they live. They discuss in detail the idea of teleconnections, in which spatiotemporal events spanning enormous geographical distances can show high degrees of correlation. Thus, extreme climate and weather events in one location can induce other potentially extreme events in a distant location. The authors emphasize the use of methods developed by network science to understand and discover such teleconnections.

Chapter 2, entitled "Structured Estimation in High Dimensions: Applications in Climate," continues the theme of making predictions based on observational data. Goncalves and Banerjee focus on two specific problems: first, the authors address the problem of making temperature and precipitation predictions over nine landmasses given measurements taken over the oceans. The authors show that by imposing known structure on the prediction variables, the prediction quality can exceed prior baselines. The second focus of the chapter is on combining predictions made by global climate models using a multitask learning approach based on structured learning yielding predictions, which significantly exceed competing methods.

Chapter 3, by McQuade and Monteleoni entitled, "Spatiotemporal Global Climate Model Tracking" describes the use of ensemble machine learning models to combine predictions of global climate models using information from spatial and temporal patterns. They explore the fixed-share algorithm and provide an extension that allows the prediction weights of the global climate models to switch based on both spatial and temporal patterns. This approach has thematic commonality with the preceding chapter, although it addresses the problem using Markov random fields.

Chapter 4 features a discussion on "Statistical Downscaling in Climate with State-of-the-Art Scalable Machine Learning" in which Vandal, Bhatia, and Ganguly explore the important problem of projecting the output of simulations made at low spatial resolution to higher spatial resolutions necessary for studying the impacts of climatic changes on socioeconomic and ecological systems. They discuss techniques such as sparse regression, Bayesian methods, kriging, neural networks, and other transfer function-based techniques to address the problem. In addition, the authors discuss deep belief networks and provide a comparison of this method against other methods on a precipitation prediction problem.

In Chapter 5, Johnson and Chawla provide an overview of methods to understand and predict the proliferation of biological species due to changes in environmental conditions. In their chapter entitled, "Large-Scale Machine Learning for Species Distributions," the authors provide an overview of this pressing problem and discuss correlative methods to predict species distributions based on environmental variables. These methods are contrasted with causal approaches that rely on significant domain knowledge regarding the species in question and the impact of environmental variations on those species. The authors provide a theoretical framework for the correlative approaches and show how this problem can be reduced to a familiar classification problem.

Chapter 6 addresses the problem of "Using Large-Scale Machine Learning to Improve Our Understanding of the Formation of Tornadoes." In this chapter, McGovern, Potvin, and Brown show how spatiotemporal relational random forests can be used to analyze massive four-dimensional data sets to detect and potentially differentiate storms that will lead to tornadoes from those that do not. The challenge that the researchers address is multifold due to the fact that the data sets are massive while the processing must be done in near real-time to yield actionable results. In addition to trying to address the actionability of the results, the authors also address the problem of providing potential explanations for tornadogenesis.

Ganguly et al., utilize a critical innovation in neural networks in Chapter 7 entitled, "Deep Learning for Very High-Resolution Imagery Classification." The authors address the problem of classifying images that have very high resolution using a class of deep learning algorithms. The authors show the performance of stacked autoencoders, deep belief networks, convolutional neural networks, and deep belief neural networks with a special set of features derived from the underlying data used as inputs. The study is performed on a modern cloud-based computing platform and the NASA Earth Exchange (NEX) platform and reveals that the feature extraction method coupled with the deep belief neural network has a significantly higher classification performance compared to the other methods.

Chapter 8 by Kumar et al., addresses the critical problem of unmixing spectral signals in remote sensing images of land cover. By the inherent nature of remote sensing data, any given image pixel, regardless of the imaging resolution can be composed of many underlying sources—grass, trees, cement, water, etc. Land cover classification techniques can be improved significantly through the application of algorithms that use principled methods to decompose the spectral signals to reveal the underlying sources. This chapter features a detailed analysis of numerous unmixing algorithms followed by a comprehensive set of results across several data sets. Given the proliferation of medium- to high-resolution global satellite data sets, these algorithms will play a key role in the monitoring of our planet for detecting changes in forest cover, urbanization, and surface water.

Chapter 9 addresses the "Semantic Interoperability of Long-Tail Geoscience Resources over the Web." In this chapter, Elag, Kumar, Marini, and Peckham apply long-tail distributions, which are a well-known concept in business, economics, and other disciplines in the context of numerous small, highly curated data sets that an individual scientist may have on their local network. While each data set by itself has a certain scientific value, the authors present the idea that the totality of the information contained in union of all of these data would present an invaluable source of information across the Earth sciences. For such data sets to be easily accessible, the authors discuss methods to improve the interoperability of the data sets using semantic web mining. They provide concrete examples of potential approaches to embed context and other relevant information in a geosemantics framework.

The editors are thankful to the authors of this book who have given a detailed discussion of pressing research issues covering a diverse array of topics ranging from combining global climate models to new methods for creating semantic interoperability across data sets, large and small. We believe that these contributions will form an important resource for interdisciplinary research in the Earth and computer sciences. Developing a deeper understanding of the issues facing our planet through the pursuit of scientific inquiry based on observable data is more pressing now than ever before.

Ashok N. Srivastava, PhD
VP Data and Artificial Intelligence Systems
Chief Data Scientist, Verizon

Ramakrishna Nemani, PhD
Senior Earth Scientist
NASA Ames Research Center

Karsten Steinhaeuser, PhD
Research Scientist, Computer Science & Engineering
University of Minnesota
Data Scientist, Progeny Systems Corporation

1

Network Science Perspectives on Engineering Adaptation to Climate Change and Weather Extremes

1.1	Introduction	1
1.2	Motivation	2
1.3	Network Science in Climate Risk Management	3
	New Resilience Paradigm	
1.4	Network Technology Stack	6
1.5	Case Studies: Telescoping Systems of Systems	8
1.6	Conclusion	9

Udit Bhatia

Auroop R. Ganguly

1.1 Introduction

Weather and hydrological extremes, which may be exacerbated by climate variability and change [1,2], severely stress natural, engineered, and human systems. What makes these climate hazards particularly worrisome is their rapidly changing nature, along with our lack of understanding of the hazards attributes that may matter the most for impact analyses. With climate change as threat multiplier, the weather extremes have devastating impacts on essential functionality of infrastructure networks, specifically, lifelines including transportation networks, power-grid, water distribution networks, and communication networks [3]. The failure of one of the lifeline systems may cause massive social disruption, and localized damage may trigger a cascade of failure within the network or across network multiplex [4,5]. Environmental impacts of such infrastructures also span spatial scales from regional to national to global. For example, fossil fuel combustion in power generation facilities can result in heat island effects, all of which can have profuse impact on humans and biodiversity in the natural systems. The impacts of climate hazards, and hence preparedness and management of natural hazards as well as climate adaptation [6] and to a great extent mitigation, crucially depend upon our understanding of the interdependencies of the complex systems. While component-based risk management frameworks have long been used to inform the disaster mitigation and management, there is a growing realization that the conventional risk management framework should be coupled with a system-based resilience approach that can confront the ever-growing complexity and interdependencies of infrastructure systems. Challenges stem from the attributes of the data, the systems under consideration, and the nature of the problems. One concern is the ability to deal with "Big Data," typically defined through attributes such as volume, variety, and velocity [7]. The interest in extremes or large changes, often unprecedented, requires the extraction of signatures or precursors of rare

1

events and anomalies. Interrelationships among systems motivate the need for correlative and predictive analyses [8], to understand processes such as spatiotemporal complexity, cascading failures [9], and flow or spreading phenomena [10]. A combination of the best-available methods in data sciences (e.g., machine learning), process models (e.g., agent-based systems), and physics models (e.g., global climate models) is required to address these challenges. Machine learning in particular needs to be able to look for extremes, anomalies, and change from massive and diverse ("Big") data while being aware of complex dependence structures. This has motivated recent advances in the machine learning literature in areas such as rare event analytics and extreme value theory, network science, and graphical models and algorithms specifically designed for massive spatiotemporal data. Complex networks, including their theory and efficient computational implementations [11], are among the tools that are increasingly being used by theoretical and applied machine learning researchers to solve the urgent societal problems discussed in this chapter.

1.2 Motivation

Unprecedented end correlated weather extremes, and drastic regional climate changes, are no longer surprising; legacy infrastructures are crumbling, while multilayer lifeline networks magnify stresses through percolation within and cascading failures across layers; natural environments from wetland to marine ecosystems are degrading beyond the stage when they can protect coastlines or cities, or recover from biodiversity loss, while population and assets continue to be moved in vulnerable urban or coastal regions [12]; and indirect multiplier effects can be large given increased connectivity [13]. Crucial issues that need to be understood in this context are generation and/or exacerbation of extreme stresses, resilience (robustness and recovery) of the interconnected built and natural systems to these stresses, and consequences of the stresses and systemic resilience elements across impacted natural, infrastructural, and societal systems. Key knowledge gaps include understanding the generation processes [14] and statistical attribute of the extremes, potential changes in the nature of stressors and/or stressed and impacted systems, and uncertainties (including intrinsic variability) associated with our understanding of these entities.

In the context of climate-related hazards, there been increasing recognition of spatial and spatiotemporal correlation [15] and tail dependence among extreme events. Therefore, understanding the dependence structure among extreme events (such as large-scale extremes and anomalies and intense local storms, global warming, and extreme precipitation events [16]) is important to understand impacts of these stressors on infrastructure and natural systems (stressed systems), and human lives and economies (impacted systems). Correlation and interdependency in climate systems over large spatial scales are characterized by teleconnection patterns, which refer to a recurring and persistent large-scale pressure and/or circulation pattern that spans over vast geographical area [17]. The organization of teleconnections plays a crucial role in stabilization of climate patterns, and hence delineating the global changes in architecture of these patterns could be useful to identify the signature of global climate change. Temporally, climate systems are affected by the slow responding subsystems such as oceans, and hence climate variability usually exhibits long-term memory, which means the present climate states may have long-term influences on the states in far future [18]. Thus, the dominance of long-term processes suggests that any future exceedance in threshold of sea-level rise regionally might be exceeded earlier or later than from anthropogenic change alone as a result of melting glaciers and snowcaps [19]. Also, the simultaneous occurrence of extreme events, such as simultaneous storms and floods at different locations, has serious impact on risk assessment and mitigation strategies.

A strong feedback that exists between the weather extreme-induced stresses, and built and natural systems severely impacts these systems [20], resulting in loss of their essential functionality. These impacts are further exacerbated by aging of critical infrastructure systems and lifelines, habitat loss and mass extinctions in ecosystems as a result of increasing urbanization, and high population pressure. Infrastructure systems, which are "backbone" of prosperity and quality of life in modern societies, are highly networked in nature, and failure of some specific components can result in the disruption of entire system's functionality via cascading failure [21]. Hence, failure in one system can percolate across the multiple systems, resulting in the extreme state of disruptions by "triggering a disruptive avalanche of cascading and

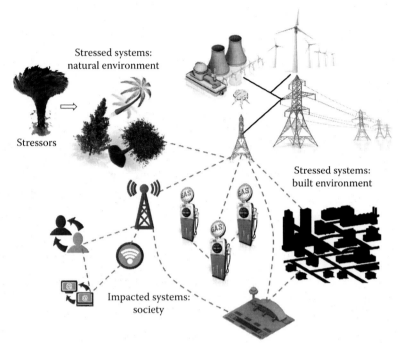

Cascading interdependencies across natural and built environments

FIGURE 1.1 Strong interdependencies exist between the stressors (e.g., hurricane, shown in the figure), stressed systems (e.g., ecosystems and lifeline networks for coastal megacities), and the impacted systems (e.g., urban communities and financial institutions).

escalating failures" [4,5]. Similarly, loss of a small fraction of species from a natural ecosystem causes a series of extinction of other dependent species, putting these systems at the risk of ecological community collapse [22].

The climate- and weather-related extremes not only expose built environments to multiple hazards but also bring about myriad consequences for humans and society. Community structure and connectivity in the society are the key factors that determine how different shocks and disruptions, both natural and man-made, impact the society. With the arrival of new tide of "soft technologies" [23], the societies are getting less connected physically and strongly connected virtually. While closely operating societies may be more adaptable and resilient to these shocks [24], the vulnerability of distant people is strongly correlated through global environmental changes, global market linkages, and flows of information as shown in Figure 1.1.

Past few decades have witnessed the steep rise in population in coastal regions, megacities, and adjoining regions, which is expected to continue for coming decades. These populations are highly sensitive to many hazards and risks: flood, climate change-induced sea-level rise, and disease outbreaks. Also, the severe population stress in these cities as a result of mass migration and urban-centric growth puts extreme stress on already crumbling infrastructure systems of these megacities. The combined effect of over-virtualization of social networks and mass migrations coupled with evolving nature of hazards, specifically in the backdrop of climate change, has made these social networks dynamic in space and time, thus contributing to the "nonstationarity" in social networks [25]. Table 1.1 summarizes the role of "nonstationarity" and "deep uncertainty" across the three systems under consideration.

1.3 Network Science in Climate Risk Management

Given the strong correlation that exists within and across the stressors, stressed, and impacted systems, network representation can be a plausible unified framework to represent knowledge in the aforementioned

TABLE 1.1   Common Attributes of the Stressors, Stressed, and Impacted Systems

Attributes → and systems ↓	Correlation	Extremes	Nonstationarity	Deep uncertainty
Climate change and weather extremes (stressors)	Teleconnections [17]; multivariate and multiscale dependence; long-term memory [8]	More intense heat waves [26]; persisting cold extremes [27]; intensifying precipitation extremes [16]	Trends in extremes [28]	Climate chaos [29]; predictive surprises [30]; climate variability [31]
Lifeline and natural systems (stressed)	Interdependent systems [32]; mutualistic ecological networks [33]	Cascades [34]; first-order transition [35]; ecological community collapse [36]	Time-evolving networks [37]	Topological diversity [38]; parameter uncertainty [39]; chaos in ecosystems [29]
Social systems (impacted)	Social network analysis [40]	Extreme stress on communities [40]	Mass migrations [40]	Uncertain data [25]; unanticipated policies [41]

systems. However, knowledge acquired from network-based representation and analyses should be supplemented and complemented with the information that aids our understanding about individual systems. For example, in case of climate-related stressors, while network can give dynamical understanding of climate oscillators and teleconnections, data sciences, statistics, and local-scale physics [14] still play a pivotal role in generation of actionable insights. In case of infrastructure systems, both built and natural, the network-based knowledge needs to be complemented and supplemented with data acquisition, data representation, information management, real-time data ingestion, and offline data analyses. Finally, in impacted systems, while network-based solution framework can bridge the key knowledge gaps in our understanding of societal resilience, this knowledge may not be sufficient enough to inform the policy, econometrics, and incentive structures for decision making.

The three systems alluded to above and represented in Figure 1.2, specifically, stressors (climate change and weather extremes), stressed (lifelines and natural environment), and impacted systems (societal and human) share two important attributes. First, the systems are dependent, often in a complex manner, and further, they may be either subject to or cause/exacerbate extremes dependent in space and time, termed as "correlated extremes." Second, recent evolution of these system lead to what has been called "nonstationarity with deep uncertainty", or considerable changes in patterns (giving rise to situations where history is no longer a sufficient guide to future) and unwieldy uncertainty (a mix of known unknowns, unknown unknowns, and intrinsic systemic variability).

Despite the increasing complexity in threats and interdependencies in technical, social, and economical systems, conventional risk management frameworks have focused on strengthening these specific components to withstand the identified threats to an acceptable level and hence preventing the overall system failure. As outlined in the commentary in *Nature Climate Change* [6], enabling these capacities requires specific methods to define and measure resilience and modeling and simulation techniques for highly complex systems to develop resilience engineering. Figure 1.3 shows the resilience management framework with risk as a central component. Risk of climate-related hazards results from the interaction of climate-related hazards with the vulnerability and exposure of natural, built, and human systems [42].

It is noted that analogy can be drawn between the three elements of risks (Figure 1.3) with the three layers shown in Figure 1.2. That is, threats are analogous to climate stressors, vulnerabilities to the stressed systems, and exposed systems to the impacted systems, which are social systems in the present context. However, given the strong bidirectional feedback among the three systems referred to in Figure 1.2, this analogy may not always hold true since specific system can assume any of the three roles in risk management framework.

1.3.1 New Resilience Paradigm

While conventional risk management framework focuses on strengthening one specific component at a time, resilience management framework adopts a holistic system-level approach to identify system-level

Network view to stressors, stressed, and impacted systems

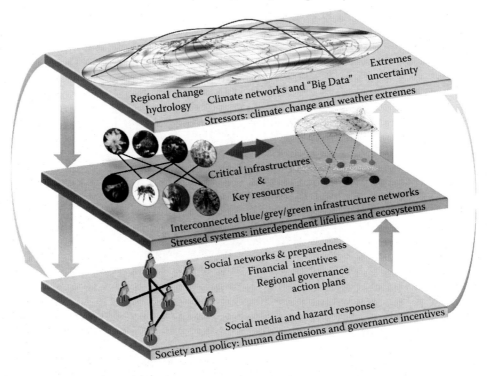

FIGURE 1.2 (See color insert.) A multilayered and unified framework can represent the coupled system where (a) climate change and weather or hydrological extremes are the stressors; (b) interdependent lifeline and environmental systems are the stressed systems; and (c) social networks of communities and regions are the impacted systems. This representation enables a unified quantitative framework, but particular instantiations of this network are necessary and sufficient to answer specific questions.

Resilience management framework

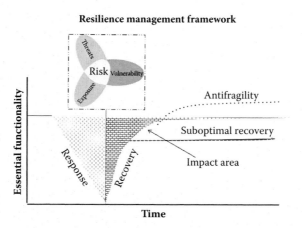

FIGURE 1.3 In system resilience framework, risk can be interpreted as the total reduction in critical functionality, while resilience is related to absorption of the stress, rate of reduction in essential functionality, and recovery of essential functionality after adversity. The postrecovery state of the system—measured in terms of essential functionality—can be better, equal to, or worse than prehazard state of the system.

functionalities that are crucial, understand and improve the interconnectivity of interdependent networks, enhance social networks, and bridge the gaps in policy shortfalls to mitigate and recover from the unforeseeable impacts the evolving stressors, including climate change, can produce. It should be noted that resilience is not an alternate to either risk management or engineering design principles for designing specific components. Rather, resilience is a complementary attribute to improve the traditional risk management framework, which is useful and widely accepted tool in mitigating the foreseeable and anticipated stress situations.

A roadmap to enable resilient design practices requires (1) understanding the correlations and evolving nature of adverse events that can impact the functionality of infrastructure systems and/or society; (2) approaches for computing robustness of the highly complex systems; (3) identifying the most important (or influential) components within individual systems; and (4) development of new simulation techniques to understand the resilience (response and recovery) of isolated as well as interdependent systems. The next section discusses how network science-based technology Stack (Net tech Stack) can serve as a unified framework to understand and hence quantify resilience for highly complex and interdependent systems. However, the instantiations of each of these systems, represented by networks, are necessary to address specific challenges. The representation and model can be combined with sensor and data engineering, computational process models, data science methods, and uncertainty propagation in forward or feedback processes to develop intervention strategies for monitoring and early warning, managing cascading failures and graceful recovery, proactive designing, and developing vulnerability and risk models.

1.4 Network Technology Stack

As discussed in previous sections, the three systems under consideration, stressors, stressed, and impact, are highly complex, interconnected, and interdependent in nature. Hence, understanding the complicacy and control of these systems is one of the key challenges to enable resilient design practices. Network science-based technology stack (Figure 1.4) in combination with data diagnostics, online (real-time) data acquisition, offline data analyses, and process modeling can help in understanding the complexity and interdependency, thus enabling the embedment of resilient engineering practices into design of the networked systems. Network science-based methods have been used for change detection, correlative and predictive analyses in weather and climate [8], robustness and resilience of infrastructures [3] and ecosystems, including percolations and cascades of failures, as well as for characterization of social systems. Thus, network science methods may act as a connective technology across the stressors, stressed, and the impacted systems, even though a comprehensive treatment of each system will require other approaches as well. The full technology stack that enables network science methods and their adaptation to climate change and resilient engineering is described next.

Network science finds its roots in graph theory and statistical physics. What distinguishes the network science from graph theory is its empirical and data-driven nature. Tools of networks derive the formalism to deal with graphs from graph theory and the conceptual framework to deal with universal organizing principles and randomness from statistical physics [43]. Since real-world networks usually have massive volume of data behind them, this field actively borrows algorithms from data sciences, including database management, and data mining. From modeling perspective, a network comprises of nodes and links. Links between a pair of nodes are placed to reproduce the complexity and randomness of individual systems. The key attribute of each node in a given network is degree, representing the number of links it has to other nodes with degree distribution representing the probability distribution of these degrees over the entire network. Along with the topological complexity, these systems display a large heterogeneity in the capacity and intensity of links, which is measured by the weight of each link. Analogous to the degree of the node, strength of a node is measure of the sum of weight of all links incident upon a node.

Since processes that shape individual systems differ greatly, the network architecture behind individual systems may vary greatly. Yet a key discovery of network science is universality, which means that evolution

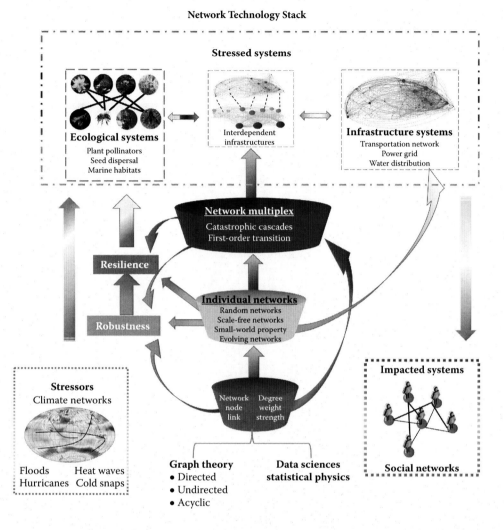

FIGURE 1.4 (See color insert.) Complex network-based technology stack in combination with sensor and data engineering, computational process models, and data science methods can potentially serve as a unified framework to meet these challenges. Tools to model single and interdependent networks come from network science and contribute to insights for the lifelines and ecosystems. Since all the three systems—stressors, stressed, and impacted systems—share common attributes of correlation and interdependence, fundamental network science breakthroughs and state of the art leads to novel adaptations or (in some cases) customization and then onward to new science or engineering insights.

of networks emerging in various domains of science and technology are similar to each other, enabling the applications of common set of mathematical and machine learning-based tools to explore these similarities making network science applicable to the heterogeneous systems. For example, many real-life systems such as transportation networks [44] and communication networks [45] exhibit a exponentially decaying tail in degree distribution and hence characterized as "scale-free" networks [43].

Although real-world infrastructure systems are highly interdependent and interconnected in nature, large number of studies have modeled and analysed the networks as single networks with underlying assumption that these networks can be isolated from each other. However, in interdependent networks, relatively localized failure in one system can percolate through the interconnected web or network of networks [5], triggering a "disruptive avalanche of cascades" [4]. Despite the fact that network science-based methods have advanced our understanding of individual and interdependent systems, percolation-based models are a very stylized representation of the network's response to the damage and hence lack the realism to capture many of the features that contribute to the resilience and robustness of real-world systems [4]. Moreover, given the strong bidirectional feedback that exists between the climate-induced stressors, stressed, and impacted systems, enabling resilient design practices requires a coherent understanding of the feedbacks within and across these systems, network science-based frameworks, in combination with sensor and data engineering and computational process models, which can lead to new scientific and engineering insights.

1.5 Case Studies: Telescoping Systems of Systems

Given this universality and applicability of network science, complex networks have been used to understand the structure, dynamics, and robustness of individual systems such as climate, Internet, transportation networks, and ecological and social systems. Network science analysis can delve into complex systems ranging from climate networks, infrastructure systems, to ecology and human systems. While understanding the network characteristics is crucial to understand the fundamental network properties (such as degree and strength distribution, network centrality measures, community structure, identification of hubs), enabling resilience into the system design may require information at higher spatial resolution (e.g., vulnerabilities and interdependencies of lifeline networks at facility level). In case of natural systems, network science has been used to understand the large-scale interdependencies among climate variables, regional analysis of spatial and temporal analysis of extreme rainfall patterns over South East Asia [46], sub-system level analysis using spatial pattern analysis [52], and phase synchronization analysis to track the climate oscillators in different geographical regions [47].

In context of built systems, complex network-based framework has been used to understand the connectivity patterns in infrastructure systems ranging from global (e.g., world air transportation network, global cargo ship networks) to national (e.g., Chinese [44] and Indian railroad networks [53]) to system scale (e.g., Boston subway) [51].

Similarly, social networks have been studied at multiple spatial scales to understand the network characteristics at global scales (e.g., robustness and search in global social networks), national scale (e.g., impact of climate-related hazards on social networks in Uruguay), and to answer specific questions (e.g., identifying key players in the key network).

Table 1.2 shows the applicability of network science to stressors, stressed, and impacted systems at spatial scales ranging from global to regional to local. Despite the generalizability of network science-based tools to a broad spectrum of systems, the insights obtained at a particular spatial scale may not be generalized at multiple spatial scales. Hence, development of network science-based "telescoping of systems-of-systems" (Figure 1.5), analogous to nested systems in climate (e.g., regional climate models [RCMs] that sit within

TABLE 1.2 Conceptual Representation of "Telescoping Systems of Systems" for Infrastructure Systems Operating at Spatial Scales Ranging from System to Regional to National and Global Scales

Scale → and systems ↓	Global	Regional/national	System level
Climate change and weather extremes (stressors)	Multiscale dependence in climate systems [8]	Analysis of spatiotemporal monsoon rainfalls over South Asia [46]	Phase synchronization analysis to track El-Niño [47]
Lifeline and natural systems (stressed)	Air transportation network [48]; global cargo networks [49]	National Railways system [44]; resilience of urban infrastructure [50]	Small world property in Boston subway [51]; nestedness in ecosystems [33]
Social systems (impacted)	Search in global social networks	National social networks; Anatomy of urban–social networks	Key players in social networks

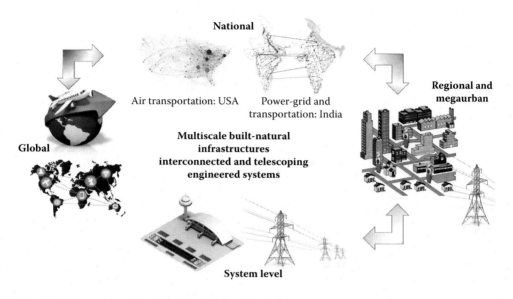

National

Air transportation: USA   Power-grid and
                                 transportation: India

**Multiscale built-natural
infrastructures
interconnected and telescoping
engineered systems**

Global

**Regional and
megaurban**

System level

FIGURE 1.5   "Telescoping systems of systems" for multiscale infrastructure systems.

global circulation models [GCMs]), can provide more specific detailed simulations for systems operating at diverse spatial scales.

1.6 Conclusion

The challenges in the coupled system-of-systems comprised of the stressors (climate change and extreme weather), stressed (built-natural lifelines), and impacted (human and social) systems are twofold: correlated and unprecedented extremes on highly interconnected systems as well as nonstationarity and deep uncertainty in the systems. Nonstationarity in this context goes beyond statistical definitions to large changes where history is no longer a sufficient guide to the future and engineering and planning principles need to change accordingly, while deep uncertainty refers to situations where translation to likelihood-based risks and hence risk-based design may not be feasible, thus motivating flexible design principles. While complex network-based frameworks, including network science technology stack and "telescoping systems of systems," could potentially help enable the resilient design practices in changing climate scenarios. Given the promise these climate networks have exhibited to understand the characteristics of highly complex systems, it may not be even necessary and sufficient to represent the systems' state. For example, in context of social networks, though network science-based methodologies have delved crucial information about architecture of these networks, it does not take into behavioral, social, and cultural attributes of these systems into account. Similarly, in context of climate-related stressors, while understanding of coupling in climate systems may yield new insights about stability of climate systems, processes that generate extremes may be synoptic or mesoscale. This means that the network science should not serve as "the hammer and the nail" solution to the challenges discussed for each of the three systems. Enabling the resilient design and engineering practices, specifically in the context of climate change and weather-related hazards, needs to integrate the system-level visualization with sensor and data engineering, computational process models, data science methods, and uncertainty propagation in forward or feedback processes. A 2014 survey at the World Economic Forum (WEF) identified "failure of climate change mitigation and adaptation" and "greater incidence of extreme weather events (e.g., floods, storms, fires)" as two of the top ten global risks of highest concern. The U.S. National Academy of Engineering (NAE) lists "restore and improve urban infrastructures" and "engineer the tools for scientific discovery" as two of fourteen grand

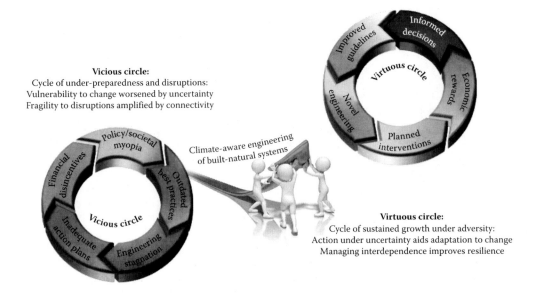

FIGURE 1.6   Motivating climate-aware engineering to meet the uncertain disasters.

challenges in engineering. The broad vision for future research is to create a consensus for addressing the challenges described in this chapter. A critical momentum is needed (Figure 1.6) to prevent from plunging into a vicious cycle of economic disincentives, policy myopia, and social inertia, resulting in stagnation of engineering and science, to a virtuous cycle of resilient infrastructures, climate preparedness, sustained economic growth, social justice, and greater security.

References

1. Climate change: Cold spells in a warm world. *Nature*, 472(7342):139–139, April 2011. Available at http://www.nature.com/nature/journal/v472/n7342/full/472139d.html.
2. Aiguo Dai. Increasing drought under global warming in observations and models. *Nature Climate Change*, 3(1):52–58, January 2013.
3. Udit Bhatia, Devashish Kumar, Evan Kodra et al. Network science based quantification of resilience demonstrated on the Indian railways network. *PLOS ONE*, 10(11):e0141890, 2015.
4. Alessandro Vespignani. Complex networks: The fragility of interdependency. *Nature*, 464(7291):984–985, April 2010.
5. Jianxi Gao, Sergey V. Buldyrev, H. et al. Percolation of a general network of networks. *Physical Review E*, 88(6):062816, December 2013.
6. Igor Linkov, Todd Bridges, Felix Creutzig et al. Changing the resilience paradigm. *Nature Climate Change*, 4(6):407–409, June 2014
7. Amit Sheth. Transforming Big Data into smart data: Deriving value via harnessing volume, variety, and velocity using semantic techniques and technologies. In *2014 IEEE 30th International Conference on Data Engineering (ICDE)*, pp. 2–2, March 2014.
8. Karsten Steinhaeuser, Auroop R. Ganguly, and Nitesh V. Chawla. Multivariate and multiscale dependence in the global climate system revealed through complex networks. *Climate Dynamics*, 39(3–4):889–895, June 2011.
9. Zhenning Kong and E.M. Yeh. Resilience to degree-dependent and cascading node failures in random geometric networks. *IEEE Transactions on Information Theory*, 56(11):5533–5546, November 2010.

10. Linyun Yu, Peng Cui, Fei Wang et al. From micro to macro: Uncovering and predicting information cascading process with behavioral dynamics. *arXiv:1505.07193 [physics]*, May 2015. arXiv: 1505.07193.

11. Mario Ventresca and Dionne Aleman. Efficiently identifying critical nodes in large complex networks. *Computational Social Networks*, 2(1):6, March 2015.

12. Jeroen C. J. H. Aerts, W. J. Wouter Botzen et al. Evaluating flood resilience strategies for coastal megacities. *Science*, 344(6183):473–475, May 2014.

13. S.M. Rinaldi, J.P. Peerenboom, and T.K. Kelly. Identifying, understanding, and analyzing critical infrastructure interdependencies. *IEEE Control Systems*, 21(6):11–25, December 2001.

14. Paul A. O'Gorman and Tapio Schneider. The physical basis for increases in precipitation extremes in simulations of 21st-century climate change. *Proceedings of the National Academy of Sciences*, 106(35):14773–14777, September 2009.

15. Gabriel Kuhn, Shiraj Khan, Auroop R. Ganguly et al. Geospatialtemporal dependence among weekly precipitation extremes with applications to observations and climate model simulations in South America. *Advances in Water Resources*, 30(12):2401–2423, December 2007.

16. Seung-Ki Min, Xuebin Zhang, Francis W. Zwiers et al. Human contribution to more-intense precipitation extremes. *Nature*, 470(7334):378–381, February 2011.

17. Anastasios A. Tsonis, Kyle L. Swanson et al. On the role of atmospheric teleconnections in climate. *Journal of Climate*, 21(12):2990–3001, June 2008.

18. Naiming Yuan, Zuntao Fu, and Shida Liu. Extracting climate memory using fractional integrated statistical model: A new perspective on climate prediction. *Scientific Reports*, 4, October 2014.

19. Sönke Dangendorf, Diego Rybski, Christoph Mudersbach et al. Evidence for long-term memory in sea level. *Geophysical Research Letters*, 41(15):2014GL060538, August 2014.

20. D. J. Rapport, H. A. Regier, and T. C. Hutchinson. Ecosystem behavior under stress. *The American Naturalist*, 125(5):617–640, May 1985.

21. Sergey V. Buldyrev, Roni Parshani, Gerald Paul et al. Catastrophic cascade of failures in interdependent networks. *Nature*, 464(7291):1025–1028, April 2010.

22. Mikael Lytzau Forup, Kate S. E. Henson, Paul G. Craze et al. The restoration of ecological interactions: Plantpollinator networks on ancient and restored heathlands. *Journal of Applied Ecology*, 45(3):742–752, June 2008.

23. Zhouying Jin. *Global Technological Change: From Hard Technology to Soft Technology*. Chicago: Intellect Books, 2011.

24. W Neil Adger, Hallie Eakin, and Alexandra Winkels. Nested and teleconnected vulnerabilities to environmental change. *Frontiers in Ecology and the Environment*, 7(3):150–157, May 2008.

25. Eytan Adar and Christopher Ré. Managing uncertainty in social networks. *IEEE Data Engineering Bulletin*, 30:2007, 2007.

26. Gerald A. Meehl and Claudia Tebaldi. More intense, more frequent, and longer lasting heat waves in the 21st century. *Science*, 305(5686):994–997, August 2004.

27. Evan Kodra, Karsten Steinhaeuser, and Auroop R. Ganguly. Persisting cold extremes under 21st-century warming scenarios. *Geophysical Research Letters*, 38(8):L08705, April 2011.

28. Kenneth E. Kunkel, Karen Andsager, and David R. Easterling. Long-term trends in extreme precipitation events over the conterminous United States and Canada. *Journal of Climate*, 12(8):2515–2527, August 1999.

29. William M. Schaffer. Order and chaos in ecological systems. *Ecology*, 66(1):93–106, February 1985.

30. Ning Lin and Kerry Emanuel. Grey swan tropical cyclones. *Nature Climate Change*, advance online publication, August 2015.

31. Richard W. Katz and Barbara G. Brown. Extreme events in a changing climate: Variability is more important than averages. *Climatic Change*, 21(3):289–302, July 1992.

32. Jianxi Gao, Sergey V. Buldyrev, H. Eugene Stanley et al. Networks formed from interdependent networks. *Nature Physics*, 8(1):40–48, January 2012.

33. Jordi Bascompte, Pedro Jordano, Carlos J. Melin et al. The nested assembly of plant animal mutualistic networks. *Proceedings of the National Academy of Sciences*, 100(16):9383–9387, August 2003.

34. Leonardo Dueñas-Osorio and Srivishnu Mohan Vemuru. Cascading failures in complex infrastructure systems. *Structural Safety*, 31(2):157–167, March 2009.

35. Jianxi Gao, Sergey V. Buldyrev, Shlomo Havlin et al. Robustness of a network of networks. *Physical Review Letters*, 107(19):195701, November 2011.

36. Thomas LaBar, Colin Campbell, Suann Yang et al. Restoration of plant-pollinator interaction networks via species translocation. *Theoretical Ecology*, 7(2):209–220, January 2014.

37. Reka Albert and Albert-Laszlo Barabasi. Topology of evolving networks: Local events and universality. *Physical Review Letters*, 85(24):5234–5237, December 2000.

38. James Winkler, Leonardo Dueñas-Osorio, Robert Stein et al. Performance assessment of topologically diverse power systems subjected to hurricane events. *Reliability Engineering and System Safety*, 95(4):323–336, 2010.

39. Gangquan Si, Zhiyong Sun, Hongying Zhang et al. Parameter estimation and topology identification of uncertain fractional order complex networks. *Communications in Nonlinear Science and Numerical Simulation*, 17(12):5158–5171, December 2012.

40. Berman Rachel, Claire Quinn, and Jouni Paavola. The impact of climatic hazards on social network structure: Insights from community support networks in Western Uganda. Working paper, Sustainability Research Institute, Leeds, United Kingdom, June 2014.

41. Vincent A. W. J. Marchau and Warren E. Walker. Addressing deep uncertainty using adaptive policies: Introduction to section 2. *Technological Forecasting and Social Change*, 77(6):917–923, 2010.

42. IPCC. *Climate Change 2014: Impacts, Adaptation, and Vulnerability. Part B: Regional Aspects. Contribution of Working Group II to the Fifth Assessment Report of the Intergovernmental Panel on Climate Change*. Barros, V.R., C.B. Field, D.J. Dokken, M.D. Mastrandrea, K.J. Mach, T.E. Bilir, M. Chatterjee, K.L. Ebi, Y.O. Estrada, R.C. Genova, B. Girma, E.S. Kissel, A.N. Levy, S. MacCracken, P.R. Mastrandrea, and L.L. White (eds.). Cambridge and New York: Cambridge University Press, 2014.

43. Reka Albert and Albert-Laszlo Barabsi. Statistical mechanics of complex networks. *Reviews of Modern Physics*, 74(1):47–97, January 2002.

44. W. Li and X. Cai. Empirical analysis of a scale-free railway network in China. *Physica A: Statistical Mechanics and its Applications*, 382(2):693–703, August 2007.

45. Reka Albert, Hawoong Jeong, and Albert-Laszlo Barabasi. The Internet's Achilles' heel: Error and attack tolerance of complex networks. *Nature*, 406:200–0, 2000.

46. Nishant Malik, Bodo Bookhagen, Norbert Marwan et al. Analysis of spatial and temporal extreme monsoonal rainfall over South Asia using complex networks. *Climate Dynamics*, 39(3-4):971–987, August 2011.

47. A. Gozolchiani, S. Havlin, and K. Yamasaki. Emergence of El Niño as an Autonomous Component in the Climate Network. *Physical Review Letters*, 107(14):148501, September 2011.

48. R. Guimerà, S. Mossa, A. Turtschi et al. Amaral. The worldwide air transportation network: Anomalous centrality, community structure, and cities' global roles. *Proceedings of the National Academy of Sciences of the United States of America*, 102(22):7794–7799, May 2005.

49. Pablo Kaluza, Andrea Kölzsch, Michael T. Gastner et al. The complex network of global cargo ship movements. *Journal of the Royal Society Interface*, 7(48), July 2010.

50. Min Ouyang, Leonardo Dueñas-Osorio, and Xing Min. A three-stage resilience analysis framework for urban infrastructure systems. *Structural Safety*, 3637:23–31, May 2012.

51. Vito Latora and Massimo Marchiori. Is the Boston subway a small-world network? *Physica A: Statistical Mechanics and its Applications*, 314(14):109–113, November 2002.

52. Karsten Steinhaeuser and Anastasios A. Tsonis. A climate model inter-comparison at the dynamics level. *Climate Dynamics*, 42(5-6):1665–1670, April 2013.

53. Parongama Sen, Subinay Dasgupta, Arnab Chatterjee et al. Small-world properties of the Indian railway network. *Physical Review E*, 67(3):036106, March 2003.

2

Structured Estimation in High Dimensions: Applications in Climate

André R Goncalves

Arindam Banerjee

Vidyashankar
Sivakumar

Soumyadeep
Chatterjee

2.1   Introduction.. 13
2.2   Sparse Structured Estimation and Structure Learning.............. 14
2.3   Sparse Group Lasso and Climate Applications 16
      SGL and Hierarchical Norms • Experiments
2.4   Multitask Sparse Structure Learning and Climate Applications . 20
      Multitask Sparse Structure Learning • Experiments on GCMs
      Combination
2.5   Conclusions 30

2.1 Introduction

One of the central challenges of data analysis in climate science is understanding complex dependencies between multiple spatiotemporal climate variables. The data are typically high dimensional with each climate variable in each spatial grid or time period denoting a separate dimension. In fact, in many climate problems, the dimensionality, that is, the number of possible features or factors potentially affecting a response variable, is usually much larger than the number of samples that are typically reanalysis data sets over the past few decades. For example, in one of the problems considered in this chapter, one wants to predict climate variables like monthly temperature, precipitable water, etc. over land locations using information from six climate variables over oceans. We formulate it as a regression problem with the climate variable over a land location as the response variable. We consider 439 locations on oceans, so that there are a total of $6 \times 439 = 2634$ covariates in our regression problem. The data are the monthly means of the climate variables for 1948–2007 from the National Centers for Environmental Prediction/National Center for Atmospheric Research (NCEP/NCAR) Reanalysis 1 data set [1], so that we have a total of $60 \times 12 = 720$ data samples. Traditional statistical methods like least squares regression do not work in such high-dimensional, low-sample scenarios.

Another key feature in these data sets is the presence of spatial and temporal patterns and relationships. Capturing these relationships in statistical models often improves estimation and prediction performance. As an example, consider the problem of multimodel global climate model (GCM) ensemble combination. GCMs are complex mathematical models of climate systems which are run as computer simulations, to predict climate variables such as temperature, pressure, and precipitation over multiple centuries [2]. Several such GCMs are reported to the Intergovernmental Panel on Climate Change (IPCC). Due to the varied nature of the GCM assumptions, future projections of GCMs show high variability. To minimize the variability and prediction error, the predictions from the various GCMs are suitably combined. One of the popular approaches is to perform a weighted combination of GCM predictions, where the weights signify model skills. GCM model skills are estimated for each spatial location based on their performance in reproducing historical observations at that spatial location. For each spatial location, GCM skill estimation is posed as a regression problem. Reanalysis data provide historical observations of the response variable, while the GCM predictions are the covariates, and the model skills are the parameter vectors that are to be estimated. The model skills at different spatial locations are related to each other. This can be represented as a spatial graph, where the spatial locations are the nodes and edges represent a relationship between the model skills for the two spatial locations connected by the edge. For example, [3] shows that, for modeling temperature, imposing a constraint that encourages neighboring locations to have similar weights not only improves prediction performance but also gives interpretable results for model skill evaluation. But in many problems, we do not know the structure of the spatial graph, that is, we do not know which spatial locations are related, and thus one has to simultaneously learn the graph structure while estimating model skills. Learning the structure from the data is challenging because of the large number of spatial locations. For example, for our problem of predicting temperature over South America, the data set has 250 spatial locations.

Similar to the climate domain, the above highlighted issues of dealing with high-dimensional data sets are increasingly faced in domains such as astronomy, finance, biotechnology, medicine, ecology, etc. with the rapid development of data collection and storage technologies [4–6]. Recent advances in the field of high-dimensional statistics have developed a suite of tools to analyze such high-dimensional data sets. All of these methods are based on imposing additional constraints on the problem, which enforces some prior structure known from the problem domain. As an example, consider the climate variable prediction problem described earlier. We assume that only a few ocean locations are relevant for the prediction problem. It is also evident that the climate variables at a particular location are naturally grouped. Similarly, for the GCM multimodel ensemble combination problem, we impose the constraint that the spatial graph is sparse, that is, each node is connected to only a few other nodes in the graph. Such models have the additional advantage of learning a parsimonious and compact model which is more interpretable. Generally, estimation involves solving a convex optimization problem, which can be efficiently solved using well-known convex optimization techniques.

The rest of this chapter is organized as follows. We give a brief introduction to methods in high-dimensional statistics in Section 2.2. We elaborate on the two problems, which we briefly discussed above, in Sections 2.3 and 2.4 before concluding in Section 2.5.

2.2 Sparse Structured Estimation and Structure Learning

Estimating dependencies in the context of climate sciences needs to consider a few key aspects which pose serious challenges to the state-of-the-art. Climate phenomena of interest such as change of surface air temperature (SAT), can be considered manifestations of a complex nonlinear dynamical system. From such a perspective, finding the dependency of a variable y on another variable x is difficult, since the number of possible native and derived features can easily run into millions, often with strong spatial and temporal correlations. Recent work on sparse regression methods provides a solution to such challenges, even in the case when the sample size is less than the total number of features.

Structured Regression: Given a set of p covariates $(X_1, \ldots, X_p)$ and a predictand Y, where $E[Y|X_1, \ldots, X_p] = \sum_{i=1}^{p} \theta_i^* X_i$, linear regression aims to estimate the (unknown) statistical parameter θ^*, from n samples (x_i, y_i), $i = 1, \ldots, n$. Note that a coefficient θ_i^* is zero if and only if the predictand y is *conditionally* independent of X_i, given all other variables $X_{\setminus i}$. Thus, regression also provides an estimate of the set of conditional dependencies between Y and the covariates. For example, for Gaussian conditional distributions $\mathcal{P}(Y|X_1, \ldots, X_p)$, linear regression amounts to maximizing the conditional log-likelihood.

Regression in high-dimensional scenarios faces a key issue due to the dimensionality (p) of the problem being larger than the sample-size (n) [7]. For a fixed sample size, the matrix of covariates, called the design matrix, becomes ill-conditioned, or even singular when $p > n$ [8]. In such scenarios, the estimation of the parameter θ is difficult without additional constraints. Recent work on sparse regression focuses on the scenarios where the model parameter θ^* is *sparse*, meaning that many of its entries are zero [8–10], which also signifies a *low-complexity* model, since most covariates are conditionally independent of the predictand y. Estimation proceeds by maximizing the *penalized* log-likelihood of data, which leads to a regularized regression problem of the form

$$\min_{\theta} \mathcal{L}((\mathbf{y}, \mathbf{X}), \theta) + \lambda \mathcal{R}(\theta), \tag{2.1}$$

where $(\mathbf{y}, \mathbf{X})$ constitute the predictands and predictors, and $\mathcal{R}(\theta)$ is a regularizer. For example, the popular Lasso problem [9] considers an L_1 norm regularizer, and least squares loss function, and requires optimization of the following objective:

$$\hat{\theta}_n = \operatorname*{argmin}_{\theta} \left\{ \frac{1}{2n} \|\mathbf{y} - \mathbf{X}\theta\|^2 + \lambda_n \|\theta\|_1 \right\}, \tag{2.2}$$

where $\lambda_n > 0$ is a penalty parameter. Recent literature in machine learning and statistics has proposed several other regularizers, such as the Group Lasso [11], which considers $R(\theta) = \sum_G \|\theta_G\|_2$ and enforces sparsity of groups, the Fused Lasso [12], which considers $R(\theta) = \|\theta\|_1 + \sum_{j=2}^{p} |\theta_j - \theta_{j-1}|$ and ensures that non zero coefficients have similar magnitudes, etc. The Lasso and related methods have been applied to a variety of problems and domains [11,13–15], and have provided encouraging empirical results. A generalization of the Lasso is the notion of structured sparsity, wherein the model parameter θ^* is allowed to follow a certain hierarchical structure in the sparsity induced by the regularizer $\mathcal{R}(\cdot)$. In Section 2.3, we discuss one such hierarchical sparse regularizer, called the Sparse Group Lasso [16–18]. The SGL has been proposed in [17,18], and has been found to outperform unstructured sparse regularizer, such as Lasso in various scientific domains [16,18].

Multitask Learning: A multitask learning (MTL) problem consists of learning, in a unified framework, a set of tasks simultaneously rather than independently. By exploiting the fact that some of the tasks might be related, it is possible to transfer information (data samples, learned parameters, etc.) between related tasks so that it may improve the performance of individual tasks. This transfer capability is particularly interesting when a limited amount of data is available for each task.

In the unified framework, an internal shared representation is needed to allow information exchange between tasks. Examples of such representations include: shared hidden nodes in neural networks [19], common prior in hierarchical Bayesian model [20], shared parameters of Gaussian process [21], etc. Other MTL methods do not have an explicit shared representation, but instead use regularizers to impose certain structure about task relationship, leading to a minimization problem of the form

$$\min_{\Theta} \sum_{k=1}^{m} \mathcal{L}\left((\mathbf{y}_k, \mathbf{X}_k), \theta_k\right) + \lambda \mathcal{R}(\Theta) \tag{2.3}$$

where $\lambda > 0$ is a penalization parameter and $\mathcal{L}(\cdot)$ is the loss function corresponding to the task being dealt with, and includes squared loss, logistic loss, and hinge loss as examples; Θ is a matrix whose columns are the parameter vectors θ_k for each task $k = 1, ..., m$. The goal is to learn K parameter vectors $\theta_1, ..., \theta_m \in \mathbb{R}^d$ such that $f(\mathbf{X}_k, \theta_k) \approx \mathbf{y}_k$. Examples of structure assumption include task parameter vectors of all tasks are close to each other [22], tasks are related in a group structure [23], all models share a common set of features [24], and task relatedness via a shared low-rank structure [25].

In many problems, only a high level or sometimes no understanding about the task relationship is available. The structure of the tasks relationship thus needs to be learned from the data. Only a few methods with such capability have been proposed in the literature [26,27]. In Section 2.4, we discuss in detail our methodology for GCM combination which belongs to this class of algorithms. We also show that our method provides a better estimation of future temperature in South America when compared with traditional methods, while capturing the underlying dependencies between geographical locations.

2.3 Sparse Group Lasso and Climate Applications

In order to motivate this section, let us consider a particular predictive modeling task. The task is to predict climate variables over *target* regions on land by using information from multiple climate variables over oceans. From a climate science point of view, it is often of importance to understand how oceanic phenomena affect atmospheric variables over land. For example, it is well known that Pacific SST affects much of climate across the world due to the El Niño oscillation.

The predictands of interest in this problem are (1) SAT (in °C) and (2) precipitable water (in kg/m^2) over nine "target regions" on land, viz., Brazil, Peru, Western and Eastern United States, Western Europe, Sahel, South Africa, Central India, and Southeast Asia, as shown in Figure 2.1. Prediction was done for SAT and precipitable water at each of these 9 locations and, in total, we considered 18 response variables. These regions were chosen following [28] because of their diverse geological properties and their impact on human interests. The covariates in our problem are six climate variables measured at ocean locations over the entire globe: (1) SST, (2) sea level pressure, (3) precipitable water, (4) relative humidity, (5) horizontal wind speed, and (6) vertical wind speed. From the NCEP Reanalysis 1 data set [1], one can obtain monthly measurements of these variables from 1948 to 2007. At a 10° × 10° resolution, the data set has $T = 439$ locations on the oceans, so that we have $p = 6 \times T = 2634$ covariates in our regression model. Further, we only have $N = 720$ samples of monthly values over 60 years, and therefore one encounters a high-dimensional regression problem. In such scenarios, using *sparse* regularizers is essential in order to control model complexity.

Assuming a linear model, we can denote

$$\mathbf{y} \sim \mathbf{X}\theta^* + w, \tag{2.4}$$

where $\mathbf{y} \in \mathbb{R}^n$ is the n-dimensional vector of observations of a climate variable at a target region, $\theta^* \in \mathbb{R}^p$ is the coefficient associated with all p variables at all locations, $\mathbf{X} \in \mathbb{R}^{n \times p}$ is the covariate matrix and $w \in \mathbb{R}^n$ is the noise vector. Our goal is twofold:

1. Understand which covariates are relevant/important for predicting the target variable.
2. Build a suitable regressor based on these relevant variable(s).

Note that the spatial structure of the data indicates a natural "grouping" of the variables at each ocean location. Simple sparse regularizers, such as the L_1 norm used in Lasso, do not respect this structure inherent in the data. Structured regularizers, which impose *structured sparsity* that respects this spatial nature, tend to be more useful in this scenario. The *SGL* (in the sequel) is a regularized regression method, a generalization of sparse regression methods, such as the Lasso [9,10], or group Lasso [11,13].

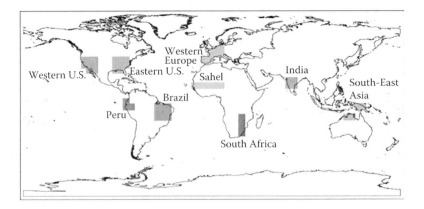

FIGURE 2.1   Land regions chosen for predictions (From Steinhaeuser et al. *Statistical Analysis and Data Mining*, 4(5), 2011).

2.3.1 SGL and Hierarchical Norms

The geographic grouping of variables gives rise to a natural sparsity structure among covariates, viz., if a particular location on oceans is *irrelevant*, then coefficients of all variables at that location should be zero. For example, if a particular location is deemed "relevant" then we should be able to select the "most important" variable(s) at that location to be considered for prediction. Note that Group Lasso or Overlapping Group Lasso will only enforce sparsity over groups of locations, and cannot select the most relevant variable(s) at any geographical location. Therefore, SGL regularization is better suited to this task than only group sparsity.

Formally, recall that we have $T = 439$ locations over oceans, and consider that the total number of variables as covariates in the regression problem is $p = 6 \times T$. Then, for a penalty parameter λ_n, the SGL estimator [16] is given by

$$\hat{\theta}_{\text{SGL}} = \underset{\theta \in \mathbb{R}^p}{\text{argmin}} \left\{ \frac{1}{2N} \|\mathbf{y} - \mathbf{X}\theta\|_2^2 + \lambda_n r(\theta) \right\}, \tag{2.5}$$

where r is the SGL regularizer given by

$$r(\theta) := r_{(1, \mathcal{G}_2, \alpha)} = \alpha\|\theta\|_1 + (1 - \alpha)\|\theta\|_{1, \mathcal{G}}, \tag{2.6}$$

where

$$\|\theta\|_1 = \sum_{i=1}^p |\theta_i|, \ \ \|\theta\|_{1, \mathcal{G}} = \sum_{k=1}^T \|\theta_{G_k}\|_2 \tag{2.7}$$

and $\mathcal{G} = \{G_1, \ldots, G_T\}$ are the groups of variables at the T locations considered. The mixed norm $\|\theta\|_{1, \mathcal{G}}$ penalizes groups of variables at irrelevant locations, while the L_1 norm $\|\theta\|_1$ promotes sparsity among variables chosen at selected locations.

The SGL regularizer is a tree-structured hierarchical norm regularizer, when the height of the tree is 2. The first level of the tree contains nodes corresponding to the T disjoint groups $\mathcal{G} = \{G_1, \ldots, G_T\}$, while the second level contains the singletons (Figure 2.2). It combines a group-structured norm with an element-wise norm (Equation 2.6).

Theoretically, we have illustrated in [16] that when the covariates form a hierarchy as in Figure 2.2, SGL provides statistically consistent estimate of θ^*, that is, the estimation error $\|\hat{\theta}_{\text{SGL}} - \theta^*\|_2$ decreases to zero with high probability with increasing sample size. Theoretically, the benefit of using structured sparse

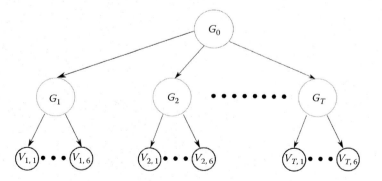

FIGURE 2.2 The hierarchical structure of the SGL norm. $G_1, \ldots, G_T$ are groups of variables at T ocean locations, while $V_{k,1}, \ldots, V_{k,6}$ are groups of variables at the kth location.

regularizers over Lasso, when presence of such hierarchies are known, is twofold. First, SGL requires lower model complexity than Lasso for similar statistical performance since the model is more constrained than simple sparsity. Second, the error in estimation for a given (finite) sample size is lower since entire groups of irrelevant variables are forced to have zero coefficients, as opposed to Lasso, which does not consider any group sparsity.

2.3.2 Experiments

Experiments were conducted on the prediction problem discussed earlier, viz. to predict climate variables over certain land regions using climate variables over oceans. We considered the data from January 1948 to December 1997 as the training data and from January 1998 to December 2007 as the test data in our experiments. Therefore, the training set contained $n_{\text{train}} = 600$ samples and the test set $n_{\text{test}} = 120$ samples. Last, we used the Sparse Learning with Efficient Projections (SLEP) package [29] for MATLAB® to run SGL on our data set. It may be noted that we do not take into account temporal relationships that exist in climate data. Moreover, since we consider monthly means, temporal lags of less than a month are typically not present in the data. However, the data does allow us to capture more long-term dependencies present in climate.

Removing Seasonality and Trend: As illustrated in [28], seasonality and autocorrelation within climate data at different time points often dominate the signal present in it. Hence, when trying to utilize such data to capture dependency, we look at a series of *anomaly* values, that is, the deviation at a location from the "normal" value. First, we remove the seasonal component present in the data by subtracting the monthly mean from each data point and then normalize by dividing it by the monthly standard deviation. At each location, we calculate the monthly mean μ_m and standard deviation σ_m for each month $m = 1, \ldots, 12$ (i.e., separately for January, February, etc.), for the entire time series. Finally, we obtain the anomaly series for location k as the z-score of the variable at location k for month m over the time series.

Further, we need to detrend the data to remove any trend components in the time series, which might also dominate the signal present in it and bias our regression estimate. Therefore, we fit a linear trend to the anomaly series at each location over the entire time period 1948–2010 and take the residuals by subtracting the trend. We use this deseasonalized and detrended residuals as the data set for all our subsequent experiments.

2.3.2.1 Prediction Accuracy

Evaluation of our predictions was done by computing the root mean square errors (RMSEs) on the test data and comparing the results against those obtained in [28], using ordinary least squares (OLS) estimates, and using Lasso. Note that the problem is high-dimensional, since the number of samples (~ 600) for training is much less than the problem dimensionality (~ 2400). [28] uses a correlation-based approach to separately cluster each ocean variable into *regions* using a k-means clustering algorithm. The *regions* (78 clusters in total) for all ocean variables are used as covariates for doing linear regression on response variables. Their model is referred to as the *Network Clusters* model. RMSE values were computed, as mentioned earlier, by predicting monthly mean anomaly for each response variable over the test set for 10 years. The RMSE scores are summarized in Table 2.1. We observed that SGL consistently performs better than both the *Network Clusters* method and the OLS method.

The higher prediction accuracy might be explained through the model parsimony that SGL provides. Applying SGL, only the most relevant predictor variables are given nonzero coefficients and any irrelevant variable is considered as noise and suppressed. Since such parsimony will be absent in OLS, the noise contribution is large and therefore, the predictions are more erroneous. Further, SGL often performs better than Lasso, although Lasso also provides model parsimony and has more freedom in selecting relevant covariates. Moreover, structured variable selection provided by SGL often has greater interpretability than Lasso. We elaborate on this aspect in Section 2.3.2.2.

TABLE 2.1 RMSE Scores for Prediction of SAT (in °C) and Precipitable water (in kg/m^2) Using SGL, Lasso, Network Clusters, and OLS. The Number in Brackets Indicate Number of Covariates Selected by SGL from among the 2634 Covariates. The Lowest RMSE Value in Each Task Is Denoted as Bold

Variable	Region	SGL	Lasso	Network Clusters	OLS
Air Temperature	Brazil	**0.198** (651)	0.211	0.534	0.348
	Peru	**0.247** (589)	0.259	0.468	0.387
	West USA	**0.270** (630)	0.291	0.767	0.402
	East USA	**0.304** (752)	0.307	0.815	0.348
	W Europe	0.379 (835)	**0.367**	0.936	0.493
	Sahel	**0.320** (829)	0.322	0.685	0.413
	S Africa	0.136 (685)	**0.130**	0.726	0.267
	India	**0.205** (664)	0.206	0.649	0.3
	SE Asia	0.298 (596)	**0.277**	0.541	0.383
Precipitable Water	Brazil	**0.261** (762)	0.307	0.509	0.413
	Peru	**0.312** (739)	0.344	0.864	0.523
	West USA	**0.451** (824)	0.481	0.605	0.549
	East USA	**0.365** (133)	0.367	0.686	0.413
	W Europe	0.358 (820)	**0.321**	0.450	0.551
	Sahel	0.427 (94)	**0.413**	0.533	0.523
	S Africa	0.235 (34)	**0.215**	0.697	0.378
	India	0.146 (593)	**0.143**	0.672	0.264
	SE Asia	**0.159** (571)	0.168	0.665	0.312

Source: K. Steinhaeuser, et al. *Statistical Analysis and Data Mining*, 4(5), 2011.
Abbreviations: Lasso, Least Absolute Shrinkage and Selection Operator; OLS, ordinary least squares; RMSE, root mean square errors; SGL, Sparse Group Lasso.

2.3.2.2 Variable Selection for Brazil

The high prediction accuracy of SGL brings to light the inherent power of the model to select appropriate variables (or features) from the covariates during its training phase. To quantitatively elaborate on this aspect, we looked more closely at temperature prediction in Brazil.

In order to evaluate the covariates which consistently get selected from the set, we conduct hold out cross-validation. During the training phase, an ocean variable was considered *selected*, if it had a corresponding nonzero coefficient. So, in each run of cross-validation, some of the covariates were selected, while others were not.

We observed that there are ∼ 60 covariates among the 2634 covariates that are selected in every single run of cross-validation. In Figure 2.3, we plot the covariates which are given high-coefficient magnitudes by SGL by training on the training data set from years 1948 to 1997, in order to illustrate that SGL consistently selects *relevant* covariates. It turns out that these covariates are exactly those which were selected in every cross-validation run. Most of these covariates lie off the coast of Brazil. The influences of horizontal wind speed and pressure are captured, which is consistent with the fact that the ocean currents affect land climate typically through horizontal wind. The tropical climate over Brazil is expected to be influenced by the inter tropical convergence from the north, polar fronts from the south, and disturbances in ocean currents from the west, as well as the influence of Easterlies from the east and immediate south. It is interesting to see that SGL model captures these influences, as well as the spatial autocorrelation present in climate data, without having any explicit assumptions.

In order to do a comparison, in Figure 2.4 we plot the variables selected by Lasso in every cross-validation run. There is overlap between this set of variables and the ones selected by SGL, particularly

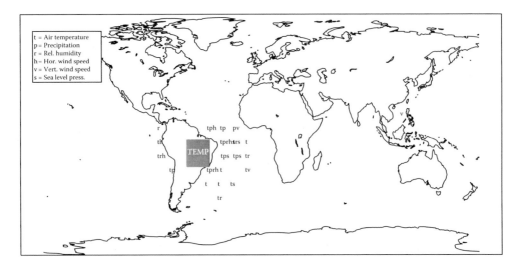

FIGURE 2.3 Temperature prediction in Brazil: SGL-selected variables. All the plotted variables are selected in every single run of cross-validation. SGL, Sparse Group Lasso.

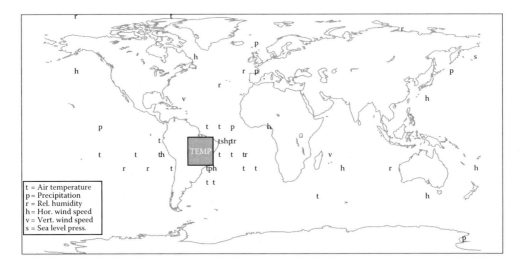

FIGURE 2.4 Temperature prediction in Brazil: Lasso-selected variables. All the plotted variables are selected in every single run of cross-validation. Lasso, Least Absolute Shrinkage and Selection Operator.

similar variables are selected off the coast of Brazil. However, Lasso has less discretion in the geographic spread of variables chosen. Thus, wind, pressure, and relative humidity at various locations around the globe are also selected, as shown in the figure. These variables are hard to interpret climatologically, and the model learnt by Lasso, as shown in Table 2.1, often performs worse than SGL.

2.4 Multitask Sparse Structure Learning and Climate Applications

GCMs are complex mathematical representations of the major climate system components (atmosphere, land surface, ocean, and sea ice) and their interactions [2]. GCMs are important tools for improving our understanding and predictability of climate behavior on seasonal, annual, decadal, and centennial timescales. Several GCMs have been proposed by climate science institutes from different countries (Table 2.2 gives a few examples).

Due to different assumptions and parameterizations adopted by the different models, the forecasts of climate variables as predicted by these models exhibit high variability [2]. This in turn introduces uncertainty in analysis based on these predictions. A common approach to deal with such uncertainty is to suitably combine the projections from the different climate models [2]. The simplest of such methods is to use the average model, that is, assigning equal weights to all models. Still, other models learn weights on models based on their skill in predicting observed climate variables. These weights are then used in a weighted average combination model to predict future climate. The fundamental hypothesis in such models is that different models have different skills in modeling the climate variables or processes at different spatial locations. More recently, Subbian and Banerjee [3] proposed a spatially regularized model, which encourages nearby geographical locations to have similar weights.

In this work, we pose the GCM combination problem as a MTL problem. Learning the model weights at each spatial location constitutes a task. Like in [3], we assume that the different tasks are related, but instead of assuming the relationship structure we model it as a sparse graph. The graph is estimated from the data using sparse estimation techniques. As will be seen from the results, the resulting graph is climatologically meaningful, and the prediction accuracy across South America significantly outperforms existing approaches.

2.4.1 Multitask Sparse Structure Learning

In this section, we describe the basic idea of our multitask sparse structure learning (MSSL) framework. We start by giving a small description of the notations used throughout the section, followed by a short introduction to the structure estimation problem. It will be good to remember that for this problem, learning the weights at each spatial location constitutes a task, the GCM model's predictions for the locations are the covariates and the samples are the GCM model's monthly predictions across a few decades. So, for a particular location k, the model is

$$\mathbf{y}_k = \mathbf{X}_k \mathbf{w}_k + \mathbf{e}_k \tag{2.8}$$

where $\mathbf{y}_k$ represents the observed values for a climate variable for example, temperature, $\mathbf{X}_k$ is the GCM model output, $\mathbf{w}_k$ is the parameter vector, which is the model skill for location k and $\mathbf{e}_k$ represents the error. Our aim is to learn the $\mathbf{w}_k$s, given a few decades of data of $(\mathbf{y}_k, \mathbf{X}_k)$. The idea behind MTL is that there is a relationship between the m tasks, and we can perform better in learning the $\mathbf{w}_k$'s if we use this relationship in our model than by learning the $\mathbf{w}_k$'s separately for each spatial location.

2.4.1.1 Notation and Preliminaries

Let m be the number of tasks, d be the number of covariates which are shared across all tasks, and n_k, $k \in [1, ..., m]$ be the number of samples for each task. Also $\mathbf{X}_k \in \mathbb{R}^{n_k \times d}$ be the covariates matrix (input data) for the kth task, $\mathbf{y}_k \in \mathbb{R}^{n_k \times 1}$ be the response for the kth task, and $\mathbf{W} \in \mathbb{R}^{d \times m}$ be the parameter matrix, where columns are parameters $\mathbf{w}_k$ for each task and $\hat{\mathbf{w}}_j$ denotes the j-th row of $\mathbf{W}$.

We consider a linear model $\mathbf{y}_k = \mathbf{X}_k \mathbf{w}_k + \mathbf{e}_k$ for each task where $\mathbf{e}_k$ is the residual error for the kth task. When we consider task relatedness based on the residual errors $\mathbf{e}_k$, we assume that all tasks have the same number of training samples n and we define the matrix $\mathbf{E} = (\mathbf{e}_1, \mathbf{e}_2, ..., \mathbf{e}_m) \in \mathbb{R}^{n \times m}$. We denote the j-th row of $\mathbf{E}$ as $\hat{\mathbf{e}}_j$. Also, $\mathbf{\Omega} \in \mathbb{R}^{m \times m}$ is a symmetric matrix that captures the task relationship structure. $\|\mathbf{A}\|_1$ and $\|\mathbf{A}\|_F$ are the ℓ_1 and Frobenius norm of the matrix $\mathbf{A}$, respectively.

2.4.1.2 Structure Estimation

We show how to model the task relationship structure using $\mathbf{\Omega} \in \mathbb{R}^{m \times m}$. Let $\mathbf{s} = (s_1, ..., s_m)$ be an m-variate random vector with joint distribution $\mathcal{P}$. Any such distribution can be characterized by an undirected graph $G = (V, E)$, where the vertex set V represents the m covariates of $\mathbf{s}$ and edge set E represents the conditional dependence relations between the covariates of $\mathbf{s}$. If s_i is conditionally independent of s_j given the other variables, then the edge (i, j) is not in E. Assuming $\mathbf{s} \sim \mathcal{N}(\mathbf{0}, \boldsymbol{\Sigma})$, the missing edges correspond to zeros in the inverse covariance matrix or *precision* matrix given by $\boldsymbol{\Sigma}^{-1} = \boldsymbol{\Omega}$, that is, $(\boldsymbol{\Sigma}^{-1})_{ij} = 0 \ \forall (i, j) \notin E$.

Classical estimation approaches [30] work well when m is small. Given n independent and identically distributed (i.i.d.) samples $\mathbf{s}_1, \ldots, \mathbf{s}_n$ from the distribution, the empirical covariance matrix is $\hat{\boldsymbol{\Sigma}} = \frac{1}{n} \sum_{i=1}^{n} (\mathbf{s}_i - \bar{\mathbf{s}})^\top (\mathbf{s}_i - \bar{\mathbf{s}})$, where $\bar{\mathbf{s}} = \frac{1}{n} \sum_{i=1}^{n} \mathbf{s}_i$. However, when $m \gg n$, $\hat{\boldsymbol{\Sigma}}$ is rank-deficient and its inverse cannot be used to estimate the precision matrix $\boldsymbol{\Omega}$. However, for a sparse graph where most of the entries in the precision matrix are zero, several methods exist that can estimate $\boldsymbol{\Omega}$ [8,31].

2.4.1.3 General MSSL Formulation

Consider the linear model, $\mathbf{y}_k = \mathbf{X}_k \mathbf{w}_k + \mathbf{e}_k$, for each task. MSSL will learn both the task parameters $\mathbf{w}_k$ for all tasks and the structure which will be estimated based on some information from each task. Further, the structure is used as inductive bias in the model parameter learning process so that it may improve tasks generalization capability. So, the learning of one task is biased by other related tasks [32].

We investigate and formalize two ways of learning the relationship structure, represented by $\boldsymbol{\Omega}$: (1) modeling $\boldsymbol{\Omega}$ from the task-specific parameters $\mathbf{w}_k, \forall k = 1, .., m$, and (2) modeling $\boldsymbol{\Omega}$ from the residual errors $\mathbf{e}_k, \forall k = 1, .., m$. Based on how we model $\boldsymbol{\Omega}$, we propose p-MSSL (from tasks parameters) and r-MSSL (from residual error).

At a high level, the estimation problem in r-MSSL is of the form

$$\min_{\mathbf{W}, \boldsymbol{\Omega} > 0} \ \sum_{k=1}^{m} \mathcal{L}\Big((\mathbf{y}_k, \mathbf{X}_k), \mathbf{w}_k\Big) + \mathcal{B}(\mathbf{W}, \boldsymbol{\Omega}) + \mathcal{R}_1(\mathbf{W}) + \mathcal{R}_2(\boldsymbol{\Omega}), \tag{2.9}$$

where $\mathcal{L}(\cdot)$ denotes a suitable task-specific loss function, $\mathcal{B}(\cdot)$ is the inductive bias term, and $\mathcal{R}_1(\cdot), \mathcal{R}_2(\cdot)$ denote suitable sparsity inducing regularization terms. The interaction between parameters $\mathbf{w}_k$ and the relationship matrix $\boldsymbol{\Omega}$ is captured by the $\mathcal{B}(\cdot)$ term. When $\boldsymbol{\Omega}_{k,k'} = 0$, the parameters $\mathbf{w}_k$ and $\mathbf{w}_{k'}$ have no influence on each other.

2.4.1.4 Parameter Precision Structure

If the tasks are unrelated, one can learn the columns $\mathbf{w}_k$ of the parameter matrix $\mathbf{W}$ independently for each of the m tasks. However, if there exist relationships between the m tasks, learning the columns of $\mathbf{W}$ independently fails to take advantage of these dependencies. In such a scenario, we propose to utilize the precision matrix $\boldsymbol{\Omega} \in \mathbb{R}^{m \times m}$ among the tasks in order to capture pairwise partial correlations.

In the p-MSSL model, we assume that each *row* $\hat{\mathbf{w}}_j$ of the matrix $\mathbf{W}$ follows a multivariate Gaussian distribution with zero mean and covariance matrix $\boldsymbol{\Sigma}$, that is, $\hat{\mathbf{w}}_j \sim \mathcal{N}(\mathbf{0}, \boldsymbol{\Sigma}) \ \forall j = 1, \ldots, d$, where $\boldsymbol{\Sigma}^{-1} = \boldsymbol{\Omega}$. The problem then is to estimate both the parameters $\mathbf{w}_1, \ldots, \mathbf{w}_m$ and the precision matrix $\boldsymbol{\Omega}$. By imposing such a prior over the *rows* of $\mathbf{W}$, we are capable of explicitly estimating the dependence structure between the tasks via the precision matrix $\boldsymbol{\Omega}$.

Having a multivariate Gaussian prior over the *rows* of $\mathbf{W}$, its posterior can be written as

$$P(\mathbf{W}|(\mathbf{X}, \mathbf{Y}), \boldsymbol{\Omega}) \propto \prod_{k=1}^{m} \prod_{i=1}^{n_k} P\left(y_k^{(i)} \Big| \mathbf{x}_k^{(i)}, \mathbf{w}_k\right) \prod_{j=1}^{d} P\left(\hat{\mathbf{w}}_j | \boldsymbol{\Omega}\right), \tag{2.10}$$

where the first term in the right-hand side denotes the conditional distribution of the response given the input and parameters, and the second term denotes the prior over *rows* of $\mathbf{W}$. We consider the *penalized maximization* of Equation 2.10, assuming that the parameter matrix $\mathbf{W}$ and the precision matrix $\boldsymbol{\Omega}$ are sparse.

2.4.1.5 Least Squares Regression

With a Gaussian conditional distribution we have

$$
\mathcal{P}\left(y_k^{(i)}\Big|\mathbf{x}_k^{(i)}, \mathbf{w}_k\right) = \mathcal{N}\left(y_k^{(i)}\Big|\mathbf{w}_k^\top \mathbf{x}_k^{(i)}, \sigma^2\right), \tag{2.11}
$$

where we assume $\sigma^2 = 1$ and consider penalized maximization of Equation 2.10. We can write this optimization problem as minimization of the negative logarithm of Equation 2.10, which corresponds to a linear regression problem [33]

$$
\min_{\mathbf{W}, \mathbf{\Omega} > 0} \left\{ \frac{1}{2} \sum_{k=1}^m \|\mathbf{X}_k \mathbf{w}_k - \mathbf{y}_k\|_2^2 - \frac{d}{2} \log |\mathbf{\Omega}| + \frac{1}{2} \mathrm{tr}(\mathbf{W} \mathbf{\Omega} \mathbf{W}^\top) \right\}. \tag{2.12}
$$

In order to avoid the task imbalance problem, which arises when one task has so many data points that it dominates the joint empirical loss, we weigh each task empirical loss inversely to its number of data samples. Further, assuming $\mathbf{\Omega}$ and $\mathbf{W}$ are sparse, we add ℓ_1-norm regularizers over both parameters and obtain a regularized regression problem [33]. Therefore, the new formulation is

$$
\min_{\mathbf{W}, \mathbf{\Omega} > 0} \left\{ \frac{1}{2} \sum_{k=1}^m \frac{1}{n_k} \|\mathbf{X}_k \mathbf{w}_k - \mathbf{y}_k\|_2^2 - \frac{d}{2} \log |\mathbf{\Omega}| + \frac{1}{2} \mathrm{tr}(\mathbf{W} \mathbf{\Omega} \mathbf{W}^\top) + \lambda \|\mathbf{\Omega}\|_1 + \gamma \|\mathbf{W}\|_1 \right\}, \tag{2.13}
$$

where $\lambda, \gamma > 0$ are penalty parameters. The sparseness inducing ℓ_1 regularization in the weight matrix $\mathbf{W}$ is to appropriately select the important features for the learning task, by setting to zero the nonrelevant ones. From the optimization perspective, such penalization adds a nonsmooth term into the problem, which, in principle, makes it more challenging. However, plenty of efficient algorithms for large-scale ℓ_1-regularized convex problems have been proposed in the recent past [34].

In this formulation, the term involving the trace outer product $\mathrm{tr}(\mathbf{W} \mathbf{\Omega} \mathbf{W}^\top)$ affects the *rows* of $\mathbf{W}$, such that if $\mathbf{\Omega}_{ij} \neq 0$, then the columns $\mathbf{w}_i$ and $\mathbf{w}_j$ are constrained to be similar. We also observe that for a fixed $\mathbf{\Omega}$, since we have the inductive bias term, the regularization over $\mathbf{W}$ is more like the elastic-net penalty, since it has both ℓ_1 and ℓ_2 regularization on $\mathbf{W}$. Such elastic-net type penalties have the added advantage of picking up correlated relevant covariates (unlike Lasso), but have the downside that they are hard to interpret statistically. The objective function is nevertheless well defined from an optimization perspective. Also, we note that modeling the relationship as a multivariate Gaussian distribution is not necessarily a restrictive assumption. Recently there has been work in developing the Gaussian copula family of models [35], which addresses a large class of problems commonly encountered. Our framework can be extended to such models.

To solve the optimization problem (Equation 2.13), we propose an iterative optimization algorithm which alternates between updating $\mathbf{W}$ (with fixed $\mathbf{\Omega}$) and $\mathbf{\Omega}$ (with fixed $\mathbf{W}$). This results in alternating between solving two nonsmooth convex optimization problems in each iteration.

Alternating Optimization: The alternating minimization algorithm proceeds as follows:

1. Initialize $\mathbf{\Omega}^0 = I_m$ and $\mathbf{W}^0 = 0_{d \times m}$
2. For $t = 1, 2, \ldots$ **do**

$$
\mathbf{W}^{(t+1)} | \mathbf{\Omega}^{(t)} = \underset{\mathbf{W}}{\mathrm{argmin}} \left\{ \frac{1}{2} \sum_{k=1}^m \frac{1}{n_k} \|\mathbf{X}_k \mathbf{w}_k - \mathbf{y}_k\|_2^2 + \frac{1}{2} \mathrm{tr}(\mathbf{W} \mathbf{\Omega}^{(t)} \mathbf{W}^\top) + \gamma \|\mathbf{W}\|_1 \right\} \tag{2.14}
$$

$$
\mathbf{\Omega}^{(t+1)} | \mathbf{W}^{(t+1)} = \underset{\mathbf{\Omega} > 0}{\mathrm{argmin}} \left\{ \frac{1}{2} \mathrm{tr}(\mathbf{W}^{(t+1)} \mathbf{\Omega} \mathbf{W}^{\top(t+1)}) - \frac{d}{2} \log |\mathbf{\Omega}| + \lambda \|\mathbf{\Omega}\|_1 \right\} \tag{2.15}
$$

Note that this alternating minimization procedure is guaranteed to converge to a minima, since the original problem (Equation 2.13) is convex in each argument Ω and $\mathbf{W}$ [36].

Update for W: The update step (Equation 2.14) is a general case of the formulation proposed by Subbian and Banerjee [3] in the context of climate model combination, where in our work, Ω is any positive definite precision matrix, rather than a fixed Laplacian matrix.

Using vec() notation, we can re-write the optimization problem in (Equation 2.14) as

$$\min_{\mathbf{W}} \left\{ \frac{1}{2} \sum_{i=1}^{\tilde{n}} a_{ik} \left(\bar{\mathbf{x}}_i \, \text{vec}(\mathbf{W}) - \text{vec}(\mathbf{Y})_i \right)^2 + \frac{1}{2} \text{vec}(\mathbf{W})^{\top} \mathbf{P}(\Omega \otimes I_d) \mathbf{P}^{\top} \text{vec}(\mathbf{W}) + \gamma \|\mathbf{W}\|_1 \right\} \qquad (2.16)$$

where $\tilde{n} = \left(\sum_{k=1}^{m} n_k \right)$ is total number of samples from all tasks, a_{ik} is a weight corresponding to $\frac{1}{n_k}$ when the i-th data sample is from the kth task, $\bar{\mathbf{x}}_i$ is the i-th row of the matrix $\bar{\mathbf{X}}$, $\otimes$ is the Kronecker product, I_d is a $d \times d$ identity matrix, and $\mathbf{P}$ is a permutation matrix that converts the column stacked arrangement of $\mathbf{W}$ to a row stacked arrangement. $\bar{\mathbf{X}}$ is a block diagonal matrix where the main diagonal blocks are the task data matrices $\mathbf{X}_k, \forall k = \{1, .., m\}$ and the off-diagonal blocks are zero matrices. Therefore, the problem is a ℓ_1 penalized quadratic optimization program, which we solve using established proximal gradient descent methods such as fast iterative shrinkage thresholding algorithm (FISTA) [37].

We note that in the MTL formulation, the multiple original learning tasks in Equation 2.8 are put together in a larger regularized joint task, Equation 2.16. Therefore, even if the original tasks are in a low-dimensional space, when dealing with hundreds of tasks, the joint task lies in a high-dimensional space. This will be clearer from the experiments.

Note that the optimization of Equation 2.14 can be scaled up considerably by using an alternating method of multipliers (ADMM) [38] algorithm, which decouples the non smooth ℓ_1 term from the smooth convex terms in the objective of Equation 2.14 ADMMs [38] is a strategy that is intended to blend the benefits of dual decomposition and augmented Lagrangian methods for constrained optimization. It takes the form of a *decomposition–coordination* procedure in which the solutions to small local problems are coordinated to find a solution to a large global problem [38].

Update for Ω: The update step for Ω, given in Equation 2.15, is known as *sparse inverse covariance selection problem* [39], which can be efficiently solved using ADMM [38]. We refer interested readers to Section 6.5 of [38] for details on derivation of the updates.

The problem (Equation 2.15) can be simplified as

$$\min_{\mathbf{W}} \quad \text{tr}(\mathbf{S}\Omega) - \log |\Omega| + \bar{\lambda} \|\Omega\|_1 \qquad (2.17)$$

where $\mathbf{S}$ is the sample covariance matrix of the task parameters $\mathbf{W}$, that is, $\mathbf{S} = \frac{1}{d} \mathbf{W}^{\top} \mathbf{W}$ and $\bar{\lambda} = \frac{2}{d} \lambda$. As $\bar{\lambda}$ is a user defined penalty parameter; we embed the scaling factor into λ without any effect on the optimization problem.

We start by forming the augmented Lagrangian function of the problem (Equation 2.17)

$$L_{\rho}(\Omega, \mathbf{Z}, \mathbf{U}) = \text{tr}(\mathbf{S}\Omega) - \log |\Omega| + \lambda \|\mathbf{Z}\|_1 + \frac{\rho}{2} \|\Omega - \mathbf{Z} + \mathbf{U}\|_F^2 - \frac{\rho}{2} \|\mathbf{U}\|_F^2 \qquad (2.18)$$

where $\mathbf{U}$ is the scaled dual variable. Note that the nonsmooth convex function (Equation 2.15) is divided into two functions by adding an auxiliary variable $\mathbf{Z}$, besides a linear constraint $\Omega = \mathbf{Z}$. The ADMM update steps for the problem (Equation 2.17) is presented below:

1. Initialize $\Theta^0 = \Omega^{(t)}$, $\mathbf{Z}^0 = 0_{m \times m}$, $\mathbf{U}^0 = 0_{m \times m}$.
2. For $l = 1, 2, \ldots$ **do**

$$\Theta^{(l+1)} := \underset{\Theta > 0}{\text{argmin}} \left\{ \text{tr}(\mathbf{S}\Theta) - \log |\Theta| + \frac{\rho}{2} \|\Theta - \mathbf{Z}^{(l)} + \mathbf{U}^{(l)}\|_F^2 \right\} \qquad (2.19)$$

$$Z^{(l+1)} := \underset{Z}{\text{argmin}} \left\{ \lambda \|Z\|_1 + \frac{\rho}{2} \|\Theta^{(l+1)} - Z + U^{(l)}\|_F^2 \right\} \tag{2.20}$$

$$U^{(l+1)} := U^{(l)} + \Theta^{(l+1)} - Z^{(l+1)}. \tag{2.21}$$

3. Output $\Omega^{(t+1)} = \Theta^l$, where l is the number of steps for convergence.

Each ADMM step can be solved efficiently. For the Θ-update, we can observe from the first-order optimality condition of Equation 2.19 and the implicit constraint $\Theta \succ 0$ that

$$\rho\Theta - \Theta^{-1} = \rho(Z^{(l)} - U^{(l)}) - S. \tag{2.22}$$

Next, we take the eigenvalue decomposition of $\rho(Z^{(l)} - U^{(l)}) - S = Q\Lambda Q^\top$, where $\Lambda = \text{diag}(\lambda_1, \cdots, \lambda_m)$ and $Q^\top Q = QQ^\top = I$. We now multiply Equation 2.22 by $Q^\top$ on the left and by Q on the right to get $\rho\widetilde{\Theta} - \widetilde{\Theta}^{-1} = \Lambda$, where $\widetilde{\Theta} = Q^\top \Theta Q$.

$$\widetilde{\Theta}_{ii} = \frac{\lambda_i + \sqrt{\lambda_i^2 + 4\rho}}{2\rho} \tag{2.23}$$

Now $\Theta = Q\widetilde{\Theta}Q^\top$ satisfies Equation 2.22.

The Z-update (Equation 2.20) can be computed in closed form

$$Z^{(l+1)} = S_{\lambda/\rho}\left(\Theta^{(l+1)} + U^{(l)}\right), \tag{2.24}$$

where $S_{\lambda/\rho}(\cdot)$ is an element-wise soft-thresholding operator [38].

2.4.1.6 Residual Precision Structure

In r-MSSL, we assume that the rows of the residual error matrix E, $\hat{e}_j \sim \mathcal{N}(0, \Sigma), \forall j = 1, ..., n$, where $\Sigma^{-1} = \Omega$, and Ω is the precision matrix among the tasks. In contrast to p-MSSL, the relationship among tasks is modeled in terms of partial correlations among the errors $e_1, \ldots, e_m$ instead of considering explicit dependencies between the parameters $w_1, \ldots, w_m$ for the different tasks.

Finding the dependency structure among the tasks now amounts to estimating the precision matrix Ω. Such models are commonly used in spatial statistics [40] in order to capture spatial autocorrelation between geographical locations. For example, in domains such as climate or remote sensing, there often exist noise autocorrelations over the spatial domain under consideration. Incorporating this dependency through the precision matrix of the residual errors is then more interpretable than explicitly modeling the dependency among the parameters W.

We assume that the parameter matrix W is fixed, but unknown. Since the rows of E follow a Gaussian distribution, maximizing the likelihood of the data, penalized with a sparse regularizer over Ω, reduces to the following optimization problem:

$$\min_{W, \Omega \succ 0} \left\{ \sum_{k=1}^{m} \frac{1}{n_k} \|y_k - X_k w_k\|_2^2 - \frac{d}{2} \log |\Omega| + \frac{1}{2} \text{tr}(E\Omega E^\top) + \lambda \|\Omega\|_1 + \gamma \|W\|_1 \right\} \tag{2.25}$$

subject to $\quad e_k = X_k w_k - Y_k, \; k = 1, \ldots, m.$

We use the alternating minimization scheme illustrated in previous sections to optimize the above objective. Note that since the objective is convex in each of its arguments W and Ω, it is thus guaranteed to reach a local minima through alternating minimization [36]. Further, the model can be extended to losses other than the squared loss, which we obtain by assuming that the columns of E are i.i.d. Gaussian.

Similar to most earlier work on GCMs combination, both p-MSSL and r-MSSL formulations assume that the GCMs' skill, characterized by the tasks' weights, do not change over time. Ways to handle time-varying GCMs' skill in an MTL setting is an ongoing work.

2.4.2 Experiments on GCMs Combination

In this analysis, we consider the problem of GCM outputs combination for land surface temperature prediction in South America. Being the world's fourth-largest continent, covering approximately 12% of the Earth's land area, the climate of South America varies considerably. The Amazon river basin in the north has the typical hot wet climate suitable for the growth of rain forests. The Andes Mountains, on the other hand, remain cold throughout the year. The desert regions of Chile is the driest part of South America.

We use 10 GCMs from the CMIP5 data set [41]. The details about the origin and location of the data sets are listed in Table 2.2.

The global observation data for surface temperature is obtained from the Climate Research Unit (CRU, http://www.cru.uea.ac.uk). We align the data from the GCMs and CRU observations to have the same spatial and temporal resolution, using publicly available climate data operators (CDO, https://code.zmaw.de/projects/cdo). For all the experiments, we used a 2.5° × 2.5° grid over latitudes and longitudes in South America, and monthly mean temperature data for 100 years, 1901–2000, with records starting from January 16, 1901. In total, we consider 1200 time points (monthly data) and 250 spatial locations over land (Data set is available at http://www.cs.umn.edu/~andre). For the MTL framework, each geographical location forms a task (regression problem). Note that the original regression problems are in a 10-dimensional space. However, as discussed earlier, when solving all the 250 regression problems in an MTL formulation, the resulting joint regression task is in a 2500-dimensional space.

Baselines and Evaluation: We consider the following four baselines for comparison and evaluation of MSSL performance. We will refer to these baselines and MSSL as the "models" in the sequel and the constituent GCMs as "submodels," the four baselines are **(1) Average model** is the current technique used by IPCC (http://www.ipcc.ch), which assigns equal weight to all GCMs at every location; **(2) Best GCM**, which uses the predicted outputs of the best GCM in the training phase (lowest RMSE), this baseline is not a combination of models, but a single GCM instead; **(3) Linear Regression** is an OLS regression for each geographic location; and **(4) Multimodel Regression with Spatial Smoothing (S^2M^2R)** is the model recently proposed by [3], which is a special case of MSSL with pre defined dependence matrix Ω equal to the Laplacian matrix.

TABLE 2.2 Earth System Models Description

Global Climate Model (GCM)	Origin	Ref.
BCC_CSM1.1	Beijing Climate Center, China	[42]
CCSM4	National Center for Atmospheric Res., USA	[43]
CESM1	National Science Foundation, Department of Energy, NCAR, USA	[44]
CSIRO	Commonwealth Scientific and Industrial Research Organisation, Australia	[45]
HadGEM2	Met Office Hadley Centre, UK	[46]
IPSL	Institut Pierre-Simon Laplace, France	[47]
MIROC5	Atmosphere and Ocean Research Inst., Japan	[48]
MPI-ESM	Max Planck Inst. for Meteorology, Germany	[49]
MRI-CGCM3	Meteorological Research Inst., Japan	[50]
NorESM	Norwegian Climate Centre, Norway	[51]

For our experiments, we considered a moving window of 50 years of data for training and the next 10 years for testing. This was done over 100 years of data, resulting in five train/test sets. Therefore, the results are reported as the average RMSE over these test sets.

Table 2.3 reports the average and standard deviation RMSE for all 250 geographical locations. While average model has the highest RMSE, r-MSSL has the smallest RMSE in comparison to the baselines. We note the similarity of OLS and S^2M^2R performances. r-MSSL performance is slightly better than p-MSSL, but it is not statisticallly significant. Next, we show a detailed discussion focusing only on r-MSSL results.

Figure 2.5 shows the precision matrix estimated by r-MSSL algorithm and the Laplacian matrix assumed by S^2M^2R. Not only is the precision matrix for r-MSSL, able to capture the relationship between a geographical locations' immediate neighbors (as in a grid graph), but it also recovers relationships between locations that are not immediate neighbors.

The RMSE per geographical location for average model and r-MSSL is shown in Figure 2.6. As previously mentioned, South America has a diverse climate and not all of the GCMs are designed to take into account and capture this. Hence, averaging the model outputs as done by IPCC, reduces prediction accuracy. On the other hand, r-MSSL performs better because it learns the right weight combination on the model outputs and incorporates spatial smoothing by learning the task relatedness.

Figure 2.7 presents the relatedness structure estimated by r-MSSL among the geographical locations. The regions connected by blue lines are dependent on each other. We immediately observe that locations in the northwest part of South America are densely connected. This area has a typical tropical climate and comprises the Amazon rainforest which is known for having a hot and humid climate throughout the year with low-temperature variation [52].

TABLE 2.3 Mean and Standard Deviation of RMSE over All Locations for r-MSSL and the Baseline Algorithms. r-MSSL Performs Best in Predicting Temperature over South America

Average	Best GCM	OLS	S^2M^2R	p-MSSL	r-MSSL
1.621	1.410	0.866	0.863	**0.783**	**0.780**
(±0.020)	(±0.037)	(±0.037)	(±0.067)	**(±0.040)**	**(±0.039)**

Abbreviations: MSSL, multitask sparse structure learning; S2M2R, Multi Model Regression with Spatial Smoothing.

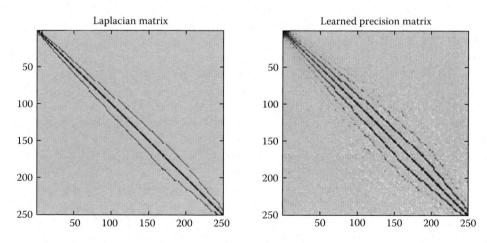

FIGURE 2.5 The Laplacian matrix (on grid graph) assumed by S^2M^2R (left) and the precision matrix learned by r-MSSL (right). r-MSSL can capture spatial relations beyond immediate neighbors. MSSL, multitask sparse structure learning; S^2M^2R, Multi-Model Regression with Spatial Smoothing.

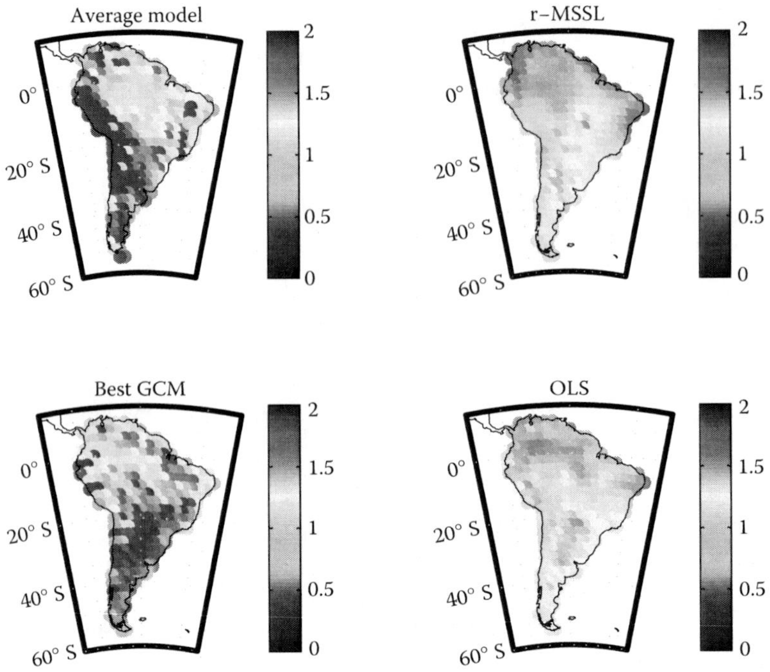

FIGURE 2.6 [Best viewed in color] RMSE per location for each one of the four approaches. Since S^2M^2R produced almost the same RMSE than OLS, it is not shown in the figure. OLS, ordinary least squares; RMSE, root mean square errors; S^2M^2R, Multi-Model Regression with Spatial Smoothing.

South America

FIGURE 2.7 Relationships between geographical locations estimated by r-MSSL algorithm. The blue lines indicate that connected locations are dependent on each other. Also see Figure 2.8 for a chord diagram of the connectivity graph. MSSL, multitask sparse structure learning.

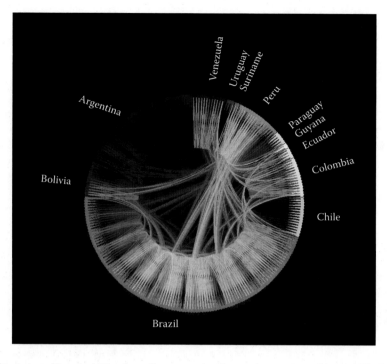

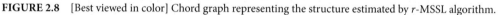

FIGURE 2.8 [Best viewed in color] Chord graph representing the structure estimated by r-MSSL algorithm.

The cold climates which occur in the southernmost parts of Argentina and Chile are clearly highlighted. Such areas have low temperatures throughout the year, but there are large daily variations [52].

An important observation can be made about the South American west coast, ranging from central Chile to Venezuela passing through Peru, which has one of the driest deserts in the world. These areas are located to the left side of the Andes mountains and are known for their arid climate. The average model is not performing well on this region compared with r-MSSL. We can see the long lines connecting these coastal regions, which probably explains the improvement in terms of RMSE reduction achieved by r-MSSL. The algorithm uses information from related locations to enhance its performance on these areas.

The lack of connecting lines in central Argentina can be explained by the temperate climate which presents greater range of temperatures than the tropical climates and may have extreme climatic variations. It is a transition region between the very cold southernmost region and the hot and humid central area of South America. It also comprises Patagonia, a semiarid scrub plateau that covers nearly all of the southern portion of the mainland, whose climate is strongly influenced by the South Pacific air current which increases the region's temperature variability [52]. Due to such high variability, it becomes harder to provide accurate temperature predictions.

Figure 2.8 presents the dependency structure using a chord diagram. Each point on the periphery of the circle is a location in South America and represents the task of learning to predict temperature at that location. The locations are arranged serially on the periphery according to the respective countries. We immediately observe that the locations in Brazil are heavily connected to parts of Peru, Colombia, and parts of Bolivia. These connections are interesting as these parts of South America comprise the Amazon rainforest. We also observe that locations within Chile and Argentina are less densely connected to other parts of South America. A possible explanation could be that while Chile, which includes the Atacama Desert is a dry region located to the west of the Andes, Argentina, especially the southern part experiences heavy snowfall which is different from the hot and humid rain forests or the dry and arid deserts on the west coast. Both these regions experience climatic conditions which are disparate from the northern rain forests and from each other. The task dependencies estimated from the data reflect this disparity.

2.5 Conclusions

Structured regression methods provide powerful tools for high-dimensional data analysis problems encountered in climate science. In this chapter, we considered two such climate problems, both of which can benefit from the recent developments in high-dimensional regression methods. First, we considered the task of predicting climate variables over land regions using climate variables measured over oceans. We illustrated that the SGL can encode sparsity arising from the natural grouping within covariates, using a hierarchical norm regularizer. Experiments prove improved prediction accuracy, and interpretable model selection by SGL compared to ordinary least squares. Second, we considered the problem of GCM model combination, in order to improve predictions. We formulated the problem as an MTL problem and provided a framework for learning relationships within tasks using structure learning methods, while estimating the model skills. Experimental results showed that the proposed MSSL method vastly outperforms traditional methods such as independent regression and even state-of-the-art MTL techniques in terms of RMSE of prediction and learns task dependencies interpretable in terms of the varied topography and vegetations of the region. Thus, structured estimation methods hold enormous promise for applications to statistical modeling tasks in varied climate science problems.

References

1. NCEP/NCAR Reanalysis 1: Surface air temperature (0.995 sigma level). http://www.esrl.noaa.gov/psd/data/gridded/data.ncep.reanalysis.surface.html.
2. IPCC: *Climate Change 2014: Synthesis Report.* Contribution of Working Groups I, II and III to the Fifth Assessment Report of the Intergovernmental Panel on Climate Change [Core Writing Team, R.K. Pachauri and L.A. Meyer (eds.)]. IPCC, Geneva, Switzerland, 151 pp, 2014.
3. K. Subbian and A. Banerjee. Climate multimodel regression using spatial smoothing. In *SIAM International Conference on Data Mining (SDM)*, pp. 324–332. SMD, 2013.
4. P. J. Bickel, J. B. Brown, H. Huang, and Q. Li. An overview of recent developments in genomics and associated statistical methods. *Philosophical Transactions of the Royal Society of London Scripta A Mathematical Physical Engineering Sciences*, 367:4313–4337, 2009.
5. D. Landgrebe. Hyperspectral image data analysis as a high dimensional signal processing problem. *IEEE Signal Processing Magazine*, 19:17–28, 2008.
6. M. Lustig, D. Donoho, J. Santos, and J. Pauly. Compressed sensing MRI. *IEEE Signal Processing Magazine*, 27:72–82, 2008.
7. S. N. Negahban, P. Ravikumar, M. J.Wainwright, and B. Yu. A unified framework for high-dimensional analysis of m-estimators with decomposable regularizers. *Statistical Science*, 27(4):538–557, 2012.
8. N. Meinshausen and P. Buhlmann. High-dimensional graphs and variable selection with the lasso. *Annals of Statistics*, 34(3):1436–1462, 2006.
9. R. Tibshirani. Regression shrinkage and selection via the lasso. *Journal of the Royal Statistical Society, Series B*, 58:267–288, 1996.
10. P. Zhao and B. Yu. On model selection consistency of lasso. *Journal of Machine Learning Research*, 7:2541–2563, 2006.
11. M. Yuan and Y. Lin. Model selection and estimation in regression with grouped variables. *Journal of Royal Statistical Society, Series B*, 68 (1):49–67, 2006.
12. R. Tibshirani, M.Saunders, S. Rosset, J. Zhu, and K. Knight. Sparsity and smoothness via the fused lasso. *Journal of the Royal Statistical Society, Series B (Statistical Methodology)*, 67(1):91–108, 2005.
13. F. Bach. Consistency of the group lasso and multiple kernel learning. *Journal of Machine Learning Research*, 9:1179–1225, 2008.
14. J. Liu and J. Ye. Moreau-Yosida regularization for grouped tree structure learning. In *Advances in Neural Information Processing Systems (NIPS)*, pp. 1459–1467, 2010. NIPS, 2010.

15. L. Jacob, G. Obozinski, and J. Vert. Group lasso with overlap and graph lasso. In *International Conference on Machine Learning (ICML)*, pp. 433–440. ACM, 2009.

16. S. Chatterjee, K. Steinhaeuser, A. Banerjee, S. Chatterjee, and A. R. Ganguly. Sparse group lasso: Consistency and climate applications. In *SIAM International Conference on Data Mining (SDM)*, pp. 47–58. SIAM, 2012.

17. J. Friedman, T. Hastie, and R. Tibshirani. A note on the group lasso and a sparse group lasso. Preprint arXiv:1001:0736v1, 2010.

18. P. Sprechmann, I. Ramirez, G. Sapiro, and Y. C. Eldar. C-HiLasso: A collaborative hierarchical sparse modeling framework. *IEEE Transactions on Signal Processing*, 59(9):4183–4198, 2011.

19. R. Caruana. Multitask learning. *Machine Learning*, 28(1):41–75, 1997.

20. J. Zhang, Z. Ghahramani, and Y. Yang. Learning multiple related tasks using latent independent component analysis. In *Advances in Neural Information Processing Systems (NIPS)*, pp. 1585–1592, 2005.

21. N. D. Lawrence and J. C. Platt. Learning to learn with the informative vector machine. In *International Conference on Machine Learning (ICML)*, 2004.

22. T. Evgeniou and M. Pontil. Regularized multitask learning. In *Knowledge Discovery and Data Mining (KDD)*: Proceedings of the ACM SIGKDD international conference on Knowledge discovery and data mining, pp. 109–117, 2004.

23. J. Zhou, J. Chen, and J. Ye. Clustered multitask learning via alternating structure optimization. In *Advances in Neural Information Processing Systems (NIPS)*, 2011.

24. G. Obozinski, B. Taskar, and M. I. Jordan. Joint covariate selection and joint subspace selection for multiple classification problems. *Statistics and Computing*, 20:231–252, 2010.

25. S. Ji and J. Ye. An accelerated gradient method for trace norm minimization. In *International Conference on Machine Learning (ICML)*, pp. 457–464, 2009.

26. Y. Zhang and D. Yeung. A convex formulation for learning task relationships in multitask learning. In *UAI Proceedings of the Twenty-Sixth Conference on Uncertainty in Artificial Intelligence (UAI)*, pp. 733–742, 2010.

27. A. R. Gonçalves, P. Das, S. Chatterjee, V. Sivakumar, F. J. Von Zuben, and A. Banerjee. Multitask sparse structure learning. In *Conference on Information and Knowledge Management (CIKM)*, 2014.

28. K. Steinhaeuser, N. Chawla, and A. Ganguly. Complex networks as a unified framework for descriptive analysis and predictive modeling in climate science. *Statistical Analysis and Data Mining*, 4(5):497–511, 2011.

29. J. Liu, S. Ji, and J. Ye. *SLEP: Sparse Learning with Efficient Projections*. Arizona State University, 2009. http://www.public.asu.edu/~jye02/Software/SLEP.

30. A. P. Dempster. Covariance selection. *Biometrics*, 157–175, 1972.

31. J. Friedman, T. Hastie, and R. Tibshiran. Sparse inverse covariance estimation with the graphical lasso. *Biostatistics*, 9(3):432–441, 2008.

32. R. Caruana. Multitask learning: A knowledge-based source of inductive bias. In *Proceedings of the 10th International Conference on Machine Learning (ICML)*, pp. 41–48. Morgan Kaufmann, 1993.

33. T. Hastie, R. Tibshirani, and J. Friedman. *The Elements of Statistical Learning; Data mining, Inference and Prediction*. New York, NY: Springer Verlag, 2001.

34. G.X. Yuan, K. W. Chang, C. J. Hsieh, and C. J Lin. A comparison of optimization methods and software for large-scale l1-regularized linear classification. *Journal of Machine Learning Research*, 11:3183–3234, 2010.

35. H. Liu, F. Han, M. Yuan, J. Lafferty, and L. Wasserman. High dimensional semiparametric Gaussian copula graphical models. *The Annals of Statistics*, 40(40):2293–2326, 2012.

36. A. Gunawardana and W. Byrne. Convergence theorems for generalized alternating minimization procedures. *Journal of Machine Learning Research*, 6:2049–2073, 2005.

37. A. Beck and M. Teboulle. A fast iterative shrinkage-thresholding algorithm for linear inverse problems. *SIAM Journal on Imaging Sciences*, 2(1):183–202, 2009.

38. S. Boyd, N. Parikh, E. Chu, B. Peleato, and J. Eckstein. Distributed optimization and statistical learning via the alternating direction method of multipliers. *Foundations and Trends in Machine Learning*, 3(1):1–122, 2011.

39. O. Banerjee, L. El Ghaoui, and A. d'Aspremont. Model selection through sparse maximum likelihood estimation for multivariate Gaussian or binary data. *Journal of Machine Learning Research*, 9:485–516, 2008.

40. K. V. Mardia and R. J. Marshall. Maximum likelihood estimation of models for residual covariance in spatial regression. *Biometrika*, 71(1):135–146, 1984.

41. K. E. Taylor, R. J. Stouffer, and G. A. Meehl. An overview of CMIP5 and the experiment design. *Bulletin of the American Meteorological Society*, 93(4):485, 2012.

42. L. Zhang, T. Wu, X. Xin, M. Dong, and Z. Wang. Projections of annual mean air temperature and precipitation over the globe and in China during the 21st century by the BCC Climate System Model BCC CSM1. 0. *Acta Metallurgica Sinica*, 26(3):362–375, 2012.

43. W. M. Washington et al. The use of the Climate-science Computational End Station (CCES) development and grand challenge team for the next IPCC assessment: An operational plan. *Journal of Physics*, 125(1), 2008.

44. Z. M. Subin, L. N. Murphy, F. Li, C. Bonfils, and W. J. Riley. Boreal lakes moderate seasonal and diurnal temperature variation and perturb atmospheric circulation: Analyses in the Community Earth System Model 1 (CESM1). *Tellus, Series A*, 64, 2012.

45. H. B. Gordon et al. *The CSIRO Mk3 Climate System Model*, volume 130. CSIRO Atmospheric Research, 2002.

46. W.J. Collins et al. Development and evaluation of an Earth-system model–HadGEM2. *Geoscience Model Development Discuss*, 4:997–1062, 2011.

47. J. L. Dufresne et al. Climate change projections using the IPSLCM5 Earth System Model: From CMIP3 to CMIP5. *Climate Dynamics*, 2012.

48. M.Watanabe et al. Improved climate simulation by MIROC5: Mean states, variability, and climate sensitivity. *Journal of Climate*, 23:6312–6335, 2010.

49. V. Brovkin, L. Boysen, T. Raddatz, V. Gayler, A. Loew, and M. Claussen. Evaluation of vegetation cover and land-surface albedo in MPI-ESM CMIP5 simulations. *Journal of Advances in Modeling Earth Systems*, 2013.

50. S. Yukimoto, Y. Adachi, and M. Hosaka. A new global climate model of the meteorological research institute: MRI-CGCM3: Model description and basic performance. *Journal of the Meteorological Society of Japan*, 90:23–64, 2012.

51. M. Bentsen et al. The Norwegian Earth System Model, NorESM1-M-Part 1: Description and basic evaluation. *Geoscience Model Development Discuss*, 5:2843–2931, 2012.

52. V. A. Ramos. South America. In *Encyclopaedia Britannica Online Academic Edition*. 2014.

3

Spatiotemporal Global Climate Model Tracking

3.1 Introduction.. 33
    Related Work

3.2 Technical Approach 34
    Background • Fixed-Share Represented as an MRF • Extension to
    the Spatial Lattice • Learning Parameters • Addressing Challenges in
    Climate Data • Inference in the MRF • Summary of the MRF-based
    Method • NTCM: An Approximate Online Method

3.3 Experiments.. 43
    Simulated Data • Climate Model Data

3.4 Summary 52

Scott McQuade

Claire Monteleoni

3.1 Introduction

We seek to combine the predictions of global climate models in the spatiotemporal setting by incorporating spatial influence into the problem of learning with expert advice. Climate models, in particular general circulation models (GCMs), are mathematical models that simulate processes in the atmosphere and ocean such as cloud formation, rainfall, wind, ocean currents, radiative transfer through the atmosphere and so on and are used to make climate forecasts. GCMs designed by different laboratories will often produce significantly different predictions as a result of the varying assumptions and principles that are used to derive the models. The Intergovernmental Panel on Climate Change (IPCC) is informed by different GCMs from a number of different laboratories and groups. Due to the high variance between the different models, climate scientists are currently interested in methods to combine the predictions of this "multimodel ensemble," as indicated at the IPCC Expert Meeting on Assessing and Combining Multi-Model Climate Projections in 2010.

In the usual setting for online learning with expert advice (see [1]), an algorithm maintains a set of weights over the experts and outputs a combined prediction at each time iteration. Here an expert can be any "black-box predictor," and the learning algorithm does not directly incorporate any domain knowledge of a particular set of experts. After outputting the combined prediction, the true state of the environment is revealed and then the algorithm may update the weighting over experts. This problem of online learning with expert advice has been well-studied, and a number of theoretical bounds on the performance of proposed algorithms have been established (see [1] for a thorough overview). Our contribution extends several of these algorithms to learn with expert advice in the spatiotemporal setting.

We first show that several existing algorithms for online learning with expert advice are equivalent to performing marginal inference on an appropriately defined Markov random field (MRF). We are concerned with the setting where the best expert may change over time and space. The fixed-share algorithm of [2] models the setting where the best expert may vary over time, fixed-share can be represented as a chain MRF (or simply a hidden Markov model [HMM] with respect to time) where there is a separate latent variable representing the identity of the best expert at each time iteration. These latent variables are chained together in the MRF by links that represent the transition dynamics of the best expert over time. We extend this paradigm from the single temporal MRF chain to a lattice over space and time, developing a framework that allows the best expert to "switch" over both space and time.

3.1.1 Related Work

A number of studies have looked at how a multimodel ensemble of climate models can be used to enhance information over and above the information available from just one model. For example, [3,4] showed that the average of the models' output gives a better estimate of the real world than any single model. Additionally, there has been recent work on developing and applying more sophisticated ensemble methods [5–12]. For example, Richard et al. [11], proposed univariate and multivariate Bayesian approaches to combine the predictions over a variety of locations of a multimodel ensemble, in the batch setting. In the case of regional climate models, Sain and Furrer [12] proposed ensemble methods involving Bayesian hierarchical models.

Previous work provided techniques to combine the predictions of the multimodel ensemble, at various geographic scales, when considering each geospatial region as an independent problem [4,13]. However, since climate patterns can vary significantly and concurrently across the globe, this assumption is unrealistic and could therefore limit the performance of these previous approaches. Our work relates most directly to the tracking climate models (TCMs) approach of Claire Monteleoni et al. [13]. To our knowledge, our work is the first extension of the TCM to model spatial influence.

MRF-based methods have been used recently to analyze climate data. In [14], an MRF-based approach was used to spatially and temporally detect drought states throughout the twentieth century. Our proposed method differs from [14] in that (1) we are primarily interested in making an inference about the most recent time slice in the MRF cube (representing best models at the current time) as the MRF grows and evolves over time, and (2) we attempt to learn the optimal parameters for the MRF over time.

3.2 Technical Approach

3.2.1 Background

The fixed-share algorithm of [2] is one of the several "share update" algorithms from the literature on online learning with expert advice [2,15,16] that incorporate switching dynamics to model situations where the best expert may switch over time. Fixed-share uses the following update rule:

$$p_t(i) = \frac{1}{Z_t}\left[(1-\alpha)p_{t-1}(i)e^{-L_{t-1}(i)} + \frac{\alpha}{n-1}\sum_{j\neq i}p_{t-1}(j)e^{-L_{t-1}(j)}\right] \qquad (3.1)$$

Here $p_t(i)$ is the weight of expert i (the probability of being the best predictor) at time t, $L_t(i)$ is the loss of expert i at time t (in this paper we use the squared loss: the squared difference between a prediction and the observation), n is the number of experts, Z_t is a normalization constant to ensure that the weights sum to 1, and α is a parameter that captures how frequently the identity of the best expert switches. The second term in Equation 3.1 determines, via the α parameter, how much weight an expert "shares" with the other experts during an update. In the case where $\alpha = 0$, this update rule reduces to the multiplicative updates of the static-expert algorithm of [2], which models situations where the best expert does not change over time.

In [17], it was shown that a family of algorithms for online learning with experts, including fixed-share, could be derived as Bayesian updates of the appropriately defined generalized* HMM, where the identity of the current best expert is the latent variable. This family of algorithms generalizes the fixed-share algorithm due to [2] to arbitrary transition dynamics among experts.

3.2.2 Fixed-Share Represented as an MRF

Figure 3.1 shows a standard HMM that can be used to represent fixed-share, expressed as an MRF. The black nodes are observed variables or evidence: L_t represents the losses we observed for each GCM at time t. The white nodes are the latent variables: p_t represents the identity of the GCM that is the best predictor at time t. Here we are primarily interested in inferences about the latent variables conditioned on the values of the observed variables, so we are not concerned with any dependencies between the observed variables (this can be thought of as a simplification of the generalized HMM of [17], which does model dependencies between the observed variables).

In an MRF, the joint probability distribution of all variables is expressed as a product of factors, where each factor is a function of a subset of the variables. For pairwise MRFs, such as Figure 3.1, all factors are a function of only two variables, and each edge in the graph represents a factor involving the two variables connected by the edge. For every neighboring (i.e., connected by an edge) pair of variables, x_i and x_j, there is an associated "energy" function $E(x_i, x_j)$ that defines the dependency between the two variables, with lower energy indicating a more likely state assignment for the pair. The energy between an observed variable y_i and a latent variable x_i can be encoded as a function of latent variable, $E(x_i | y_i)$ (for simplicity we will represent this as $E(x_i)$ with an implied conditioning on the observed variables y). The factor associated with each energy function is simply the exponential of the negative energy function, so the joint distribution of all latent variables (conditioned on the observed variables) becomes

$$p(x_1, x_2, \ldots x_{\max}) = \frac{1}{Z} \exp \left\{ - \left(\sum_{(i,j)} E(x_i, x_j) + \sum_i E(x_i) \right) \right\}$$

The indices (i, j) are over the set of all neighboring variables, and Z is a normalization factor (also known as the partition function) [18]. The energy functions in Figure 3.1 can be defined to accommodate arbitrary transition dynamics among the experts; however, in this paper, we focus specifically on the transition

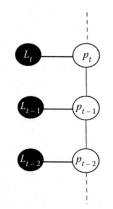

FIGURE 3.1 Fixed-share expressed as an MRF.

---

* This HMM generalization allows arbitrary dependence among the observations.

dynamics that correspond to fixed-share, which are defined using the α parameter. To make the marginal inference of p_t in this MRF equivalent to the fixed-share updates, we define the energy functions as

$$E_L(p_{t,r} = i) = L_{t,r}(i) \tag{3.2}$$

$$E_{\text{time}}(p_{t,r} = i, p_{t-1,r} = j) = \begin{cases} -\log(1 - \alpha_{\text{time}}) & \text{if } i = j \\ -\log(\frac{\alpha_{\text{time}}}{n-1}) & \text{if } i \neq j \end{cases} \tag{3.3}$$

E_L (represented by the black links in Figure 3.1) defines the energy between the loss and best expert at time t (where the loss is the observed variable), and E_{time} (represented by the gray links in Figure 3.1) defines the energy between the best experts at two sequential time iterations (both latent variables). This is equivalent to an HMM transition matrix consisting of $(1 - \alpha)$ diagonal terms and $\frac{\alpha}{n-1}$ off-diagonal terms. Here we also index the variables by the region (or location) r, which will be used in our extension to the spatial lattice. Similarly, we specify α_{time} to indicate that this parameter represents the temporal (as opposed to spatial) switching rate. At time $t = T$, before we have observed L_t, we can calculate the marginal probability of p_T in Figure 3.1 using Equations 3.2 and 3.3 to confirm that it is equivalent to the fixed-share update rule in Equation 3.1:

$$p_T(i) = \frac{1}{Z_T} \sum_{p_1 \cdots p_{T-1}} \exp\left\{-\left(\sum_{t=1}^{T-1} E(p_{t+1}, p_t) + \sum_{t=1}^{T-1} E(p_t)\right)\right\}$$

Let p_{T-1}^* be the marginal inference of p_{T-1}^* at time $t = T - 1$

$$p_{T-1}^*(i) = \frac{1}{Z_{T-1}} \sum_{p_1 \cdots p_{T-2}} \exp\left\{-\left(\sum_{t=1}^{T-2} E(p_{t+1}, p_t) + \sum_{t=1}^{T-2} E(p_t)\right)\right\}$$

Then

$$p_T(i) = \frac{1}{Z_T'} \sum_j \exp\left\{-\left(E(p_T(i), p_{T-1}(j)) + E(p_{T-1}(j))\right)\right\} p_{T-1}^*(j)$$

Here Z_T' normalizes this form of p_T. Applying Equation 3.2

$$p_T(i) = \frac{1}{Z_T'} \sum_j e^{-E(p_T(i), p_{T-1}(j))} e^{-L_{T-1}(j)} p_{T-1}^*(j)$$

Applying Equation 3.3

$$p_T(i) = \frac{1}{Z_T'} \left[(1 - \alpha) e^{-L_{T-1}(i)} p_{T-1}^*(i) + \sum_{j \neq i} \frac{\alpha}{n-1} e^{-L_{T-1}(j)} p_{T-1}^*(j)\right]$$

which is equivalent to Equation 3.1.

3.2.3 Extension to the Spatial Lattice

While the fixed-share MRF in Figure 3.1 models temporal dependencies between the best expert, we also allow the best expert to vary spatially in this work. Since an MRF is an undirected graph, this "switching" energy function can be naturally applied to spatial links between variables as well.

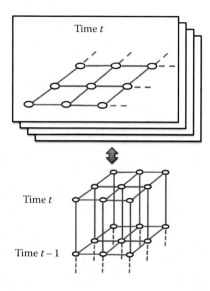

FIGURE 3.2 (See color insert.) The MRF extended over space.

Figure 3.2 illustrates how we construct a spatial lattice MRF by using a fixed-share MRF at each location. The p_t node for each region is linked to each of its four adjacent spatial neighbors (the horizontal links in Figure 3.2), with the energy of each link (E_{space}) taking the same form as the energy of the temporal switching dynamics (E_{time}) but with a different switching rate parameter α_{space}:

$$E_{\text{space}}(p_{t,r_1} = i, p_{t,r_2} = j) = \begin{cases} -\log(1 - \alpha_{\text{space}}) & \text{if } i = j \\ -\log\left(\frac{\alpha_{\text{space}}}{n-1}\right) & \text{if } i \neq j \end{cases} \tag{3.4}$$

where r_1 and r_2 are neighboring regions. Naturally, we would not expect the spatial and temporal switching rates to be the same due to the arbitrary scales for time and space, and so we have separate α_{space} and α_{time} parameters. Note that since we now have loops in the MRF, we can no longer simply interpret these switching rates as the probability of the best expert switching, as we could in the loopless fixed-share model. These links represent factors of the joint distribution and in the case of a loopy MRF the strength of the links will generally not be equivalent strength of the marginal dependence between the two variables (e.g., in our case, if we marginalize out all but two latent variables associated with a link, the resulting marginal dependence will typically be stronger than the link). This does not turn out to be a problem in our framework as we attempt to directly learn the optimal values for these factorized switching rates. In this work, we assume a constant α_{space} value over all locations, however, extensions to allow α_{space} to vary over different locations could be explored in future work.

3.2.4 Learning Parameters

Since we have restricted the energy functions in the MRF to be of the form of Equations 3.2 through 3.4, we have only two parameters that define the MRF: α_{space} and α_{time}. This simplified structure allows us to utilize a variant of the Learn-α algorithm of [17]. Learn-α, depicted in Figure 3.3 [19], is a hierarchical learner that treats each possible value of a parameter (using a discretization in the case of a continuous parameter) as a separate expert running an instantiation of the base algorithm. Learn-α uses an exponentially weighted update scheme to track and combine the predictions of these meta-experts. Since we have two parameters in our MRF, and independence between the two would be too strong of an assumption, we must learn these

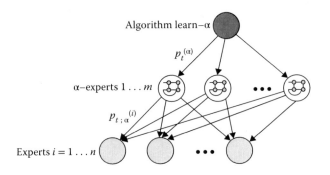

FIGURE 3.3 The Learn-α hierarchy.

parameters jointly—each possible pair of values (given a discretization) is treated as a separate meta-expert by Learn-α.

3.2.5 Addressing Challenges in Climate Data

As described below, our proposed framework addresses two common challenges with climate data—grid distortion and incomplete observation data.

3.2.5.1 Grid Distortion

GCMs produce predictions on a geographic latitude and longitude coordinate system. This type of grid suffers from spherical distortions toward the poles (i.e., the spacing between longitudinal lines decreases towards the poles). Since we assume that our α_{space} switching rate is constant over the Earth, we need to make an adjustment for these compressed distances. Each spatial link can be interpreted as a single-iteration Markov chain as described in Section 3.2.1, with a transition matrix consisting of $(1 - \alpha)$ diagonal terms and $\frac{\alpha}{n-1}$ off-diagonal terms. For transition dynamics defined by α_{space} over distance d_0, we need to find $\alpha^*_{\text{space}}(d)$ that corresponds to appropriately adjusted switching dynamics over some shorter distance d. Given an initial set of weights p_0, it can be shown algebraically that after k iterations of this Markov chain (defining only the switching dynamics and not incorporating the loss updates), the weights will be

$$p_k(i) = \frac{1}{n} + \left(1 - \frac{\alpha n}{n-1}\right)^k \left(p_0(i) - \frac{1}{n}\right) \tag{3.5}$$

The weights converge toward their steady-state uniform distribution at a rate of $(1 - \frac{\alpha n}{n-1})$. Indeed, $(1 - \frac{\alpha n}{n-1})$ is the only nonunitary eigenvalue of this transition matrix.

The compressed transition dynamics (over d) applied d_0 times should be equivalent to our normal transition dynamics (over d_0) applied d times, since in both cases the resulting dynamics should correspond to a distance of $d_0 d$. Evaluating this in the context of Equation 3.5 yields

$$\alpha^*_{\text{space}}(d) = \frac{n-1}{n}\left[1 - \left(1 - \frac{\alpha n}{n-1}\right)^{\frac{d}{d_0}}\right] \tag{3.6}$$

In the case of the latitude and longitude grid over a sphere, at a latitude of ϕ degrees, the distance between the longitudinal lines will be compressed by a factor of

$$\frac{d}{d_0} = \cos\left(\frac{\phi \pi}{180}\right) \tag{3.7}$$

3.2.5.2 Handling Missing Observations

In many cases, observed data are not available over the entire Earth for all time iterations. Our MRF-based framework proposed in this paper can easily handle partially or completely missing observed data. When observed data are not available, there is no observed "loss" node attached to the hidden variable for that location and time because loss is a function of the observation (in other words, $E(x_i)$ can be treated as a constant). This allows for both spatial and temporal influence to continue passing through that hidden node without creating discontinuities in the grid, and an inference can still be made for that node based on the spatial and more distant temporal neighbors.

3.2.6 Inference in the MRF

In the case of the temporal fixed-share MRF with no loops, the exact marginal probabilities of the hidden variable can be quickly calculated using simple algorithms such as belief propagation [18], producing results equivalent to the fixed-share algorithm. However, in our spatiotemporal MRF, the presence of loops makes both maximum a posteriori (MAP) inference and marginal inference much more difficult. There has been much progress in developing efficient approximate MAP solvers in recent years; see [20] for an overview of recent methods. These MAP solvers can be useful in our framework for predicting with the most likely expert at each location, however, in many cases we will want to form a combined weighted prediction using all of the experts. In these situations, we need to estimate the full marginal distributions of the hidden variables, which is a different problem than MAP estimation.

For the majority of our experimental results, we utilize Gibbs sampling to perform inference on the MRF. Gibbs sampling is a Markov chain Monte Carlo (MCMC) technique for generating unbiased samples from the probability distribution [21]. By accumulating Gibbs samples, we obtain an unbiased estimate for the marginal distribution of each latent variable. We use this estimated marginal distribution to form a final prediction for each location and time, either by taking a weighted average of the GCM predictions (using the marginal distribution as the weights) or by taking the prediction of the most likely (max-marginal) GCM.

While Gibbs sampling does produce unbiased estimates, it can be computationally expensive to achieve sufficient convergence, particularly with MRFs with stronger (more deterministic) links between variables. Furthermore, our MRF cube grows at each time iteration, and we must restart the Gibbs sampling at each time iteration on the new larger MRF. These challenges motivate us to develop a fast approximate algorithm that can be run in a true online setting, which we describe in Section 3.2.8.

3.2.7 Summary of the MRF-based Method

The entire MRF-based procedure including the calculation of the losses, the construction of the MRF, inference in the MRF, and hierarchical learning of the α parameters is summarized in Box 3.1. The MRF-based procedure takes a set of q geospatial regions indexed by r as input, along with weights w_r. These user-specified weights are used to calculate a weighted global prediction and could capture information such as the number of data points in each region or the area of each region. There are n global climate models that are indexed by i; here the ith model's prediction at time t, $M_{i,r}(t)$ is additionally indexed by r, the geospatial region. Similarly, j and k index the hierarchical meta-experts (in the context of the Learn-α algorithm), each MRF inference algorithm running with different values of the parameters α_{time} and α_{space}, computed via the discretization procedure in [17]. The weight of each meta-expert indexed by (j, k) is represented by pAlpha$[t, j, k, r]$ at region r and time t. The weight of each GCM i is represented by $p_{t,j,k,r}(i)$ at time t and for α_{time} and α_{space} values indexed by j and k, respectively.

Box 3.1 MRF-Based Tracking of Climate Models

Input:

 Set of geographic subregions, indexed by $r \in \{1, \cdots, q\}$ that collectively
  span the globe, with weights w_r.

 Set of climate models, $M_{i,r}, i \in \{1, \cdots, n\}$ that output predictions $M_{i,r}(t)$ at
  each time t and each region r.

 Set of $\alpha_{\text{time},j} \in [0,1], j \in \{1, \cdots, m_{\text{time}}\}$ //Discretization of α_{time} parameter.

 Set of $\alpha_{\text{space},k} \in [0,1], k \in \{1, \cdots, m_{\text{space}}\}$ //Discretization of α_{space} parameter.

Initialization:

 $\forall j, k, \ \text{MRF}_{j,k} \leftarrow \emptyset$ //Initialize the MRF to an empty graph.

 $\forall j, k, r \ \text{pAlpha}_{t=1,r}(j,k) \leftarrow \frac{1}{m_{\text{time}}m_{\text{space}}}$ //Initial weight of each meta-expert.

Upon tth observation:

 For each $r \in \{1 \dots q\}, i \in \{1 \dots n\}$:

 $L_{t,r}(i) \leftarrow (y_{t,r} - M_{i,r}(t))^2$

 For each $j \in \{1 \dots m_{\text{time}}\}, k \in \{1 \dots m_{\text{space}}\}$:

 Append a new time layer to $\text{MRF}_{j,k}$ using Equations 3.2 through 3.4

 $\forall t, r, \ p_{t,j,k,r} \leftarrow \text{Marginals}(\text{MRF}_{j,k})$ //Inference to assign weights to each GCM.

 For each $r \in \{1 \dots q\}$:

 $\text{LossPerAlpha}[j,k,r] \leftarrow -\log \sum_{i=1}^n p_{t,j,k,r}(i) e^{-L_{t,r}(i)}$

 $\text{pAlpha}_{t+1,r}(j,k) \leftarrow \text{pAlpha}_{t,r}(j,k) e^{-\text{LossPerAlpha}[j,k,r]}$

 //Weighted or Max Prediction Combination:

 $\text{PredictionPerAlpha}[j,k,r] \leftarrow \sum_{i=1}^n p_{t+1,j,k,r}(i) M_{i,r}(t+1)$

 Normalize pAlpha over j, k.

 //Weighted or Max Prediction Combination:

 $\forall r, \text{Prediction}[r] \leftarrow \sum_{j,k} \text{pAlpha}_{t+1,r}(j,k) \text{PredictionPerAlpha}[j,k,r]$

 $\text{GlobalPrediction} \leftarrow \sum_{r=1}^q w_r \text{Prediction}[r]$

3.2.8 NTCM: An Approximate Online Method

We now address some of the computational issues in the MRF-based framework and propose an alternative online algorithm, Neighborhood-augmented Tracking Climate Models (NTCM), which we introduced in [19], to approximate the spatial influence of our MRF in a more efficient manner. NTCM can be viewed as a spatial extension to the nonspatial TCMs algorithm of [13]. TCM maintains a set of weights over climate models in a single region using a Learn-α hierarchy (see Section 3.2.4) of fixed-share algorithms (see Section 3.2.1). The TCM algorithm is equivalent to our MRF-based framework run on only one region (i.e., no spatial switching dynamics). In this situation, the MRF becomes a chain without loops and exact inference is computationally simple. With NTCM, we attempt to preserve the computational efficiency of TCM, while incorporating some notion of spatial influence between neighboring regions. We accomplish this in NTCM by running the TCM algorithm at multiple locations and modifying the transition dynamics over time (corresponding to a nonhomogeneous HMM) to favor transitions to experts that are performing well in neighboring regions.

$$P(i \mid k; \alpha, \beta, r, t) = \begin{cases} (1 - \alpha) & \text{if i=k} \\ \frac{1}{Z}\left[(1 - \beta) + \beta\frac{1}{|S(r)|}\sum_{s \in S(r)} P_{\text{expert}}(i, t, s)\right] & \text{if i}\neq\text{k} \end{cases} \quad (3.8)$$

$$\text{where } Z = \frac{1}{\alpha} \sum_{\substack{i \in \{1 \cdots n\} \\ \text{s.t. } i \neq k}} \left[(1 - \beta) + \beta\frac{1}{|S(r)|}\sum_{s \in S(r)} P_{\text{expert}}(i, t, s)\right]$$

Box 3.2 shows our NTCM algorithm, using similar notation to Box 3.1. Note that in the context of NTCM there is only one α parameter (governing the temporal switching rate), so we simply refer to this

---

Box 3.2 Neighborhood-Augmented Tracking Climate Models

Input:
  Set of geographic subregions, indexed by $r \in \{1, \cdots, q\}$ that collectively
  span the globe, with weights w_r.
  Set of climate models, $M_{i,r}, i \in \{1, \cdots, n\}$ that output predictions $M_{i,r}(t)$ at
  each time t and each region r.
  Set of $\alpha_j \in [0, 1], j \in \{1, \cdots, m\}$ //Discretization of α parameter.
  β, a parameter for regulating the magnitude of the spatial influence
Initialization:
  $\forall j, r, \ \text{pAlpha}_{t=1,r}(j) \leftarrow \frac{1}{m}$ //Initial weight of each meta-expert.
  $\forall i, j, r, \ p_{t=1,j,r}(i) \leftarrow \frac{1}{n}$ //Initial weight of each expert.
Upon tth observation:
For each $r \in \{1 \ldots q\}$:
Set $P(i|k; \alpha_j, \beta, r, t)$ using Equation 3.8, $\forall i, k \in \{1, \ldots, n\}, j \in \{1, \ldots, m\}$.
For each $i \in \{1 \ldots n\}$:
  $L_{t,r}(i) \leftarrow (y_{t,r} - M_{i,r}(t))^2$
For each $j \in \{1 \ldots m\}$:
  $\text{LossPerAlpha}[j] \leftarrow -\log \sum_{i=1}^{n} p_{t,j,r}(i) \, e^{-L_{t,r}(i)}$
  $\text{pAlpha}_{t+1,r}(j) \leftarrow \text{pAlpha}_{t,r}(j) e^{-\text{LossPerAlpha}[j]}$
  For each $i \in \{1 \ldots n\}$:
    $p_{t+1,j,r}(i) \leftarrow \sum_{k=1}^{n} p_{t,j,r}(k) \, e^{-L_{t,r}(i)} \, P(i|k; \alpha_j, \beta, r, t)$
  Normalize $p_{t+1,j,r}$
  //Weighted or Max Prediction Combination:
  $\text{PredictionPerAlpha}[j, r] \leftarrow \sum_{i=1}^{n} p_{t+1,j,r}(i) \, M_{i,r}(t + 1)$
Normalize $\text{pAlpha}_{t+1,r}$
//Weighted or Max Prediction Combination:
Prediction$[r] \leftarrow \sum_{j=1}^{m} \text{pAlpha}_{t+1,r}(j) \, \text{PredictionPerAlpha}[j, r]$
GlobalPrediction $\leftarrow \sum_{r=1}^{q} w_r \, \text{Prediction}[r]$

parameter as α, whereas in context of the MRF framework we have two separate α parameters and usually specify α_{space} or α_{time}. Figure 3.4 illustrates the general concept of NTCM—separate instance of Learn-α running in different regions with spatial influence between neighbors.

For each region r, the nonhomogeneous transition matrix among experts is defined by $P(i \mid k; \alpha, \beta, r, t)$ in Equation 3.8. $S(r)$ is the set of all geographical regions that are spatial neighbors for the region r. This set is determined by the neighborhood scheme, which could be defined using a variety of shapes and sizes. In this work, we define the neighborhood as the four adjacent regions (as described in more detail in Section 3.3), however, one could also consider other options such as more circularly defined neighborhoods or a scheme that is complex enough to model teleconnections. β is a parameter regulating the magnitude of the spatial influence, $P_{expert}(i, t, s)$ (conditioned over all α values) is the current probability of expert (climate model) i, as determined by the modified Learn-α algorithm for spatial neighbor s, and Z is a normalization factor that ensures that each row of the transition matrix sums to 1 (i.e., the off-diagonal terms of each row sum to α).

The β parameter regulates the influence of the spatial neighbors. β value of 0 models no spatial influence and corresponds to the normal Learn-α algorithm (with the switching probability α being shared equally by all experts). Note, however, that our master algorithm still differs from Learn-α in that we run multiple instances of Learn-α, one per geospatial region, r. A β value of 1 would distribute the switching probability α based solely on the expert probabilities of the spatial neighbors. An interesting direction of future work is to extend the algorithm to simultaneously learn β, analogously to the learning of α.

3.2.8.1 Time Complexity of NTCM

The total running time of the global Learn-α algorithm with m values of α, a set of n models, over t time iterations, is bounded by $O(tmn^2)$, excluding all data preprocessing and anomaly calculations. Subdividing the globe into q regions adds a factor of q to this bound. The addition of the neighborhood augmentation

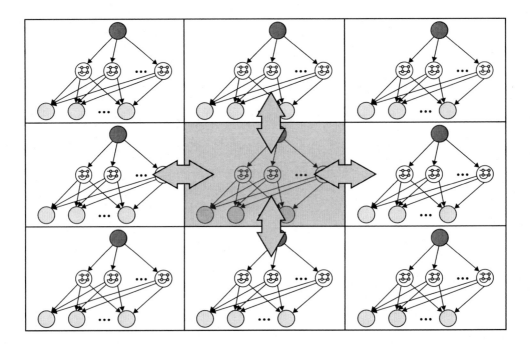

FIGURE 3.4 (See color insert.)   The NTCM algorithm with Learn-α running on each region and spatial influence between the regions.

requires $O(sn^2)$ steps each time an HMM transition matrix is updated, where s is the number of neighbors. Since the matrix would need to be updated for each time iteration, for each geographic region and for each α-value, the overall complexity would be increased by a factor of qs to $O(qstmn^2)$. Additionally, if b separate β values are used, the complexity would further increase by a factor of b. The total running time of NTCM will be bounded by $O(bqstmn^2)$, linear in all factors except the number of models.

3.3 Experiments

In this section, we describe empirical results comparing our MRF-based method and the more efficient NTCM algorithm to several existing methods. We primarily compare our results to the Learn-α algorithm [17] that was used for TCMs in [13]. Note that the TCM approach is equivalent to our NTCM approach without any spatial influence (i.e., NTCM with $\beta = 0$). Throughout our results, we refer to our proposed MRF framework as the *MRF-Based Approach* and NTCM (typically with $\beta = 1$) as *NTCM*. For comparisons with existing methods, we evaluate the performance relative to Learn-α run independently on multiple regions as *Regional Learn-α* and Learn-α run at the global level as *Global Learn-α*.

All of the methods we examine maintain a set of weights over the models in the ensemble (these weights can be interpreted as the probability of a model being the best predictor in the context of our graphical models). Given these weights, several different prediction strategies could be employed. Two examples of simple prediction strategies are the weighted mean (using the weights to combine all model predictions) and most probable expert (using the prediction of only the highest weighted expert). Since our primary objective was to identify the best model at a given time and location, we use the most probable expert strategy in most experiments to produce our results. In the context of our MRF, the most probable expert strategy is equivalent to marginal MAP estimation for all the latent variables individually, where we estimate the state of each latent variable (i.e., the identity of the best climate model) by calculating the mode of its marginal distribution. In our simulated data experiments, we use the well-studied graph cuts method (see [22]) to produce an exact MAP estimate for all latent variables in our MRF (in general, the MAP estimate is not the same as the marginal MAP estimate that corresponds to the most probable expert strategy).

We primarily report two types of loss in our results—the global loss and the mean regional loss. We use the squared loss for all algorithms and results. In this context, the global loss is the squared difference between the most probable model prediction (averaged over all valid regions) and the observed value (also averaged over all valid regions). The mean regional loss is the regional loss averaged over all regions, where the regional loss is the squared difference between the most probable model prediction and the observed value in a given region.

We are interested in the performance of our algorithms over large time periods, rather than at individual time iterations. As we are limited to a single hindcast data set for the climate model predictions, we report bootstrapped results to establish the significance between the performance of different algorithms over the entire time period. We use the block bootstrap method of [23], with a block size of 10, to help preserve the time dependencies in the sequences of losses.

In many of our results, we report the cumulative loss, which is the cumulative sum of losses up to a certain point in the hindcast (we typically plot this quantity as it increases over the entire hindcast period). Our motivation here is to analyze how performance changes over the hindcast, rather than to just summarize performance over the whole hindcast. Not that the cumulative loss at the end of the hindcast period is simply the mean loss over the hindcast multiplied by a constant factor (the number of time iterations).

Note that for these results we do not employ the correction for grid distortion or missing data described in Section 3.2.5. While the MRF-based approach is entirely capable of handling these issues, we omit the corrections for a fair and direct comparison with NTCM and the Learn-α methods, which do not have similar corrections to account for these issues. As such, we discard any region where observed data were not available at any point during the hindcast run.

In Section 3.3.1, we report results of experiments using simulated data in a simplified setting. The object of these simulated data experiments is to verify that our MRF-based approach performs well in its intended setting and that the NTCM provides a reasonable approximation. In Section 3.3.2, we report results of experiments involving surface air temperature predictions of historical climate models ensembles (hindcasts), using records of observed data to calculate the losses.

3.3.1 Simulated Data

We conducted experiments with simulated data sampled from an MRF (the generating MRF) that was structured similarly to the MRF used in our proposed algorithm. The primary goal of these experiments was to demonstrate that our proposed framework performs well when the MRF model accurately describes the data generation process. The generating MRF replicated the latent nodes and their interconnections (but not the observed nodes) from our proposed MRF model. We then ran a MCMC simulation on the generating MRF, resulting in an unbiased sample of values for all the nodes. These sampled values represent simulated identities of the best GCM at each location and time. The MCMC simulation was conducted with the following parameters: 36×72 grid of spatial locations (simulating a 5 degree latitude/longitude grid), 100 time iterations, two GCMs, $\alpha_{space} = \alpha_{time} = 0.25$, MCMC burn-in period of 1000 iterations.

After obtaining a sample of the latent nodes, the observed nodes (representing the losses for each GCM at each location and time) were simulated by sampling from a normal distribution as follows: for the "best" GCM (corresponding to the sampled value of the latent node), the loss was generated by normal (0,1) distribution truncated at 0 so that the loss would be positive; for the other GCM (not the best), the loss was generated by normal (1,1) distribution again truncated at 0.

One motivation behind this simplified setting that only includes two GCMs is that we were able to use an exact and efficient MAP estimation algorithm based on minimum graph cuts [22] to estimate the most likely configuration of the latent variables in our MRF. We integrated the UGM package of [24] to perform this MAP estimation. We only used this method for the simulated data experiment as complications arise when there are more than two GCMs; for the rest of the experiments we used the Gibbs sampling method described in Section 3.2.6 for inference in the MRF.

Figure 3.5 and Table 3.1 show the cumulative global loss results for several algorithms on the simulated data set. Figure 3.6 and Table 3.2 show the cumulative mean regional loss results for the same experiment. For our proposed MRF-based approach, we evaluated a version that was provided with the optimal α_{space}

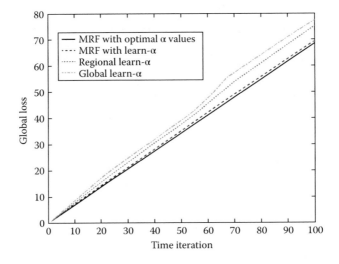

FIGURE 3.5 Cumulative global loss on simulated data.

TABLE 3.1 Mean and Cumulative Annual Global Loss on Simulated Data

	Mean Annual Loss (Bootstrapped)	Standard Deviation (Bootstrapped)	Cumulative Annual Loss
MRF w/ Optimal α	0.6832	0.0029	68.4127
MRF Learning α	0.6916	0.0039	69.3467
Regional Learn-α	0.7519	0.0121	75.2155
Global Learn-α	0.7769	0.0323	77.6634

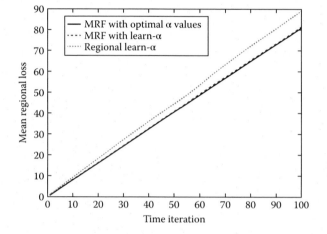

FIGURE 3.6 Cumulative mean regional loss on simulated data.

TABLE 3.2 Mean and Cumulative Annual Regional Loss on Simulated Data

	Mean Annual Loss (Bootstrapped)	Standard Deviation (Bootstrapped)	Cumulative Annual Loss
MRF w/ Optimal α	0.8081	0.0012	80.8060
MRF Learning α	0.8122	0.0012	81.2171
Regional Learn-α	0.8864	0.0015	88.6387

and α_{time} parameters (i.e., the parameter values used to simulate the data) as well the regular version that learns these parameters hierarchically from the data. As shown in both the global and regional losses, the parameter-learning version performs almost as well (within 2%) as the version with knowledge of the parameters used in the simulation. These MRF-based methods were compared to the global and regional variants of Learn-α. In both cases, the MRF-based methods showed significantly smaller loss than the Learn-α methods.

Note that for this simulated data set with only two GCMs, we did not explicitly evaluate the proposed NTCM approach since NTCM is equivalent the the regional Learn-α method when there are only two possible "best models."

3.3.2 Climate Model Data

We ran experiments with our algorithm on historical data, comparing temperature observations and GCM hindcasts (predictions of the GCMs using historical scenarios). Since the GCMs are based on first principles and not data-driven, it is valid to run them predictively on past data. The hindcast GCM predictions were obtained from the IPCC Phase 3 Coupled Model Intercomparison Project (CMIP3) archive [25]. Data from the climate of the 20th century experiment (20C3M) were used. Multiple institutions have contributed a number of different models and runs to the 20C3M archive. In this experiment, a single model from each institution, and a single run of that model, was arbitrarily selected, as this is standard practice according to [13]. Our ensemble size was 13.

We obtained historical temperature anomaly observations from across the globe and during the relevant time period from the NASA GISTEMP archive [26]. Temperature anomalies are explained in Section 3.3.2.2.

3.3.2.1 Preprocessing

The different climate model data sets had been created with different parameters, including different initialization dates and spatial resolutions. We performed several preprocessing tasks on the data sets to improve uniformity and consistency. All data sets, including both the models and the observed anomaly data, were temporally truncated to the period occurring between the beginning of 1890 and the end of 1999. The data were then averaged over entire years, forming a consistent time series of 110 points across all data sets. The data set were also spatially resampled to a geographic grid of 5 degree squares across the globe, creating a consistent spatial grid of dimensions 36x72. The data were spatially resampled using an unweighted mean of all original data points that fell within each 5 degree square. Note that these "degree square" were defined using the WGS 84 coordinate system, and as such they are neither perfect squares (in terms of their footprint on the Earth) nor of a consistent area. However, the same grid of "squares" was used for all data sets.

The observed anomaly data set contains a number of "missing" data points, where the observed temperature anomaly was not available for a variety of reasons. As described in Section 3.2.5.2, our proposed MRF framework provides a natural method to handle missing data. However, in order to directly compare results with other methods that cannot handle missing data gracefully, any 5 degree geographic square that had missing data (no observed data points within the 5 degree square) for at least 1 year in the time series was excluded from the experiment. The majority of the excluded cells are close to the Polar regions, where limited historical temperature observations are available, particularly for the early part of the data set.

3.3.2.2 Temperature Anomalies

Climate scientists often work with *temperature anomalies* as opposed to raw temperatures. A temperature anomaly is the change in temperature at a particular location from the (average) temperature at that same location during a particular benchmark period. Temperature anomalies are used by climate scientists because they tend to have lower variance when averaged across multiple locations than raw temperatures (as discussed further in [13]). Figure 3.7 shows the observed global temperature anomalies from NASA GISTEMP, as well as the anomalies for input data (GCM predictions), over the 1890–2000 period. Both the observed data and the input data had been spatially averaged over all of the valid 5 degree cells (the included regions).

The observed data from NASA GISTEMP had already been converted to temperature anomalies. The benchmark data used to create these anomalies were not provided, however, the benchmark period was noted as 1951–1980. The models provided data in absolute temperatures. To convert the absolute data to anomalies, separate benchmarks were calculated for each model by averaging the model data over the 1951–1980 period. Another motivating factor for using anomalies based on benchmarks from the respective data set was that the GCMs had been initialized at different times with different parameters. This caused a sizable and consistent offset between different models, around 4°C for some pairs of models. Using anomaly values based on a consistent benchmark period mitigates this constant offset.

In this work, we standardize all predicted and observed temperature anomalies by dividing by an estimate for the standard deviation of a temperature anomaly (this same estimated standard deviation is used to scale all anomalies at all locations). The purpose of this standardization is to make our learning

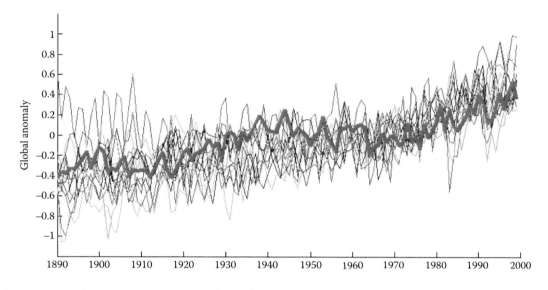

FIGURE 3.7 The observed global temperature anomaly (heavy gray), and the predicted anomalies from the 13 individual climate models.

algorithms invariant to the scale of the observations and predictions (e.g., Celsius versus Fahrenheit). This is particularly important with exponentially weighted algorithms such as NTCM, where the scale of the data is directly related to the effective learning rate.

3.3.2.3 Spatial Influence Parameters

We conducted several experiments to investigate the effect of the spatial influence parameters on the performance of the algorithms. The MRF framework and NTCM have a similar, but not equivalent, parameter that controls the amount of spatial influence between neighboring regions. In the MRF framework, the α_{space} parameter defines the link between neighboring locations in the graph. Smaller values of α_{space} represent stronger links, and thus more spatial influence. In NTCM, the β parameter defines how much a region's "switching probability" is influenced by well-performing GCMs in neighboring regions (as opposed to poorly performing GCMs). In other words, larger β values increase the spatial influence of well-performing GCMs.

3.3.2.4 Effect of α_{space} in the MRF Framework

Figure 3.8 shows our results for varying α_{space} over the hindcast. We set $\alpha_{time} = 0.05$ and ran experiments with different values for α_{space}. The right-most point on the graph corresponds to $\alpha_{space} = \frac{n-1}{n}$ (n is the GCM ensemble size) for an algorithm variant with no spatial influence, equivalent to regional Learn-α. Spatial influence increases with subsequent smaller α_{space} values. When $\alpha_{space} = 0$ the spatial influence is at a maximum, and all the hidden variables must have the same state (this case is equivalent to tracking a single set of experts over all locations without modeling any spatial variation, the same as global Learn-α). Figure 3.8 indicates that the optimal value of α_{space} is between these two extremes, as the performance initially improves with increasing spatial influence (decreasing α_{space}), but eventually diminishes with α_{space} values that are too small.

3.3.2.5 Effect of β in NTCM

Figure 3.9 shows the mean annual losses over a range of β values for several different region sizes, as well as the mean losses over all the regions. While there is not a monotonic decrease in losses with increasing β values for all of the individual region sizes, the mean loss (over all region sizes) does show a clear trend of

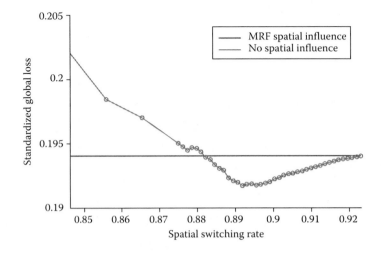

FIGURE 3.8 The performance (mean annual squared loss for global predictions) of our method versus α_{space} with $\alpha_{\text{time}} = 0.05$. The gray line represents our proposed approach. The black line is equivalent regional Learn-α with no spatial influence.

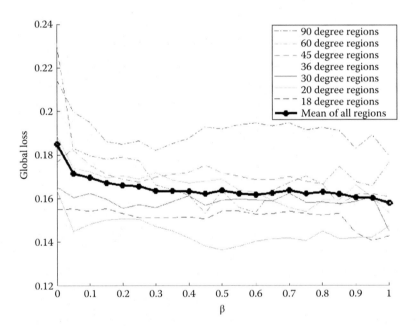

FIGURE 3.9 The performance (global loss) across different β values for seven different region sizes. The bold black line represents the average performance over all these regions, demonstrating decreased average losses as β increases to 1.

decreasing loss with increasing β values, indicating that increased influence (in the context of the β parameter) from the spatial neighbors improves the performance. Note that this is not completely analogous to the MRF situation, as the total "switching rate" in NTCM is governed by its α parameter, where β determines the fraction of that switching rate that is directed toward well-performing GCMs in neighboring cells.

3.3.2.6 Performance on the Hindcast

We compared the performance of the MRF-based approach and NTCM with the global and regional versions of Learn-α on the hindcast data set. We used 36 degree cells for all region-based methods. Figure 3.10 and Table 3.3 display the global losses while Figure 3.11 and Table 3.4 display the mean regional losses.

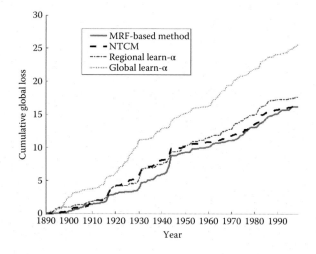

FIGURE 3.10   Cumulative global loss of the hindcast.

TABLE 3.3   Mean and Cumulative Annual Global Losses

	Mean Annual Loss (Bootstrapped)	Standard Deviation (Bootstrapped)	Cumulative Annual Loss
MRF-Based Method	0.1559	0.0244	16.2102
NTCM	0.1583	0.0225	16.5762
Regional Learn-α	0.1680	0.0215	17.6812
Global Learn-α	0.2415	0.0263	25.6658

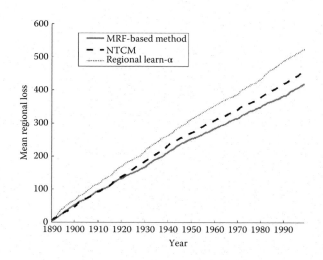

FIGURE 3.11   Cumulative mean regional loss of the hindcast.

Over the entire hindcast, the MRF-based method performed slightly better than NTCM, and both performed better than the Learn-α variants. At certain intermediate points in the hindcast, NTCM actually had a lower cumulative loss than the MRF-based method. These results support our conjecture that, in practice, NTCM is a reasonable approximation for the MRF-based method, and both of these methods that incorporate spatial influence outperform simpler methods that do not model for spatial variation and influence.

3.3.2.7 Regional Analysis

We now examine the performance of our proposed methods at the regional level, using the same losses we generated in the experiments described in Section 3.3.2.6. These experiments with 36 degree square regions split the surface of the Earth into 50 cells. Figure 3.12 shows the method that had the smallest regional loss (averaged over time) for each cell, and Table 3.5 shows the number of regions where each approach produced the smallest loss. Here NTCM actually produced the smallest loss in slightly more regions than the MRF approach, which contrasts with the total mean of all regional losses in Table 3.4 where the MRF approach produced the smallest loss. This is not entirely surprising as both measures are fairly close for NTCM and the MRF approach, and the number of regions with the smallest loss does not take into account the magnitude of the difference in performance in each region.

The regions where each of our proposed approaches produce the best results appear to be somewhat clustered in Figure 3.12. This could provide an interesting direction for future research, as the apparent clustering may indicate that each approach performs well in different areas of the Earth. So in addition to the identity of the best climate model changing over space, the best approach for TCMs may also change over space (and possibly over time as well). This observation may also motivate a location-specific α_{space} parameter (instead of the assumption that α_{space} does not change over space, as there may be different spatial transition dynamics in different areas).

TABLE 3.4 Mean and Cumulative Annual Regional Losses, Summed over All Regions

	Mean Annual Loss (Bootstrapped)	Standard Deviation (Bootstrapped)	Cumulative Annual Loss
MRF-Based Method	3.8092	0.2127	418.9585
NTCM	4.1289	0.2584	455.2941
Regional Learn-α	4.7508	0.3166	523.5269

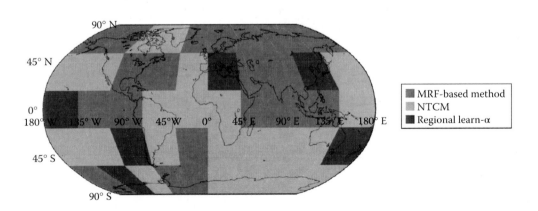

FIGURE 3.12 The method with the smallest cumulative loss in each of the 50 regions.

TABLE 3.5 Number of Regions Where Each Approach Produced the Smallest Loss

	Number of Regions
MRF-Based Method	21
NTCM	23
Regional Learn-α	6

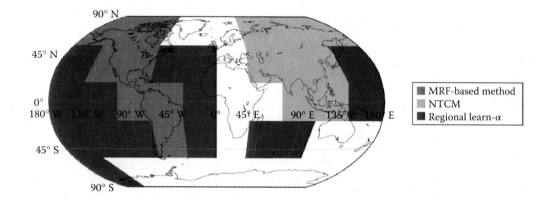

FIGURE 3.13 The three super-regions, each consisting of a set of the 36 degree regions.

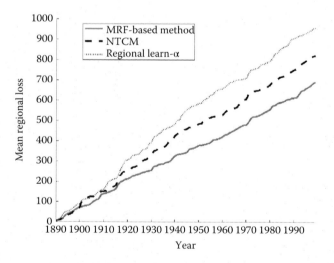

FIGURE 3.14 Mean regional losses for each method within the "Americas" region.

Continuing to investigate this idea that the best approach may depend on the geographic location, we analyzed our results within "super-regions," which are groups of individual region cells. We formed three super-regions, based on partitions loosely the Americas, Asia, and Oceans. The extents of these super-regions are displayed in Figure 3.13.

We show the cumulative regional loss (averaged over each super-region) in Figures 3.14 through 3.16 and Table 3.6. We again see that both our proposed approaches, NTCM and the MRF-based approach,

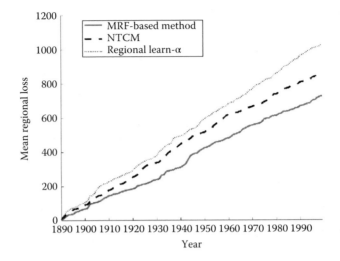

FIGURE 3.15   Mean regional losses for each method within the "Asia" region.

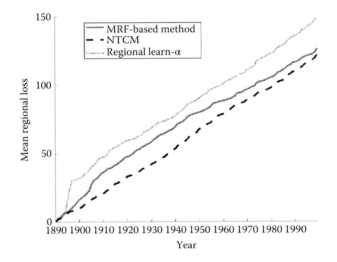

FIGURE 3.16   Mean regional losses for each method within the "Ocean" region.

outperform the simpler method of tracking the best climate model in each region independently (regional Learn-α). Also of note, NTCM actually outperforms our MRF approach by a small margin in the Ocean super-region. While the difference in this super-region is not as significant as the other super-regions, it does indicate that the strength of each approach may differ based on geographic area.

3.4 Summary

We proposed two methods for incorporating spatial influence into the multimodel ensemble problem, extending temporal learning methods into the spatial dimension. Our MRF-based framework provides a spatiotemporal probabilistic model for the best GCMs and can handle practical issues such as distorted geospatial grids and missing observation data points. Our online algorithm NTCM provides an approximation to the MRF-based framework and is more computationally efficient with large data sets. We evaluated

TABLE 3.6 Mean Regional Losses for Each of the Super-Regions

	Mean Annual Loss (Bootstrapped)	Standard Deviation (Bootstrapped)	Cumulative Annual Loss
Americas			
MRF-Based Method	6.3062	0.6863	688.8846
NTCM	7.5265	0.9636	819.9221
Regional Learn-α	8.6580	1.0253	954.9650
Asia			
MRF-Based Method	6.6316	0.6634	726.1517
NTCM	7.7689	0.7894	861.7909
Regional Learn-α	9.3655	1.0091	1026.7740
Ocean			
MRF-Based Method	1.1457	0.0854	126.3621
NTCM	1.1014	0.0862	121.7379
Regional Learn-α	1.3436	0.2042	148.1649

our algorithms on both simulated data and hindcasts of climate models. In both cases, our proposed algorithms produced predictions with smaller losses than several other existing algorithms that do not model spatial influence. We also observed that the performance of our proposed approaches varied in different geographic areas. For the simulated data, our MRF-based framework, which hierarchically learns (from the data) the parameters regulating the amount of spatial and temporal dependence, produced losses that were less than 2% larger than the losses produced when the algorithm was provided with the optimal parameter values.

There are several open questions related to our framework that could be explored in future work. We restricted our model to two parameters, one each for regulating spatial and temporal influence, in order to simplify the parameter learning process. A richer framework could accommodate links of different strengths at different locations (or at different time iterations); however, there are computational challenges with learning a large number of parameters jointly. Another area for exploration is the structure of the MRF; our MRF (lattice-based in this work) might benefit from additional links for certain applications, particularly climate patterns that exhibit teleconnections over large distances.

Acknowledgments

We thank the anonymous reviewers of AAAI 2012, the Second International Workshop on Climate Informatics, and the Third International Workshop on Climate Informatics for their comments used in revision. We acknowledge the modeling groups, the Program for Climate Model Diagnosis and Intercomparison (PCMDI) and the WCRP's Working Group on Coupled Modelling (WGCM) for their roles in making available the WCRP CMIP3 multimodel data set. Support of this data set is provided by the Office of Science, U.S. Department of Energy.

References

1. N. Cesa-Bianchi and G. Lugosi. *Prediction, Learning, and Games*, Vol. 1. Cambridge: Cambridge University Press, 2006.
2. M. Herbster and M.K. Warmuth. Tracking the best expert. *Mach Learn*, 32:151–178, 1998.
3. T. Reichler and J. Kim. How well do coupled models simulate today's climate? *Bull. Amer. Meteor. Soc.*, 89:303–311, 2008.

4. C. Reifen and R. Toumi. Climate projections: Past performance no guarantee of future skill? *Geophys. Res. Lett.*, 36, 2009.

5. A.E. Raftery, T. Gneiting, F. Balabdaoui, and M. Polakowski. Using Bayesian model averaging to calibrate forecast ensembles. *Mon. Wea. Rev.*, 133:1155–1174, 2005.

6. A.M. Greene, L. Goddard, and U. Lall. Probabilistic multimodel regional temperature change projections. *J. Climate*, 19(17):4326–4343, 2006.

7. T. DelSole. A Bayesian framework for multimodel regression. *J. Climate*, 20:2810–2826, 2007.

8. M.K. Tippett and A.G. Barnston. Skill of multimodel ENSO probability forecasts. *Mon. Wea. Rev.*, 136:3933–3946, 2008.

9. M. Pena and H. van den Dool. Consolidation of multimodel forecasts by ridge regression: Application to pacific sea surface temperature. *J. Climate*, 21:6521–6538, 2008.

10. S. Casanova and B. Ahrens. On the weighting of multimodel ensembles in seasonal and short-range weather forecasting. *Mon. Wea. Rev.*, 137:3811–3822, 2009.

11. R.L. Smith, C. Tebaldi, D. Nychka, and L.O. Mearns. Bayesian modeling of uncertainty in ensembles of climate models. *J. Am. Stat. Assoc.*, 104(485):97–116, 2009.

12. S. Sain and R. Furrer. Combining climate model output via model correlations. *Stochastic Environmental Research and Risk Assessment*, 2010.

13. C. Monteleoni, G. Schmidt, S. Saroha, and E. Asplund. Tracking climate models. *Stat. Anal. Data Min.*, 4(4):372–392, 2011.

14. Q. Fu, A. Banerjee, S. Liess, and P.K. Snyder. Drought detection of the last century: An MRF-based approach. In *SDM*, pages 24–34. SIAM, 2012.

15. N. Littlestone and M.K. Warmuth. The weighted majority algorithm. In *Proc. IEEE Symposium on Foundations of Computer Science*, pages 256–261, 1989.

16. Y. Freund and D. Ron. Learning to model sequences generated by switching distributions. In *COLT: Proceedings of the Workshop on Computational Learning Theory, Morgan Kaufmann Publishers*, 1995.

17. C. Monteleoni and T. Jaakkola. Online learning of nonstationary sequences. In *NIPS '03: Advances in Neural Information Processing Systems*, pages16, 2003.

18. J.S Yedidia, W.T Freeman, and Y. Weiss. Understanding belief propagation and its generalizations. In *Exploring Artificial Intelligence in the New Millennium*, edited by Lakemeyer G and Nebel B, Vol. 8, pp. 236–239, San Francisco, CA: Morgan Kaufmann Publishers, 2003.

19. S. McQuade and C. Monteleoni. Global climate model tracking using geospatial neighborhoods. *In Proceedings of the Twenty-Sixth AAAI Conference on Artifcial Intelligence,* July 22-26, 2012, Toronto, Ontario, Canada, pages 335–341, 2012.

20. R. Szeliski, R. Zabih, D. Scharstein, O. Veksler, V. Kolmogorov, A. Agarwala, M. Tappen, and C. Rother. A comparative study of energy minimization methods for Markov random fields. In *Computer Vision–ECCV 2006*, pages 16–29. Springer, 2006.

21. A.E Gelfand and A.F.M Smith. Sampling-based approaches to calculating marginal densities. *J Am Stat Assoc.*, 85(410):398–409, 1990.

22. Y. Boykov and V. Kolmogorov. An experimental comparison of min-cut/max-flow algorithms for energy minimization in vision. *IEEE Trans Pattern Anal Mach Intell.*, 26(9):1124–1137, 2004.

23. H.R Kunsch. The jackknife and the bootstrap for general stationary observations. *Ann Stat.*, 17:1217–1241, 1989.

24. M. Schmidt. UGM: Matlab code for undirected graphical models. http://www.cs.ubc.ca/~schmidtm/Software/UGM.html.

25. CMIP3. The World Climate Research Programme's (WCRP's) Coupled Model Intercomparison Project Phase 3 (CMIP3) multi-model dataset. http://www-pcmdi.llnl.gov/ipcc/about_ipcc.php, 2007.

26. NASA GISS. GISTEMP. http://data.giss.nasa.gov/gistemp/.

4

Statistical Downscaling in Climate with State-of-the-Art Scalable Machine Learning

	4.1	Introduction 55
		Categories of Downscaling Techniques • Uncertainty within Data
	4.2	Current State 57
	4.3	Recent Developments in SD 59
		SD Model Validation • Data Science Developments
Thomas Vandal	4.4	Bridging the Gaps 63
		Physics-Guided Data Mining • Advancements in Spatiotemporal Data
Udit Bhatia		Mining • Deep Learning for SD • Case Study: Deep Belief Networks
Auroop R. Ganguly	4.5	Conclusion 69

Thomas Vandal

Udit Bhatia

Auroop R. Ganguly

4.1 Introduction

The adaptation of infrastructure to events including hydrological and weather extremes, which may be exacerbated by climate change, largely depends on high-resolution climate projections decades and centuries into the future. Consider the damage produced by Hurricane Katrina to the city of New Orleans in the year 2005, causing an estimated 1570 deaths and $40–$50 billion in losses [1]. Climate models used to project future climate are unable to provide the necessary resolution needed to plan accordingly for such events. A wide variety of models have been suggested to increase the resolution of coarse projections, but due to the poor results and lack of consistency, stakeholders are unable to rely on the estimates. Hewitson et al. exemplified how the choice of assumptions and implementation can drastically change the projections, both in terms of magnitude and direction of change [2]. To improve our projections and give confidence to stakeholders to make important decisions, we must advance our understanding of the coarse resolution data, provide more resilient and general models, and communicate the results effectively.

Climate projections that inform stakeholders are derived from climate models that are too computationally expensive to provide outputs at a local scale but rather at scales around 100 km^2. Downscaling is used to give local climate projections by leveraging information content from lower resolution climate models (see Figure 4.1 for a graphical representation). The implicit complexity of the environment challenges the ability for downscaling models to be generalized from one region to another. Measurements of many highly complex climate and weather variables attempt to represent the environment. Each is represented at multiple elevation levels and spatial points around the globe. The combination of variables, spatial, and temporal dimensions creates high-dimensional problems. General circulation models (GCMs) are used to project

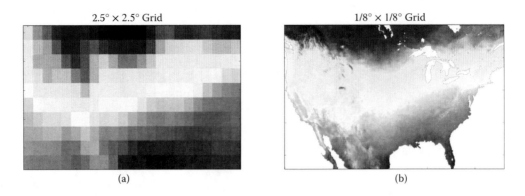

FIGURE 4.1   Two separate data sets of mean monthly temperature in January 1950. (a) A coarse data set at 2.5° by 2.5°. (b) A much higher resolution gridded data set at 1/8° by 1/8°.

future climate scenarios many years into the future by simulating the climate variables along with certain assumptions regarding greenhouse gases and initial conditions. Though GCMs are considered reliable for projecting average global and regional climates, they are coarse and fail to project high-resolution local climate. These local climate projections have the potential to inform stakeholders about the monsoons in South Asia and many other important policy decisions for years into the future.

4.1.1 Categories of Downscaling Techniques

Given that downscaling is needed to provide these local projections, two main techniques are used, dynamical downscaling and SD. Dynamical downscaling, based on regional climate models (RCMs), has been used extensively for increasing the resolution of GCMs by accounting for regional forces. This technique of downscaling allows the inclusion of sub grid parameters, such as convective schemes, soil processes, and vegetation processes, to capture the high-resolution physical processes. However, certain crucial issues concerning the use of dynamical downscaling limit its applicability. One key issue is whether this technique is capable of adding more information at different spatial scales compared to the GCMs, which impose lateral boundary conditions to the RCMs [3]. Also, sub grid-scale parameterization is not independent of the nonstationarity assumption. As the parameters and physical processes are tuned to specific regions, the models are unable to generalize from one to another. While the use of RCMs plays a pivotal role in hypothesis testing, it is much more resource intensive in comparison to the SD. The increased computational expense limits researchers' ability to produce multiple downscaled climate projections with different initial conditions and parameters. Without this ability, one is unable to provide a comprehensive evaluation of future climate scenarios. The decrease in computational cost and increased generalization make SD models preferred by many.

The SD approach uses historical data to learn a transfer function that can project local climate using GCMs. These models tend to be much more general than RCMs, which are tuned to specific regions, and are less computationally expensive. The goal is to learn a (transfer) function f with some set of coarse resolution predictors X to project a high spatial or temporal resolution variable y. Learning such a function f is challenging with the changing climate and high dimensionality. The data and method must be chosen with the knowledge of multiple assumptions: (1) X should adequately represent the variability found in y, (2) the statistical attributes to X and y must be valid outside of the time period used for fitting the model (nonstationary), and (3) X must incorporate the future climate change signal [4]. Since the climate is known to be nonstationary due to climate change, this assumption challenges many SD and machine learning models available.

4.1.2 Uncertainty within Data

As noted previously, projections of future climate are modeled using GCMs, which are then downscaled using a transfer function. The Program for Climate Model Diagnosis and Intercomparison (PCDMI) currently manages the Coupled Model Intercomparison Project Phase 5 (CMIP5) by joining a suite of highly dynamic and complex models produced by research labs all over the world. CMIP5 models attempt to simulate the climate system by encapsulating a comprehensive set of physical processes governing the system. Constraints from unknown physical processes governing the climate increase uncertainty of these models. Each CMIP5 model relies on initial conditions, boundary conditions, and parameters. The initial conditions represent the state of the climate at time 0, while the boundary conditions apply physical constraints on the model. The parameters in CMIP5 models are set to consider specific greenhouse gas and other effects that govern the climate system. Just small changes in a model's initial conditions and parameters drastically effect projections adding discrepancies between models. SD models must acknowledge such uncertainty and be applied appropriately.

To quantify uncertainty and provide valid comparisons of SD models, it is customary to use reanalysis data sets, assimilated from observed data, for modeling. Observed data sets capture the true dynamics of the climate system but contain a significant amount of uncertainty [5]. This uncertainty stems from measurement errors including failing remote sensors, sampling errors, bias errors, and others. Advanced reanalysis data sets incorporate quality control of these uncertainties by accounting for anomalies and biases, allowing researchers to be confident of their data. These data sets have high dimensionality, sometimes reaching tens of thousands of attributes, and at most 60 years of temporal observations, causing extreme overfitting. The NCEP Reanalysis I and II data sets are commonly used as predictors for SD, containing more than 50 years of global climate data [6]. Predictands also come from observational data, stemming from a mixture of stations and reanalysis data sets. When station data are used, users must be aware of possible anomalies and biases. Precipitation and temperature effect each and every one of us, making them the most common and important variables to downscale. For some applications, modeling mean precipitation and temperature provides knowledge of long-term climate effects. Other applications, like flood modeling, require the projection of extremes at finer resolutions for regional impact assessment and mitigation.

To decrease the number of dimensions in our data sets, we may consider the inclusion of climate indices. Climate indices are high-level measures of the overall climate, accounting for known physical dependencies and global trends. The El Nino Southern Oscillation (ENSO) precipitation index is one example that measures the El Niño and La Niña patterns by examining spatial averages of precipitation in the Pacific Ocean [7]. The use of such a climate index may allow for one to remove many features located far from the target variable, condensing data sets significantly.

The first step of creating SD models lies in the choice of predictors. Climate variables tend to be highly dependent both spatially and temporally, reducing information content in the data set. So even though the data set is very high dimensional, many of those variables are highly correlated, leading to multicollinearity in models. To choose the predictors, covariate selection techniques are applied, which will then be used as inputs to our model, reducing dimensionality and overfitting in the model. In theory this is a great idea but in practice it is very challenging. As Caldwell showed, obtaining statistically significant covariates through data mining alone is not sufficient and can return false dependencies [8]. This notion motivates physics-guided data mining, where covariates are chosen from a set of known physical dependencies in the environment.

4.2 Current State

SD methods are often placed into one of three possible groups including weather types, regression, and weather generators. Weather types estimate high-resolution predictions by using historical data to find the most similar scenarios that match the test data. For example, if one is predicting whether or not it will rain on a given day with certain features, a weather type would find the most nearest neighbor to the features given and return a similar precipitation value found in that nearest neighbor. The method of analogues has shown reasonable results [9] but, in most cases, fails to account for the assumption of nonstationarity by assuming similar observations in the future. Regression methods are the most common, using historical data to estimate a set of transfer functions that transform a set of features to a prediction. Multiple linear regression, neural networks, and support vector machines are widely used methods, all having the potential for further improvement—these methods will be discussed in further detail in the rest of this chapter. Weather generators, which differ significantly from weather types and regressions methods, use historical data to make predictions on future weather by using time series techniques. Stochastic weather generators have shown success in this area in the attempt to generate higher temporal resolution projections [10–12].

The downscaling of climate averages versus those focused on extreme events require their own considerations. The selection and validation of models should be carefully considered when building SD models. By definition, the number of average events greatly outweighs the number of extreme events. For example, say we want to downscale monthly mean precipitation, we are given 12 observations per year. But on the other hand, downscaling the maximum monthly mean precipitation over a year gives us just one observation per year. While the number of observations differs, so does the uncertainty in the data. GCMs are just models given a set of initial conditions; their ability to estimate extreme events, say the 95th percentile precipitation, will have a much larger confidence interval than the mean or median precipitation. The limited number of observations and uncertainty enhances the difficulties of downscaling extreme events.

Much research has been conducted in downscaling both extremes and averages. In 2010, Maraun reviewed several methods, covering applications in both mean and extreme precipitation, including linear models, additive models, method of analogues, nonlinear methods, and artificial neural networks (ANNs) [13]. Linear models are the most widely used due to their simple formulation and our theoretical understanding of variability. The IPCC 2007 Fourth Assessment report explains that linear models may be preferred over more complex nonlinear methods. Still, nonlinear methods are popular because they have the ability to gain information from the nonlinear nature of climate variables. ANNs have been shown to predict interseasonal variability but tend to underestimate heavy rainfall, also noted by the IPCC's Fourth Assessment report.

Burger et al. compared five popular methods–automated regression-based SD, bias correction spatial disaggregation (BCSD), quantile regression neural networks, TreeGen, and expanded downscaling–for downscaling extremes [14]. These methods were tested and compared on the same data sets with a broad range of predictands including both temperature and precipitation indexes. Concentrating on extremes both shrinks the sample size and reduces our confidence in GCMs. After a thorough testing procedure, expanded downscaling and BCSD provided the best performance compared to the others. The testing applied concentrated on three important keys: (1) check for sensitivity of anomalies by correlating observed and downscaled values; (2) check if the distribution of the observed matches the downscaled values; (3) lastly, a simulation of present distribution of an index from a GCM. Each of these tests concentrates on important aspects of SD, which should be carefully analyzed during validation.

BCSD has shown remarkable results in multiple studies and comparisons [15]. This method uses an important concept of quantile mapping to adjust biased data with the ability to extrapolate on data outside of the calibration range. The quantile mapping then allows for future data to be estimated in a similar distribution as the calibration set. Such quantile mapping should be considered during pre- and postprocessing for predictands as well as predictors. Similar approaches are taken in both expanded downscaling and quantile regression neural networks. This evidence suggests that future SD models should incorporate such quantile mapping to preserve the distribution but should still allow for nonstationary climate variables. This may lead to quantile mappings that change relative to future projections.

Though relatively promising methods have been proposed and a reasonable amount of data exist for data scientists to work with, there are still gaps within the observed and modeled data. As Maraun points out, data in remote locations are sparse, and sometimes errors are inherited by broken sensor, limiting validation results [13]. Along with observed data sets, GCMs are challenged with their own difficulties. The lack of understanding of the temporal variability and uncertainties between models provide valid concerns for downscaling. For example, the CMIP5 GCMs are known to have disagreement between both models and initial conditions, caused by a combination of unknown physical processes and different greenhouse gas scenarios [10]. The uncertainties in observations, climate models, and statistical models will continue to be a challenge for SD. In order to be aware of how these uncertainties effect SD, it is vital to test statistical models on a large set of climate models while differing the parameters and initial conditions. The development of SD models must stay aware of such restrictions but not be persuaded away from such research.

4.3 Recent Developments in SD

The previous sections have introduced some of the past applications and traditional methods of SD as well as challenges in observed and modeled data sets. In this section, we present recent work in climate science and data mining that have not been fully integrated with SD but may provide valuable insights in prediction and covariate selection. These developments focus on the assumptions, covariate selection, and transfer functions used for downscaling, each with possibilities of further improvements and research.

4.3.1 SD Model Validation

As mentioned previously, the assumption of nonstationarity in SD models is often overlooked and therefore compromised. Recently, Salvi et al. introduced a framework that attempts validation in a nonstationarity climate [16]. This work focused on a method to split training and test data sets by separating data with contrasting climate conditions, rather than the traditional method of chronological separation where 30 years may be in the training set and the following 10 years in the test set. For example, one may consider splitting data by hot versus cold years or ENSO versus non-ENSO years. This will provide more credible validation in the test set with the assumption that the model performs well outside of the calibration data. As radiative forcing is expected to alter climate conditions significantly in the future, it is important for researchers to consider such frameworks for validations of SD models.

Though these climate conditions are expected to change, it is unlikely that the basic relationships among atmospheric variables, governed by conservation law of physics, will be altered, allowing for assumptions regarding climate signatures, which have been shown to perform effectively [16]. Climate signatures may be viewed in both spatial and temporal dimensions to provide a higher level view of climate conditions. Such signatures can be applied using the method of analogues or other weather typing methods. Advanced approaches to quantify signatures in space and time may prove to satisfy the assumption of nonstationarity while providing improvements in predictions.

4.3.2 Data Science Developments

In parallel of developing frameworks allowing for credible validation, the statistics and machine learning communities are working on specific methods for covariate selection and prediction on various climate applications. The concentration of models utilizing advancements dimensionality reduction, sparse regression, covariate selection, and Bayesian statistics has been shown to provide advances in SD. The process of SD, as illustrated in Figure 4.2, relies on data science at each step.

Dimensionality reduction is the first step to solving important SD problems and can come in many forms. Principal component analysis (PCA), which is often used in many data science disciplines, has been shown to effectively reduce dimensionality in climate data sets. In 2010, Ghosh presented a framework which uses a PCA and support vector regression to downscale average precipitation on a monthly scale [17]. The idea here is that thousands of covariates can be described in just a few principal components, which is then followed by a nonlinear transfer function. This technique provides a relatively efficient algorithm with good results but has some downfalls that may limit future applicability. Firstly, though the principal components reduce the dimensionality, they fail to account for nonlinear dependencies. Also,

Statistical downscaling process for regression

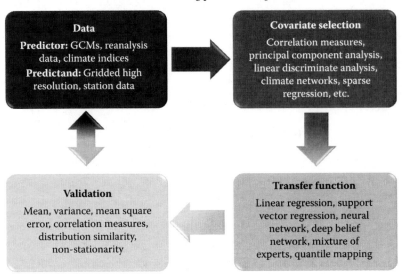

FIGURE 4.2 (See color insert.) The process of statistical downscaling for regression has four steps. (1) Data collection and splitting data into training and test sets; (2) covariate selection via various methods (no covariate selection is possible); (3) train a statistical model; (4) using the trained model and hold-out data validate using multiple statistical measures. Validation can be done on both historical and projected data.

interpretability of the components is lost during this transformation; in many cases, it would be advantageous to choose a subset of covariates which can be interpreted rather than transforming the entire data set. Lastly, as long as the feature set does not change, the principal components will stay the same for different target variables.

4.3.2.1 Sparse Regression

Recently, climate modelers have adapted sparse regression models to SD. Sparse regression models keep the interpretability of covariates and allow for the ability to update covariates depending on the target variable. This category of models is based on ordinary least squares but includes a regularization parameter to limit the number of covariates. For example, ordinary least squares estimates a parameter vector θ by solving

$$\hat{\theta}_{\mathrm{OLS}} = \operatorname*{argmin}_{\theta \in \mathbb{R}^p} \left\{ \frac{1}{2n} \|y - X\theta\|_2^2 \right\}.$$

Such a formulation is unidentifiable to solve if $n < p$ and will force all features to have nonzero coefficients. Traditional statistics techniques to solve this problem use some sort of best subset selection by comparing different sets of covariates and choosing those which return the best performing model. This process is computationally expensive, especially as the number of covariates increases, so different methods must be adapted. With the addition of a regularization term to ordinary least squares, we can guarantee a subset of coefficients to be zero, removing them as covariates, by solving the formulation

$$\hat{\theta} = \operatorname*{argmin}_{\theta \in \mathbb{R}^p} \left\{ \frac{1}{2n} \|y - X\theta\|_2^2 + \lambda r(\theta) \right\}$$

where r is a sparse regularizer.

Chatterjee et al. have applied the sparse group lasso regularizer to downscaling precipitation, demonstrating good predictive performance, robust variable selection, and interpretability of chosen covariates [18]. This work presented the model applicability in nine geographical regions at over 400 locations. Validating SD models over a variety of geographical and climatological regions around the world is valuable to the credibility of the approach and is shown to be good practice. The ability to find nonlinear covariates as well as capturing nonlinearity in the prediction may be beneficial, but could force overfitting.

4.3.2.2 Bayesian Methods

Recently, Bayesian methods in climate have been preferred by some for their ability to generalize over applications and locations as well as providing levels of uncertainty in predictions. Bayesian developments in both covariate selection and prediction are now allowing for nonlinear covariate selection. Das et al. proposed a Bayesian framework adapted from LASSO by obtaining a posterior distributions over the regression coefficients and applying it to covariate selection for extreme precipitation [19]. Lasso itself is a linear sparse regression using the L1 regularization term $\|\beta\|_1$. Applying the Poisson prior on precipitation extremes and using Guassian prior on the model's error, we can obtain a joint distribution of the frequencies and parameters. An application of a Laplace prior on β allows the implementation of sparsity. The posterior of this particular Bayesian model must be approximated using variational inference, making it very computationally expensive but is shown to outperform non-Bayesian lasso significantly while still providing sparse representation of covariates. Though this model is not directly applied to downscaling, the ability to identify nonlinear dependencies could prove to be valuable to covariate selection.

Along with covariate selection, Bayesian approaches are being applied to the predictive step of SD. An approach using Bayesian modeling averaging was recently adapted by Zhang et al. as an alternative to ordinary least squares [20]. This particular model concentrates on estimating the posterior distribution, $P(\text{Monthly Precipitation}|\text{covariates})$, over monthly mean precipitation. The idea is to lead a set of weights, w_k, for each covariate, x_k, such that

$$p(y|x_1...x_K) = \sum_{k=1}^{K} w_k p_k(y|x_k).$$

Using the set of weights found, we are able to compute the expected value of monthly precipitation as well as the uncertainty around that prediction. This particular approach performed better than ordinary least squares, just as many other models we have presented and should be tested against more advanced, sparse, linear models. Nevertheless, the framework for using Bayesian averaging is intriguing and should be researched further.

4.3.2.3 Kriging

Until now we have discussed SD approaches which have a set of covariates that can predict some observation with the hope to apply it to GCMs in the future. Such observations are assumed to be available and plentiful in space. Often times, remote sensing is used to capture these observations at stations around the world. These stations are irregularly spaced unlike the data from climate models and reanalysis data sets. The ability to transform these irregularly spaced observations to gridded data sets can be advantageous to multiple climate applications. The most popular approach is referred to as fixed kriging that uses an interpolation method to increase resolution between points. Fixed kriging focuses on computing a fixed nonstationary covariance function, computed from observed stations, to interpolate to arbitrary points in space [21].

Nychka et al. recently improved fixed kriging by introducing a Gaussian Random Markov Field to the model, while increasing the efficiency of computation over fixed kriging [22]. By accounting for the lattice structure, the model can be iterated on by increasing the resolution by a factor of two, to create multiresolution predictions, resulting in smooth and gridded data sets that may be used for further analysis. Such a method provides a strong base for the use of random processes to estimate the variability caused by downscaling.

4.3.2.4 Analogue Techniques

Previously, we had introduced multiple methods using various transfer functions for SD, but ongoing research has also continued to push the state of the art on weather typing methods. Traditionally, the nearest neighbors in the method of analogues were found by computing a distance measure between an observation and all other possibilities, treating each feature uniformly relevant. Following the theme of Bayesian hierarchical models for SD, Manor and Berkovic have developed a Bayesian inference model to aid the method of analogues [23]. Here, Manor proposed that computing the probability that two observations are nearest neighbors finds the best pair of analogues. This particular work improves two main aspects over the use of distance measures: (1) the reflection of nonlinearity between the distance and the probability is the optimal solution and (2) each feature's relevance is weighted by its statistical significance. These are significant advantages—the incorporation of nonlinearity and feature relevance has been shown to be a common theme in many of the state-of-the-art methods.

As mentioned earlier, the development of SD for climate extremes is a challenging task due a mixture of rarity and uncertainty. Weather typing has seldom been applied to the prediction of extremes, but rather daily, monthly, or seasonal means. Castellano and DeGaetono recently proposed a framework using a combination of classification and prediction [24]. To remove the bias created by most days having minimal precipitation, Castellano started by classification of extreme precipitation on a given day followed by estimation of the amount of rainfall, given extreme precipitation. In this case, all days without extreme precipitation are then removed from the training set. Though the method is simpler than the previous approaches examined, the framework for removing bias by excluding data in the training set allows for more precise predictions without the need for postprocessing bias correction.

The methods introduced provide an overview of the research currently underway in the field of SD. Each method addresses different challenges with SD, including the feature selection of many covariates, difficulties of predicting extremes, overfitting, and the quantification of uncertainty into predictions. In Table 4.1, we provide a summary of these methods along with their strengths and weaknesses. Broadly, the role of Bayesian methods is becoming more pronounced in each aspect of the development. Also, the advancements sparse regression models and nonlinear dependence measures continue to improve feature selection, leading to a reduction in overfitting.

TABLE 4.1 A Brief Summarization of the Strengths and Weakness of a Subset of Well-Known Statistical Downscaling Methods for Regression

Method	Strengths	Limitations
Automated regression-based [25]	Interpretable model, automatic feature selection	Nonlinear relationships not captured, spatial dependencies not considered
Bayesian ordinary least squares [20]	Interpretable model, inclusion of prior information, distribution of uncertainty, and parameter space	Nonlinear relationships not captured, spatial dependencies not considered, manual features selection
Bias correction spatial disaggregation [26]	Quantile mapping, considers spatial dependencies	Can only be applied to well-modeled GCMs (this excludes precipitation)
Kriging [21]	Infinite resolution, robust to irregular spatial data points	Single variable, spatial dependencies not considered, physical processes not accounted for
Quantile regression neural network [27]	Nonlinear relationships captured, quantile mapping	Not interpretable, spatial dependencies not considered
Support vector machine [17]	Nonlinear relationships captured, computationally efficient	Not interpretable, spatial dependencies not considered
Sparse group lasso [18]	Chooses informative covariates, reduces overfitting by regularization	Nonlinear relationships not captured, not interpretable, spatial dependencies not considered

4.4 Bridging the Gaps

4.4.1 Physics-Guided Data Mining

As discussed in the previous section (see dynamical downscaling), small-scale physical processes such as those related to clouds, vegetation, and soil processes are often parameterized in the GCMs. Given that not all physical information can be incorporated within these models, either due to incomplete understanding or incompatibility with present structure of the models, purely data-driven techniques without appropriate constraints based on physical principles may yield spurious insights. For example, selection of predictors or covariates is a crucial step in any of the SD approaches. Covariate selection requires a balanced understanding of knowledge of physical processes driving the predictand and data mining techniques [28]. Covariate selection based on purely data-driven techniques, including PCA and correlation measures, may result in selection of the variables whose relation with predictand cannot be interpreted physically. Also, these techniques may not be able to capture the implicit relationships between predictors and predictand, which are otherwise physically significant [29]. Therefore, inclusion of the processes, which are not directly simulated or cannot be incorporated in current GCMs, could potentially help in informing the better choice of predictors. The premise of bringing together the physical understanding which may or may not be encapsulated in the computational models with state-of-the-art data mining algorithms has been termed as physics-guided data mining (PGDM).

PGDM can be incorporated into SD in three ways: (1) pre- and postprocessing of data, (2) constraining statistical models, and (3) influencing statistical model architecture. The example above, highlighting the selection of covariates using our understanding of known physical processes, shows how PGDM can affect pre processing. While building a statistical model to downscaling precipitation, we may consider explicitly accounting for known relationships between humidity and precipitation. The use of an indicator variables representing humidity, or other known processes related to rainfall, may inform the model accordingly. A similar approach can be applied in developing a model's architecture. We previously studied an example where a classification algorithm first predicted if precipitation would occur and then given precipitation, another model was used to estimate extremes. This approach may be used to indicate which model is the most applicable to a given set of covariates of one observation. By incorporating each of these PGDM approaches, we may be able to develop more general models that can account for a changing climate.

4.4.2 Advancements in Spatiotemporal Data Mining

Interest in spatiotemporal data mining in the machine learning community has recently emerged, both in and outside the climate space. As interest continues, we are presented with an assortment of tools that may aid in SD. The use of complex networks in spatiotemporal data, which has shown to be applicable to many fields, is beginning to arise in application to climate data. With this, and the interest in causal networks, we see opportunities to use this work as a basis for covariate selection in SD. Prediction may also assist by leveraging techniques associated with multitask learning (MTL), and mixture of models may also provide greater accuracy. Accounting for the progression in spatiotemporal data mining may provide valuable results to SD and should be researched further.

4.4.2.1 Climate Networks for Covariate Selection

The application of complex networks to climate data significantly differs from most network architectures. For example, infrastructure networks examining the power grid are made up of generators connected by transmission lines. Transmission lines do not connect each and every generator, but transferring energy from one generator to the next, power can be distributed. Various statistical measures have been built to understand the robustness of these networks, which may have applicability to certain climate applications. In order to use these measures to understand climate networks, we must be able to develop networks which explain climate dependencies. The choice of nodes in climate networks largely depends on the application;

for some, high-level climate indices may be valuable, while spatial locations may be relevant to others, or even a combination of both. Here, we propose that spatial nodes could be effectively applied to covariate selection.

As each location contains temporal climate observations, dependency measures can be applied to calculate each edge. For instance, in the case of a single variate network, including just temperature, we could compute the edge as the correlation between two locations. Such a method will create an edge for each and every pair of locations in the data set; so by pruning the edges, we can eliminate weak dependencies. After removing a large portion of edges, the network is built on highly correlated nodes that can inform covariate selection by choosing only those locations relevant to the problem. Often times, these networks lead to long-range dependencies, known as teleconnections, that the user may not have previously been aware of. However, we must be aware of false dependencies in these networks simply found by chance. Donges et al. used a similar method to relate air temperature to global surface ocean currents [30]. By using climate networks, Donges was able to uncover wave-like structures of high energy flow related to global surface ocean currents. The ability to find and understand these physical dependencies shows promise in the use of climate networks.

The potential of multiple climate variables challenges many dependency measures previously explored. For instance, as previously examined, most dependency measures are built to compare two variables temporally. Such measures are able to create single variate climate networks but cannot account for the multivariate dependencies. Steinhaeuser et al. proposed a method which computes these multivariate dependencies via the Euclidean distances of cross-correlations [31]. This particular measure naively estimates the dependency by assuming that each pair is uniformly relevant and only accounts for linear dependencies. An extension of this work to allow for multivariate spatiotemporal dependency measures may not only aid the development of climate networks but also for feature selection in SD problems.

The computational costs of assembling spatial climate networks is vast due to the number of possible edges. In order to limit the computations and false dependencies found, we can consider multiscale networks by using climate indices, measuring high-level climate conditions. Joining these climate indices with locations in close proximity of our high-resolution target could greatly reduce the number of covariates. Denli et al. recently proposed a method to learn the interaction within multiscale graphical models for spatiotemporal processes [32]. The model presented has the ability to apply physically derived priors on the data, while learning sparse and rich graphical structures, accounting for both spatial and temporal dependencies. The multiscale applicability allows for high-level indices to join with lower level interactions returning physically meaningful results. This particular work, including concepts from PGDM, has the potential to provide meaningful covariate selection for SD.

4.4.2.2 Multitask Learning

Just as we have examined instances where graphical models can be used for covariate selection, they may also be used for learning related tasks in the form of MTL. MTL allows similar tasks to capture information about each other while still learning separate, hopefully high-performing, models. Similar tasks can be identified by the use of graphical models, in the form of clusters or dependency measures. In SD, we often have a larger number of spatially defined points that are similar in nature. Traditionally, SD models model each location independently without any information regarding others. Clearly, locations that are nearby are not independent from one another and could be leveraged for more informative models. In this case, MTL could be used to leverage such information.

A method called multitask sparse structure learning (MSSL) was recently proposed for a similar problem of forecasting temperature using GCMs [33]. This model is built to leverage dependencies in graphical structures while minimizing false dependencies by using a sparse precision matrix (inverse covariance). For SD, the dependencies in each model represent similar spatial dynamics, in both the feature set and target variable. The method itself assumes each location will have the same set of covariates, which as we have seen is not normally true for SD. To account for such an assumption, we may consider selecting locations that are dependent on a similar set of covariates by computing spatial distances or using methods from weather typing and clustering. For example, the Bayesian framework for covariate selection proposed by Das et al. clusters spatial regions with similar dependencies. This framework could be leveraged for the selection of spatial regions followed by MSSL to estimate downscaling.

4.4.2.3 Mixture of Experts

Just as MTL takes advantages of multiple tasks, the combination of multiple models may also contribute to better predictions. The models previously suggested all have their pitfalls and fail in certain situations. By using the collective power of many models, we may be able to outweigh these pitfalls by accounting for the better model in certain situations. Jacobs et al. originally introduced this idea in 1991 in the form of adaptive mixture of local experts (Figure 4.3), with the idea that some neural networks were better at a given task than another [34].

Adaptive mixture of experts has the potential for application in SD by using a combination of high-performing SD models as the experts. Mixture of experts has also been shown to increase the generality over many traditional models that may be able to reduce overfitting, even with a greater number of parameters. This idea has been widely studied in the form of ensembles of weather forecasting [35] and GCMs [36], further motivating the use of SD.

4.4.2.4 Downscaling Extremes

As previously mentioned, the downscaling of climate extremes is particularly challenging but is of great interest to many stakeholders. The statistical and machine learning methods outlined above are generally applicable to both downscaling means and extremes. In order to downscale extremes effectively, we must modify such methods accordingly. For example, we may consider using indicator variables that will first classify whether precipitation is likely or temperature is above some threshold to signal our transfer function to account for such a scenario, much like how Castellano and DeGaetano [24] did with the Bayesian

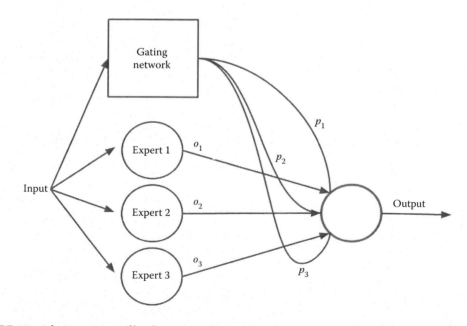

FIGURE 4.3 Adaptive mixture of local experts. Each expert is trained independently from one another. The outputs of each experts, o_i, are the same size and inputted to a basic neural network. The gating network is used to change the weights of each expert depending on the input features.

inference framework. The combination of these indicator variables within a mixture of experts type architecture may be a promising direction for researching climate extremes. The approach taken from BCSD which does quantile matching has also been shown to be effective for projecting extremes. The incorporation of quantile matching while accounting for changing distributions under nonstationarity may provide valuable projections of future extremes. Building models that can incorporate such approaches may be beneficial to downscaling climate extremes.

4.4.2.5 Computational Cost

We have introduced many models that are currently applied to SD and other which may be applicable in future work. Each of these methods requires extensive computational cost, some more than others. Along with these computationally expense contributed by training these statistical models, we must also consider the scalability of downscaling multimodel ensembles of climate models. As the number of climate models increases with more initial conditions, the greater the number of models must be downscaled. As the computational cost increases, the ability to downscale thousands of locations diminishes. The combination of both the training of statistical and machine learning models along with future downscaled projections must be considered during the process of developing models.

Some of these computations can be done in parallel, utilizing the power of distributed computing architectures. Typically, SD does not need to be applied in real time, which allows for the sacrificing of computing time. Even though these real-time computations are not needed, decreasing the cost may allow for faster iterations of experiments. Since many of these methods are directly applicable to weather forecasting, a lower cost which does allow for real-time computing may allow for a broader range of relevance. Computer vision applications, which also work with high-dimensional data, have began to use GPU computing to increase processing time with enormous success by allowing faster iterations of experiments. With this evidence, climate scientists may consider methods that can be trained on GPU processors to increase the productivity of research.

4.4.3 Deep Learning for SD

In the past years, interest in data science and machine learning has drastically increased corresponding to higher performing, more advanced, methods. The vast majority of the state-of-the-art work has been developed by researchers in the domains of computer vision, speech recognition, and natural language processing. The computer vision applications are vast, ranging from medical imaging to self-driving cars. Methods developed in this area focus on spatial dependencies within images, temporal dependencies in consecutive images (e.g., videos), dimensionality reduction, and others. Developments in speech recognition and natural language processing often build off each other, due to their temporal structure. The high dimensionality in all of these data sets continue to provide challenges, leading to solutions which may aid in climate problems.

The common theme of high dimensionality in modern data sets has led to drastic improvements in dimensionality reduction methods. PCA, which we have seen used in SD, and other traditional techniques to reduce dimensionality tend to focus on linear dependencies. The lack of nonlinearity in such models fails to take advantage of large-scale data sets. Recently, the use of unsupervised deep learning techniques have been applied to dimensionality reduction, showing large improvements over PCA. For example, when given a large number of observations, stacked autoencoders outperform PCA [37]. Autoencoders are neural networks that try to recover the input after significantly reducing the number of neurons in the hidden layers. For example, to "autoencode" a 28 by 28 pixel image, 784 input features, we could filter this through a hidden layer of just 50 units and then try to recover the image. Using such a model on high-dimensional data greatly reduces the number of dimensions while preserving information about the input. As we have seen, the high dimensionality of climate data produces challenges through the field. The use of these more complex techniques to climate data has the potential to improve prediction performance.

Autoencoders represent the most basic form of deep learning. The goal of deep learning is to allow the algorithm to learn higher level representations of the features for the given problem, rather than doing feature selection by hand. These ideas originated with basic neural networks trained through backprop-agation in the early 1990s but ended up being too computationally expensive to be practical and had difficulties locating global minimums. In recent years, the increase of computational resources as well as more advanced methods to find global minimums has renewed interest in the machine learning com-munity, known as deep learning. These deep learning methods are able to turn a complex feature space to higher level features, allowing for more generality. Many of the methods also can take advantage of unlabeled data sets within supervised models, including deep belief networks (DBNs).

DBNs are built by stacking multiple layers of restricted Boltzmann machines (RBMs) [38]. RBMs are used to create high-level feature representations by learning probability distributions over the inputs. By stacking these RBMs, using the output of one layer as the input to the next, we are able to learn reasonable weights for a regular neural network. After these weights are established, the weights can be fine-tuned by using the labels provided in the data set. We notice that the original stacking of RBMs does not depend on the labels. Since the number of unlabeled data points is often much greater than the labeled data points, we are able to gain a more general network structure. In climate data sets, it is common to have both high dimensionality and unlabeled data points, which may allow us to make use of such methods.

4.4.4 Case Study: Deep Belief Networks

As previously noted, methods such as ANNs and sparse regression models have been used extensively for SD. Here, we present a daily rainfall SD comparison between a Lasso regression model, an ANN, and a DBN. The NCEP Reanalysis I data set is used for selecting training features; a gridded data set at 2.5° by 2.5° at 6-hour and daily timeframes. We select the variables air temperature, relative humidity, precipitable water, vertical wind, horizontal wind, and pressure at 363 locations over and surrounding the continental United States, a total of 2178 features, and normalize them by subtracting the mean and dividing by the standard deviation. Our target variable, daily precipitation, is taken from the CPC Unified Gauge-based data set, which is gridded and high resolution ($1/8°$ by $1/8°$). The years 1950–1980 are used for training (11,323 days) and years 1981–1999 are used for testing. For this case study we will only train for one target location, $-71.0°$ longitude and $-42.36°$ latitude, which is directly in the middle of the continental United States.

Lasso is a linear model under the constraint of a L_1-norm, forcing many of the covariate coefficients to zero, acting as a feature selector. The ANN consists of three hidden sigmoid activation layers of sizes 500, 100, and 50 with a linear-Gaussian output layer. By decreasing the number of units in each following layer, we aim to compress the features to allow for a more general representation. The initial weights from the ANN are chosen randomly from a normal distribution. For the DBN, we take a modified approach com-pared to the ANN. The NCEP Reanalysis I data set contains data in 6-hour intervals, four times the number of observations in the daily data set, but lack precipitation data in this same interval. The pretraining step of the DBN, the stacking of RBMs, is able to take advantage of such unlabeled data. The first layer uses a Guassian RBM, allowing for continuous variable inputs, with the following two layers being plain RBMs. After the unsupervised training is complete, we are able to initialize an ANN much more intelligently, using the weights learned from the RBM. The higher level representation learned from the first layer RBM can be visualized by viewing the weighted distribution over each unit. We show an example of these distributions over the variable temperature in Figure 4.4. Hopefully, these weights will allow for the network to find a more optimal solution and maximize generality during the supervised fine-tuning stage with daily data.

After each model is trained, we validate each with the test set using the overall mean, root mean square error (RMSE), Pearson correlation, and Spearman correlation. Computing the overall mean will give us an idea of how biased each model is to low or high daily precipitation. The RMSE provides a measure of vari-ance associated relative to the observed precipitation, so the lower the RMSE the better. Both Pearson and Spearman correlations will be used to incorporate both linear and rank type measures. These results are

First layer RBM weights: Temperature

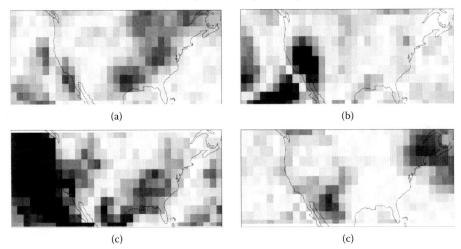

FIGURE 4.4   Each figure, (a)–(d), weights for just temperature for a random neuron in the RBM. (a) A representation of the input where the eastern United States is warmer while the western United States is colder. (b) Colder weather over the region with warmth above the Pacific Ocean. (c) Warmth over the Pacific and western United States while colder weather elsewhere. (d) A colder northeastern United States with average to warmer temperatures elsewhere.

TABLE 4.2   Estimation Results: 1981–1999

	ANN	DBN	Lasso	Observed
Mean	2.97	2.74	3.40	3.27
Pearson	0.54	0.68	0.63	–
RMSE	7.49	6.45	6.74	–
Spearman	0.50	0.66	0.66	–

DBN is biased with a mean lower than the observed but has the highest overall correlations and lowest RMSE. Lasso is biased to be above the average and also has good results on correlation and RMSE. ANN has the highest RMSE and lowest correlations.

presented in Table 4.2 and the corresponding time-series are presented in Figure 4.5. The ANN performed significantly worse than the DBN, even though the same features were used and the network architecture was the same. The only difference between these two is the way the weights within the network were initialized. By pretraining the network with stacked RBMs, we gain a better approximation of the network allowing for greater generality and the ability to locate a more likely minimum of the cost function. From other analyses, outside of this case study, we have shown this to be consistent between locations. However, the estimates for lasso are quite similar to the DBN model, if not better. In this case, the simplicity of the lasso model may outweigh the complexities of using a DBN. The large number of features relative to number of observations is likely causing overfitting in both the ANN and DBN models. With more intelligent covariate selection methods, the DBN model may perform significantly better.

This implementation of a DBN for downscaling certainly showed advancements over plain ANNs that have been used in the past, giving us hope that such a method can be improved. We also note that DBNs are able to take advantage of large numbers of observations but do not natively account for multivariate spatiotemporal data. As spatiotemporal data appear in both the predictand and predictor, we may consider using a mixture of techniques from convolutional neural networks (CNNs) and MTL. As we have

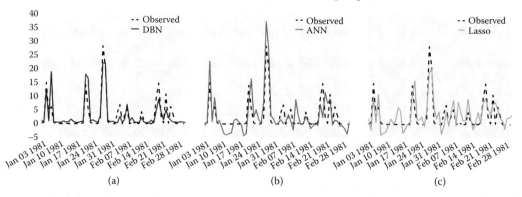

Downscaled estimates versus observed precipitation

FIGURE 4.5   Over the first 60 days of 1981, we estimate the daily precipitation with each SD model and compare to the observed data. (a) Compares the DBN with observed; we notice that the estimate learns precipitation must be greater than or equal to zero but underestimates days with large rainfall amounts. (b) An ANN compared with observed; here we see a high variance above and below the observed values. (c) The lasso estimate which is reasonably good at predicting rainfall and similar to the DBN.

already illustrated how MTL could improve the use of sparse regression models, its use in deep downscaling models could also be valuable. Spatiotemporal data sets have been successfully applied to videos for action recognition [39] where consecutive frames contain spatial information by using a CNN to create higher level representations. Though CNNs would account of spatial structure in climate data sets, it is expected hundreds of thousands of unique time-steps will be necessary to see improvements over other methods. Even with this limitation, the idea of a convolution over gridded climate data sets may be valuable to incorporating the spatial structure into SD models.

Up until now we have focused on covariate selection and regression of spatiotemporal data to minimize collinearity in our data sets. Unfortunately, this is only accounting for three dimensions of our data, latitude, longitude, and time. The fourth dimension is the multivariate nature, including variables such as wind, pressure, humidity, temperature, etc. Many of these variables will also be correlated, for instance, when there is vertical wind there is likely to be horizontal wind. Adding the fourth dimension begins to make the data set more difficult to visualize, becoming more abstract, as well as more challenging to model. A fifth dimension could also be added by incorporating different elevation levels into the data set. Most models are not built to handle this many dimensions, forcing scientists to flatten the data and sometimes, making false assumptions. Research in these high-dimensional problems is beginning, though not yet popular, which incorporate the use of tensors, *n*-dimensional matrices. Bahadori et al. proposed a low-rank tensor learning framework for multivariate spatiotemporal analysis, highlighting geospatial applications [40]. This approach is shown to work on both forecasting and cokriging (multivariate version of kriging), while providing a fast and greedy approach to learning. By accounting for both spatial and multivariate linear correlations, the method is able build a low-rank tensor to represent the predictors by assuming nearby variables are likely to be correlated. Models similar to this, those incorporating all dimensions, is yet another area where machine learning can improve SD.

4.5 Conclusion

The advancement of SD is critical to understanding the future climate and providing valuable information to stakeholders. Awareness of uncertainties in both climate models and observed data sets is essential to the calibration and validation of SD models. The downscaling of a wide range of GCMs is imperative to understanding climate change at a local scale. The choice of methods for SD should account for the non-stationarity of the climate, generalize over regions, and choose physically interpretable covariates. As work in Bayesian feature selection and network science continue to improve, they must also be applied and validated with SD models. These covariate selection models should lead to transparent end-to-end processes of SD models, allowing for replication and comparison to other models. State-of-the-art machine learning methods that are pushing the boundaries in fields such as computer vision, natural language processing, and speech recognition should be carefully considered for incorporation into SD. More specifically, the deep learning models being developed, focusing on dimensionality reduction and spatiotemporal data, have potential to improve over the current generation of SD models. Utilizing MTL to take advantage of spatially dependent observations will allow for more generalizable models. All in all, an interdisciplinary effort is crucial to improving the state of the art, allowing the interaction between climate scientists and computer scientists to effectively identify problems, use sound model validation, and thoroughly communicate results to stakeholders.

Acknowledgments

This work was funded by NSF CISE Expeditions in Computing award # 1029711, NSF CyberSEES award #1442728, and NSF BIGDATA award # 1447587. The authors thank Evan Kodra for his valuable comments.

References

1. R W Kates, C E Colten, S Laska, and S P Leatherman. Reconstruction of New Orleans after Hurricane Katrina: A research perspective. *Proceedings of the National Academy of Sciences*, 103(40):14653–14660, 2006.

2. B C Hewitson, J Daron, R G Crane, M E Zermoglio, and C Jack. Interrogating empirical-statistical downscaling. *Climatic Change*, 122(4):539–554, 2014.

3. F Giorgi, M R Marinucci, G T Bates, and G De Canio. Development of a second-generation regional climate model (REGCM2). Part ii: Convective processes and assimilation of lateral boundary conditions. *Monthly Weather Review*, 121(10):2814–2832, 1993.

4. R L Wilby, S P Charles, E Zorita, B Timbal, P Whetton, and L O Mearns. Guidelines for use of climate scenarios developed from statistical downscaling methods. *Analysis*, 27(August):1–27, 2004.

5. P Brohan, J J Kennedy, I Harris, S F B Tett, and P D Jones. Uncertainty estimates in regional and global observed temperature changes: A new data set from 1850. *Journal of Geophysical Research: Atmospheres (1984–2012)*, 111(D12), 2006.

6. E Kalnay, M Kanamitsu, R Kistler, W Collins, D Deaven, L Gandin, M Iredell, S Saha, G White, J Woollen, et al. The NCEP/NCAR 40-year reanalysis project. *Bulletin of the American Meteorological Society*, 77(3):437–471, 1996.

7. S Curtis and R Adler. ENSO indices based on patterns of satellite derived precipitation. *Journal of Climate*, 13(15):2786–2793, 2000.

8. P M Caldwell, C S Bretherton, M D Zelinka, S A Klein, B D Santer, and B M Sanderson. Statistical significance of climate sensitivity predictors obtained by data mining. *Geophysical Research Letters*, 41(5):1803–1808, 2014.

9. A J Frost, S P Charles, B Timbal, F H S Chiew, R Mehrotra, K C Nguyen, R E Chandler, J L McGregor, G Fu, D G C Kirono et al. A comparison of multi-site daily rainfall downscaling techniques under Australian conditions. *Journal of Hydrology*, 408(1):1–18, 2011.

10. R Knutti and J Sedlacek. Robustness and uncertainties in the new CMIP5 climate model projections. *Nature Climate Change*, 3(4):369–373, 2013.

11. C W Richardson. Stochastic simulation of daily precipitation, temperature, and solar radiation. *Water Resources Research*, 17(1):182–190, 1981.

12. M A Semenov and E M Barrow. Use of a stochastic weather generator in the development of climate change scenarios. *Climatic Change*, 35(4):397–414, 1997.

13. D Maraun, F Wetterhall, A M Ireson, R E Chandler, E J Kendon, M Widmann, S Brienen, H W Rust, T Sauter, M Themel et al. Precipitation downscaling under climate change: Recent developments to bridge the gap between dynamical models and the end user. *Reviews of Geophysics*, 48(2009):1–34, 2010.

14. G Burger, T Q Murdock, A T Werner, S R Sobie, and A J Cannon. Downscaling extremes—An intercomparison of multiple statistical methods for present climate. *Journal of Climate*, 25(12):4366–4388, June 2012.

15. D A Gutmann, T Pruitt, M Clark, L Brekke, J Arnold, D Raff, and R Rasmussen. An intercomparision of statistical downscaling methods used for water resource assessments in the United States. 50:7167–7186, 2014.

16. K Salvi, S Ghosh, and A R Ganguly. Credibility of statistical downscaling under nonstationary climate. *Climate Dynamics*, 46(5):1991–2023, 2016.

17. S Ghosh. SVM-PGSL coupled approach for statistical downscaling to predict rainfall from GCM output. *Journal of Geophysical Research: Atmospheres*, 115(November 2009):1–18, 2010.

18. S Chatterjee, K Steinhaeuser, A Banerjee, Sn Chatterjee, and A Ganguly. Sparse group lasso: Consistency and climate applications. In *Proceedings of the 2012 SIAM International Conference on Data Mining*, SIAM, 2012.

19. D Das, A R Ganguly, and Z Obradovic. A Bayesian sparse generalized linear model with an application to multi-scale covariate discovery for observed rainfall extremes over United States. *IEEE Transactions on Geoscience and Remote Sensing*, 53(12):6689–6702, 2014.

20. X Zhang and X Yan. A new statistical precipitation downscaling method with Bayesian model averaging: A case study in China. *Climate Dynamics*, 45(9–10), 2541–2555, 2015.

21. N Cressie and G Johannesson. Fixed rank kriging for very large spatial data sets. *Journal of the Royal Statistical Society Series B (Statistical Methodology)*, 70(1):209–226, 2008.

22. D Nychka, S Bandyopadhyay, D Hammerling, F Lindgren, and S Sain. A multi-resolution Gaussian process model for the analysis of large spatial data sets. *Journal of Computational and Graphical Statistics* (May), 24(2):579–599, 2014.

23. A Manor and S Berkovic. Bayesian Inference aided analog downscaling for near-surface winds in complex terrain. *Atmospheric Research*, 164–165:27–36, 2015.

24. C M Castellano and A T DeGaetano. A multi-step approach for downscaling daily precipitation extremes from historical analogues. *International Journal of Climatology*, 36(4):1797–1807, 2015.

25. M Hessami, P Gachon, T B M J Ouarda, and A St-Hilaire. Automated regression-based statistical downscaling tool. *Environmental Modelling & Software*, 23(6):813–834, 2008.

26. A W Wood, L R Leung, V Sridhar, and D P Lettenmaier. Hydrologic implications of dynamical and statistical approaches to downscaling climate model outputs. *Climatic Change*, 62(1–3):189–216, 2004.

27. J W Taylor. A quantile regression neural network approach to estimating the conditional density of multiperiod returns. *Journal of Forecasting*, 19(4):299–311, 2000.

28. D I Jeong, A St-Hilaire, T B M J Ouarda, and P Gachon. Comparison of transfer functions in statistical downscaling models for daily temperature and precipitation over Canada. *Stochastic Environmental Research and Risk Assessment*, 26(5):633–653, 2012.

29. P A O'Gorman and T Schneider. The physical basis for increases in precipitation extremes in simulations of 21st-century climate change. *Proceedings of the National Academy of Sciences*, 106(35):14773–14777, 2009.

30. J F Donges, Y Zou, N Marwan, and J Kurths. The backbone of the climate network. *EPL (Europhysics Letters)*, 1–6, 2009.

31. K Steinhaeuser, N V Chawla, and A R Ganguly. An exploration of climate data using complex networks. *ACM SIGKDD Explorations Newsletter*, 12(1):25, 2010.

32. H Denli, N Subrahmanya, and F Janoos. Multi-scale graphical models for spatio-temporal processes. In *Advances in Neural Information Processing Systems*, 316–324, 2014.

33. A R Goncalves, P Das, S Chatterjee, V Sivakumar, F J Von Zuben, and A Banerjee. Multi-task sparse structure learning. In *Proceedings of the 23rd ACM International Conference on Conference on Information and Knowledge Management*, pp. 451–460. ACM, 2014.

34. R A Jacobs, M I Jordan, S J Nowlan, and G E Hinton. Adaptive mixtures of local experts. *Neural Computation*, 3(1):79–87, 1991.

35. R Hagedorn, F J Doblas-Reyes, and T N Palmer. The rationale behind the success of multi-model ensembles in seasonal forecasting–i. basic concept. *Tellus*, 57(3):219–233, 2005.

36. M A Semenov and P Stratonovitch. Use of multi-model ensembles from global climate models for assessment of climate change impacts. *Climate Research (Open Access for Articles 4 Years Old and Older)*, 41(1):1, 2010.

37. G E Hinton and R R Salakhutdinov. Reducing the dimensionality of data with neural networks. *Science*, 313(July):504–507, 2006.

38. G Hinton. A practical guide to training restricted Boltzmann machines. *Momentum*, 9(1):926, 2010.

39. Q V Le, W Y Zou, S Y Yeung, and A Y Ng. Learning hierarchical invariant spatio-temporal features for action recognition with independent subspace analysis. In *Computer Vision and Pattern Recognition (CVPR), 2011 IEEE Conference on*, pp. 3361–3368. IEEE, 2011.

40. M T Bahadori, R Yu, and Y Liu. *Advances in Neural Information Processing Systems*. pp. 3491–3499, 2014.

5

Large-Scale Machine Learning for Species Distributions

5.1	Introduction	73
5.2	Theory and Concept	74
5.3	Challenges of Learning Species Distributions	76
	Spatial Extent • Spatial Resolution • Spatial Sampling • Presence-Only Data • Model Evaluation	
5.4	Modeling Methods	81
	Presence/Absence Methods • Presence/Background Methods • Presence-Only Methods	
5.5	Deploying Large-Scale Models	84
	Modeling in Practice • Modeling in the Cloud	
5.6	Future Directions	85

Reid A. Johnson

Jason
D. K. Dzurisin

Nitesh V. Chawla

5.1 Introduction

Among the greatest challenges facing biologists and ecologists today is predicting how species will respond to climate change—a challenge that necessitates an unprecedented coordination and mobilization of data, information, and knowledge [1]. Understanding the factors that determine where species live and developing predictions about their niches are important tasks for developing strategies in conservation biology and ecology. This understanding informs important actions such as setting restoration and reintroduction targets, setting conservation priorities geographically, and protecting sites against potential invasion by pest species [2–5]. The increasing availability of data and tools has facilitated the large-scale use of computer algorithms to further this understanding by predicting species' ecological niches and geographic distributions, a process broadly known as correlative niche modeling.

The development of modeling approaches has arisen in part from substantial increases in the availability of the data on individual species known as "occurrence data" or "presence data" [6–8]. The increasing availability of species occurrence data results from efforts to digitize and georeference specimens preserved in natural history collections [9,10] and to improve access to large stores of observational records accessible via Internet portals [11]. Despite these large stores of data, only a small fraction of extant species has been described in any meaningful detail, with most recent estimates suggesting that approximately 86%–91% have yet to be discovered [12]. Yet, the number of individual data points that have been collected to provide even this limited amount of information is truly enormous—and it will only get larger.

Information regarding environmental variables is now similarly plentiful and continues to grow. Extensive environmental data about climate, topography, soil, oceanographic variables, vegetation indices, land-surface reflectance, and many other factors are available across the entire planet, and at increasingly finer resolutions. These data sets are being generated by a diverse assortment of entities, such as the European and U.S. space agencies, the United Nations, university researchers, and many national institutions [13]. As these environmental data sets become larger, more accurate, and more comprehensive, their utility for the modeling of species' ecological niches and geographic distributions will only increase.

Several resources survey the field of correlative niche modeling exist (e.g., [2,3,14–17]). We offer this chapter as a synthesis of concepts relevant to the large-scale deployment of correlative niche models, marrying many useful concepts from ecology and machine learning in the development and application of these models. While the chapter is written with a focus to the machine learning community, we hope that it will be of interest to researchers in diverse aspects of ecology, biogeography, and other related fields.

We begin by describing the conceptual idea of and theory behind correlative models in Section 5.2, followed by a discussion of the domain-specific challenges that must be considered when deploying these models in Section 5.3. Next, in Section 5.4 we provide a review of commonly used correlative niche models, highlighting their theoretical and practical similarities and differences. Expanding on our discussion of modeling efforts, in Section 5.5 we provide a summary of platforms used to deploy these models and the technologies that enable their large-scale use. Finally, we outline the latest trends and future directions that we believe to hold particular promise in Section 5.6.

5.2 Theory and Concept

Impacted by the increasing availability of curated and crowdsourced data sets, the field of ecology now stands poised to benefit more than ever from the large-scale modeling of species distributions and ecological niches, two related domains. Ecological niche models (ENMs) focus on the estimation of a species' potential geographic distribution—areas that fulfill both the abiotic and biotic requisites of the species—under changed conditions and circumstances. Species distribution models (SDMs) must include further steps to transform estimated area from potential to actual distributional areas [16]. While these terms have often been used interchangeably in the literature, they entail conceptual and practical differences beyond the scope of this chapter [18]. Despite different objectives, both approaches estimate distributional areas on the basis of correlations of known species occurrences with environmental variables. Both ENM and SDM refer to what is in effect the same set of analyses and employ methods chosen from essentially the same set of mathematical algorithms, which combine similar types of species data and similar—often identical—sets of environmental variables.

Broadly, estimates of species distributions and ecological niches can be generated by two different types of models: (1) mechanistic simulation models that explicitly describe distributional processes and (2) statistical models that, by definition, describe correlative—but not necessarily causal—relationships [19]. Although mechanistic models have been proven to be powerful in capturing spatiotemporal population changes in response to environmental change, their formalism also requires the application of detailed population processes that are presently unavailable for the vast majority of species [19]. Accordingly, statistical and machine learning techniques that correlate species distribution data with environmental variables—thereby obviating any dependency on population processes—are used in many ecological applications to characterize the relationship between a species and its environment (e.g., [20,21]).

Correlative niche models are generated by combining occurrence data—locations where a species has been identified as being present or absent—with ecological and environmental variables—conditions such as temperature, precipitation, and vegetation—to model a species' ecological distribution or niche [22]. This process is illustrated in Figure 5.1. These models can forecast a species' unrealized niche (the region that is habitable for the species, but in which it has not yet been observed) and provide information regarding the potential effects of climate change, making them particularly useful where more detailed physiological data are not available for developing a process-based or mechanistic model [21,23]. In correlative niche modeling, the principal challenge is to determine, given a set of known observations (and,

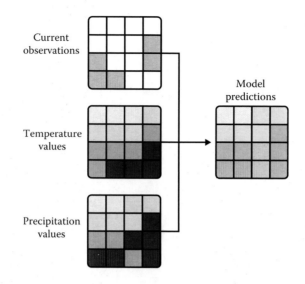

FIGURE 5.1 (See color insert.) Correlative model based on current observations in combination with temperature and precipitation levels. Observations are typically binary valued (presence or nonpresence), while environmental variables may be continuous. Correlative models identify correlations between current observations and environmental values. Darker hues indicate larger values.

potentially, absences) of a species, the ecological niche of that species within a geographic range (a set of points). To accomplish this task, correlative models need to be trained or calibrated on a set of known species observations from which they can be used to generate predictions on testing data.

By relating observational data (occurrence or abundance at known locations) of species distributions with information regarding the environmental and/or spatial characteristics of those locations, correlative models aim to identify the most influential variables explaining presences, presence/absences, abundances, or even the optimal relationships between species' distributions and these explanatory variables [2]. Figure 5.2 illustrates the correspondence between a species' geographical space and its distribution in environmental space. A correlative model identifies relationships in environmental space, with the model predictions mapped back to geographical space. These models enable future projections of species' distributions that do not depend on extensive prior knowledge of population processes, but simply on environmental and species' distribution data. The interpretation of these models can then be used to help better understand the processes underlying the correlative relationships that are inferred [24].

Fundamental to the problem of correlative modeling is the prediction of species occupancy at a given geographic location based upon the environmental characteristics of that location. This dichotomy makes correlative modeling a problem inherently amenable to binary classification. More formally, the classification task is to characterize every location within a region or study area in terms of quantitative values related to the probability of a species being present, $P(Y = 1)$, as a function of the environmental conditions presented in that location, $e \in E$, where the environment and species occurrences are linked by a model. The model, represented as $\mu(G_{\text{data}}, E)$, is a method used to estimate $f(X)$. Thus, given a set of observations, G_{data}, which has a corresponding environmental space E, the aim is to generate a model, $f(\hat{X}) = \mu(G_{\text{data}}, E)$, that approximates $f(X)$.

This formulation, while powerful, also presents significant challenges. Some of these challenges are unique to learning species' ecological niches and geographic distributions, while others manifest in a wide range of machine learning applications. In the next section, we highlight several of the focal challenges that arise when using correlative niches models and relate them to the broader context of machine learning.

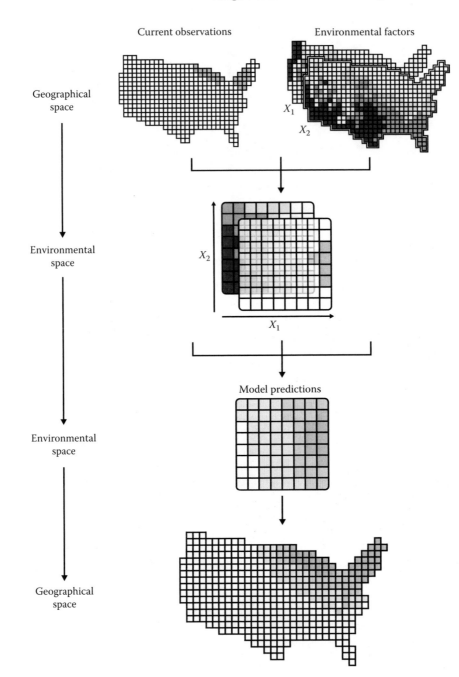

FIGURE 5.2 (See color insert.)   Correlative models use as input current observations and environmental factors in geographical space, combining them in environmental space to generate predictions. These predictions are then transformed back to geographical space for interpretation.

5.3 Challenges of Learning Species Distributions

A growth in the use of correlative niche models has followed from increased computational ability, algorithm development, and readily and freely available environmental, climatic, and species occurrence data [25–28]). Despite growth in the use and usability of these models, the methods used to analyze the dependencies of species occurrences with environmental factors still require sophisticated computational resources, advanced technical skills, and access to curated biological and climatic data. Many correlative approaches have also been criticized in the literature [29,30], mainly due to their underlying assumptions and the danger of overemphasizing results without proper acknowledgment of limitations that are often poorly understood [24]. Others recognize that although the tools are imperfect, they offer a useful way to understand potential effects of future environmental conditions for species under climate change [21,31].

An element of difficulty in using correlative niche models is the complexity of what drives species' actual and potential distributions. Despite significant progress in the development of this understanding, many challenges remain [32]. Some of these challenges are common to many applications of machine learning methods, such as questions of evaluation, interpretation, and reliability. Others arise as consequences of the spatial nature of correlative modeling, including issues of extent, scale, and bias. Often, niche modeling is hindered by a paucity of absence, or nonoccurrence, data, which are necessary for the development of presence/absence models. Further confounding the modeling task, the localities at which a species is not present can be either genuine absences or simply areas lacking occurrence information [33].

In this section, we elaborate on a variety of factors that introduce challenges into the modeling of species distributions and ecological niches. These challenges include spatial sampling, sample bias, model evaluation, skewed or imbalanced data, and model evaluation. Collectively, we believe that they characterize the primary challenges currently facing the domain of correlative niche modeling.

5.3.1 Spatial Extent

Spatial extent refers to the area over which data are collected and modeling is employed. The choice of spatial extent (e.g., state, region, continent) can have a significant effect on the output of correlative models, both conceptually and statistically. As we discuss below, performing analysis on an inappropriately expansive or restrictive spatial extents has the potential to seriously alter the interpretability and reliability of model results.

If analysis is performed over a larger spatial extent than is representative of a species' niche, the model may be prone to overfitting to environmental conditions present in the species' documented range. This can occur as a consequence of the model recognizing spurious environmental differences between the region that a species actually inhabits and other regions that, based on the available environmental features, it could inhabit but does not [34]. In the context of machine learning, this effect constitutes overfitting to the sampling bias present in the data, as the species occurrences do not represent a truly random sample from the species' potential distribution. Such overfitting leads to artificially lowered transferability of a model across space or time and should be avoided for applications that require the ability to predict new and independent data [35].

In contrast, when calibrating (i.e., training) a model using a smaller spatial extent and then applying it to a larger extent, the values for one or more environmental variables in some pixels of the larger study region may not be evaluated by the model. This effect is referred to as truncation. It typically occurs because such values do not exist in the spatial extent used for model parameterization, and instead lie outside the range of values for the corresponding variable(s) in the model [36]. This effect concerns the generalizability of the model, and the knowledge or assumptions the model makes beyond the data it has been given in order to generalize beyond it. Consequently, in order to produce predictions for pixels beyond calibration data, a model must make some assumptions [37], which may or may not hold over an unseen spatial extent.

No matter whether it is relatively larger or smaller, the extent of the study area in relation to species ranges and environmental gradients will affect how well the model describes the relationship between species distributions and environmental variables. However, the degree to which spatial extent affects correlative models is still unclear. While some have argued that including observations of species absence that are well beyond the range of environmental values where the species is present may lead to biased models [38], others contend that statistical models are robust to this type of error and may be able to overcome such limitations [39]. Restricting an analysis to the region of interest ensures that outputs are informed by data specific to the experimental region, though model results should always be interpreted with due consideration of the limitations involved [21,40].

5.3.2 Spatial Resolution

The distribution of species is usually mapped on the basis of a grid of environmental layers. Spatial scale refers to the resolution or grain size (i.e., the size of each location) of the environmental layers used in modeling. The choice of spatial scale is an important factor for correlative modeling that may affect predictions. As we discuss below, performing analysis on an inappropriately granular or coarse spatial scale has the potential to seriously affect the interpretability and reliability of model results.

Initially, one might assume that for a given extent, more granular data should always result in better predictions—and therefore better performance—than coarser data, as it should provide more information for modeling. However, a chief argument in favor of correlative models is that climatic influences on species distributions are dominant at large scales, obviating the need for modeling the small-scale interactions and processes of species distributions mechanistically [21]. While the use of coarser scales have been generally found to result in degraded model performance for fixed or locally mobile species [41], models that employ data of finer granularity are likely to have lower predictive power unless mechanistic, ecological rationales can be provided for the principal environmental predictors [25].

The effect of spatial scale also depends on the quantity and quality of the species occurrence data available, as illustrated in Figure 5.3. While larger sample sizes generally result in better models that tend to be more sensitive to scale, insufficient data will result in a model that is unreliable regardless of the granularity of the spatial scale [41]. The degree of sampling necessary to produce a reliable model is greater for more narrowly distributed species, as such models exhibit greater variability in their predictions than models for widespread species [42]. Additionally, the effect of spatial scale is only noticeable for models reaching sufficient performance or using initial data with an intrinsic error that is impacted by a coarser spatial scale [41]. If the model cannot make sufficient use of granular data, then the practical effect of highly granular data on subsequent analysis may be negligible.

Current understanding suggests that the incorporation of mechanistic effects—species migration, population dynamics, biotic interactions, community ecology, etc.—into correlative models at multiple spatial scales can improve model performance. However, to effectively make use of this information, it is important to provide a precise spatial matching between the scale at which the mechanistic effect occurs and the

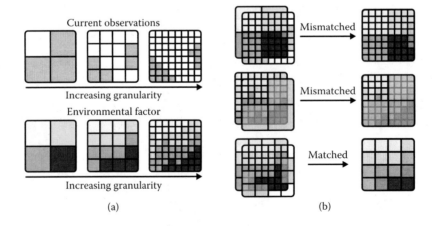

FIGURE 5.3 Correspondence between the spatial resolution of observational and environmental data. (a) Spatial resolution of observations and environmental factors. Observations and environmental factors may have different resolutions. (b) Illustration of matching observational and environmental resolutions. Environmental data that are too coarse or too granular with respect to observational data may present problems for correlative models.

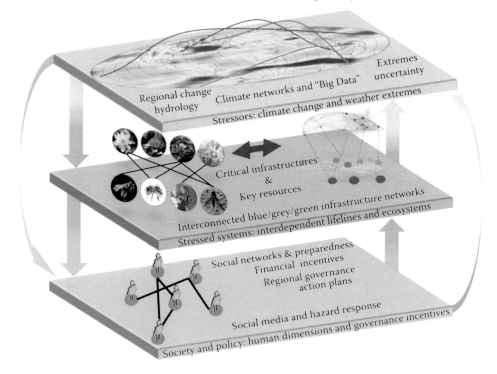

FIGURE 1.2 A multilayered and unified framework can represent the coupled system where (a) climate change and weather or hydrological extremes are the stressors; (b) interdependent lifeline and environmental systems are the stressed systems; and (c) social networks of communities and regions are the impacted systems. This representation enables a unified quantitative framework, but particular instantiations of this network are necessary and sufficient to answer specific questions.

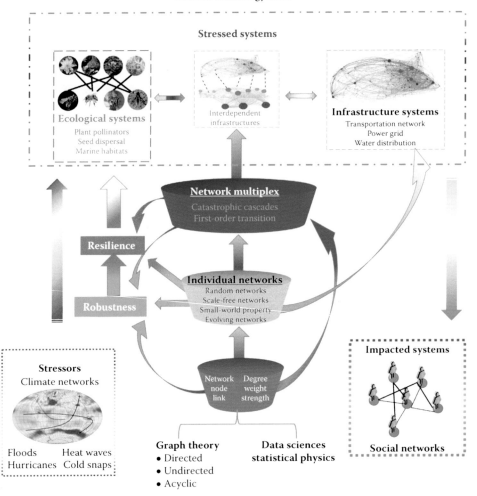

FIGURE 1.4 Complex network-based technology stack in combination with sensor and data engineering, computational process models, and data science methods can potentially serve as a unified framework to meet these challenges. Tools to model single and interdependent networks come from network science and contribute to insights for the lifelines and ecosystems. Since all the three systems—stressors, stressed, and impacted systems—share common attributes of correlation and interdependence, fundamental network science breakthroughs and state of the art leads to novel adaptations or (in some cases) customization and then onward to new science or engineering insights.

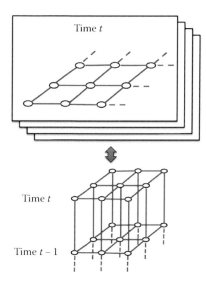

FIGURE 3.2 The MRF extended over space.

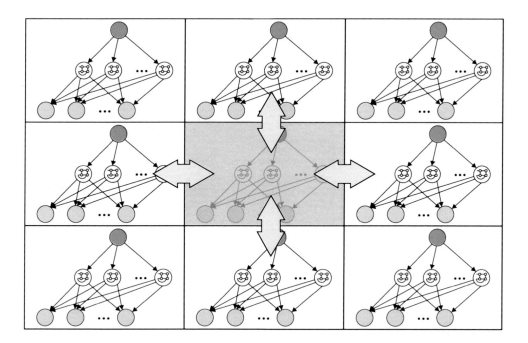

FIGURE 3.4 The NTCM algorithm with Learn-α running on each region and spatial influence between the regions.

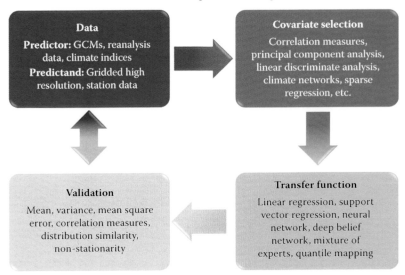

Statistical Downscaling Process for Regression

Data
Predictor: GCMs, reanalysis data, climate indices
Predictand: Gridded high resolution, station data

Covariate selection
Correlation measures, principal component analysis, linear discriminate analysis, climate networks, sparse regression, etc.

Transfer function
Linear regression, support vector regression, neural network, deep belief network, mixture of experts, quantile mapping

Validation
Mean, variance, mean square error, correlation measures, distribution similarity, non-stationarity

FIGURE 4.2 The process of statistical downscaling for regression has four steps. (1) Data collection and splitting data into training and test sets; (2) covariate selection via various methods (no covariate selection is possible); (3) train a statistical model; (4) using the trained model and hold-out data validate using multiple statistical measures. Validation can be done on both historical and projected data.

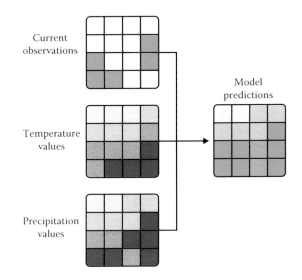

Current observations

Temperature values

Precipitation values

Model predictions

FIGURE 5.1 Correlative model based on current observations in combination with temperature and precipitation levels. Observations are typically binary valued (presence or nonpresence), while environmental variables may be continuous. Correlative models identify correlations between current observations and environmental values. Darker hues indicate larger values.

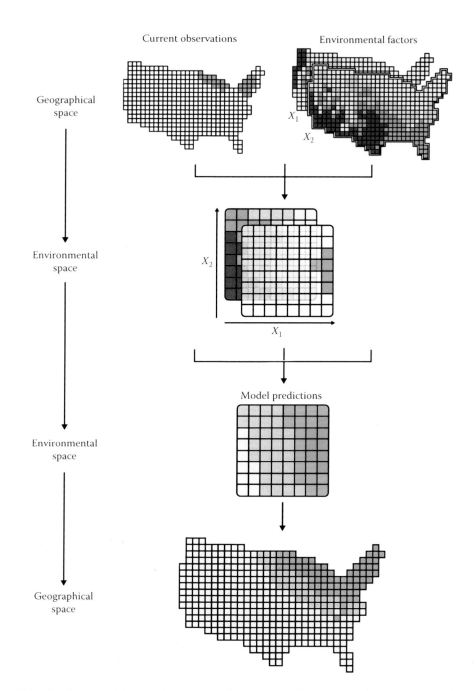

FIGURE 5.2 Correlative models use as input current observations and environmental factors in geographical space, combining them in environmental space to generate predictions. These predictions are then transformed back to geographical space for interpretation.

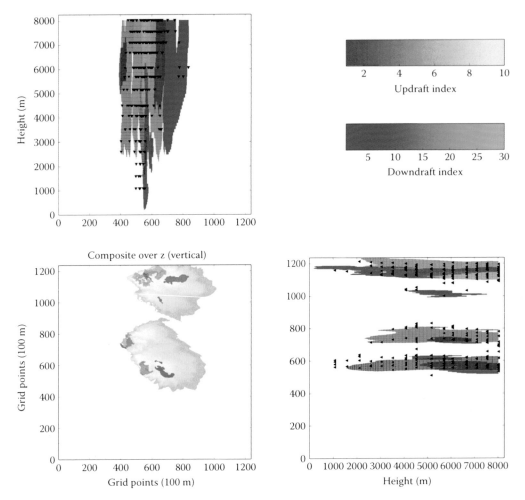

FIGURE 6.3 Visualization of the three-dimensional updraft and downdraft objects. The lower left panel shows the composite over the vertical and the top and right panels show the composites across the north–south and east–west directions, respectively. Updraft objects are shown in shades of red, and downdraft objects are shown in shades of blue. The colors correspond to the object number, not to values within the updraft or downdraft themselves. The vertical composite also includes reflectivity. The triangles in the horizontal composites indicate locations of strong vortices (described in Section 6.4.2).

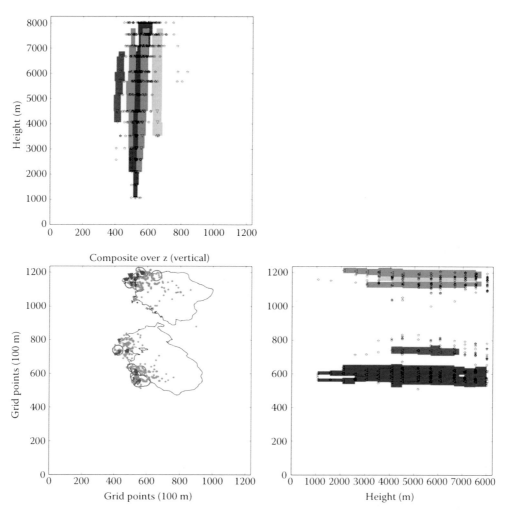

FIGURE 6.4 Mesocyclone objects at time 6000. The lower left panel shows the vertical composite with the VDAC detections shown using either a star or a triangle. Detections shown using a triangle were used in the mesocyclone objects. Positive mesocyclones are shown in red and negative in blue. There were no tornado objects at this time step.

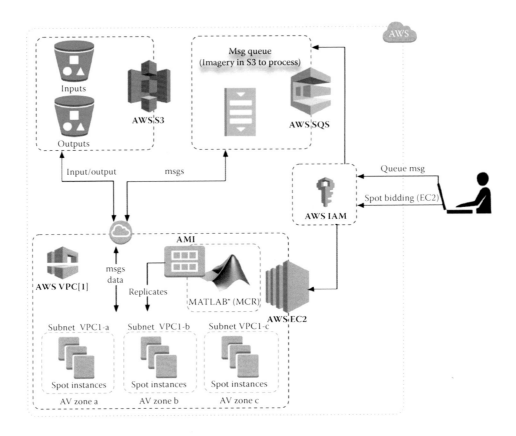

FIGURE 7.3 The AWS computing infrastructure. AMI, Amazon Machine Image; AV, availability; AWS, Amazon Web Services; EC2, Elastic Cloud Compute; IAM, AWS identity and access management; MCR, MATLAB compiler runtime; S3, Simple Storage Service; SQS, Simple Queuing Service; VPC, virtual private cloud.

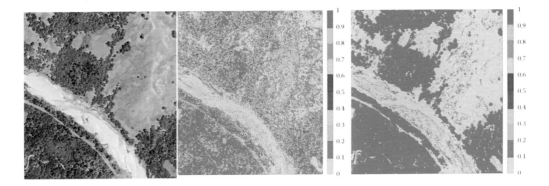

FIGURE 7.4 The output probability maps. A sample NAIP tile from the Blocksburg region in California (left), the output probability map from the DBN-based classifier (center) and the output from the CRF algorithm (right).

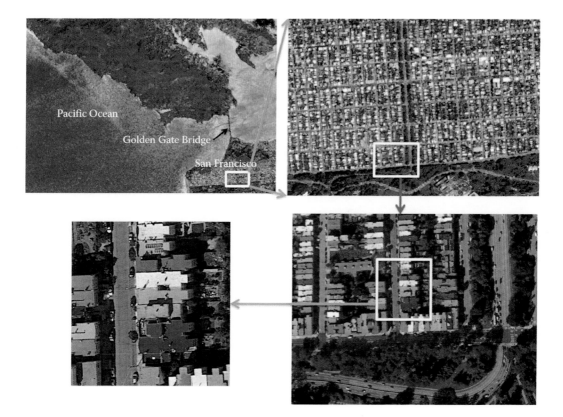

FIGURE 8.2 Study area: FCC of a part of San Francisco city. Zoomed image of the urban area (marked with rectangles in inset) shows mixing of substrate with vegetation, roads, shadow, and dark objects.

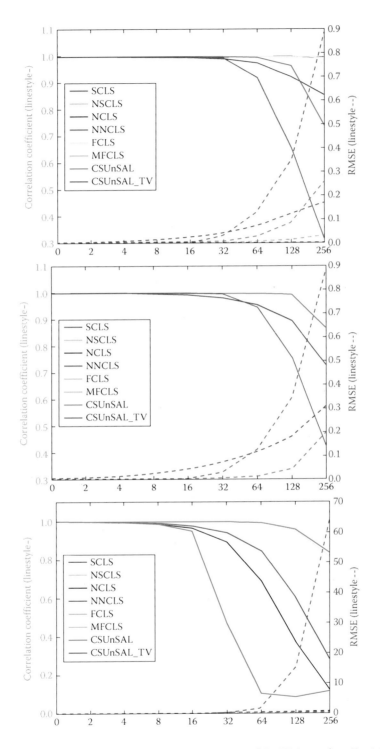

FIGURE 8.6 Pearson product–moment correlation coefficient (cc) and RMSE (secondary Y-axis) between abundance values obtained from the eight constrained algorithms and true abundances at different levels of noise. (X-axis: Noise variance.) NCLS, nonnegative constrained least squares; NCLS, normalized nonnegative constrained least squares; RMSE, root mean square error.

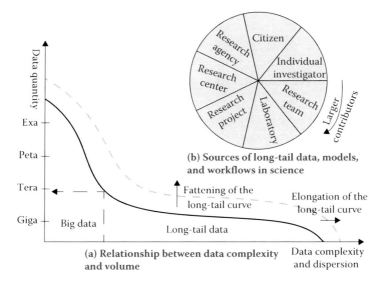

(b) Sources of long-tail data, models, and workflows in science

(a) Relationship between data complexity and volume

FIGURE 9.1 Illustration of the relationship between resource volume and its dispersion, the three forces that are re-sketching the shape of the long-tail resources curve in science, and identification of sources that are contributing to the geoscience long-tail collection.

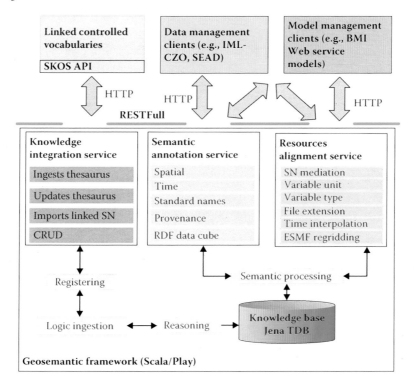

FIGURE 9.5 High-level architecture of the Geosemantic framework. The framework uses micro services architecture and provides each service as an endpoint through a RESTFul API. KIS ingests standards and can annotate them back. A unique endpoint is provided for each registered standard. The Logic Ingestion component infers the relationships among registered standards. Reasoning among ingested standard is based on Jena Reasoners and Rules Engine. Semantic Processing component handles the requests to the knowledge base and the communication among micro services.

scale at which it is modeled [25]. If only a single spatial scale is used, it is important that the granularity is appropriate for the species in question.

5.3.3 Spatial Sampling

In many cases, observing and cataloguing every individual member of a species is clearly impractical. Spatial sampling refers to the use of observations of individuals at a limited number of locations in geographic space to faithfully estimate characteristics of the larger population. A key goal of correlative modeling is to construct a model that fits well to calibration (i.e., training) data, but that does not overfit in ways such that its predictive ability is low when presented with new data, ensuring the generalizability of our learned model. As we discuss below, the manner in which training data is selected can seriously impact the quality of inferences derived from correlative niche models.

Meaningful evaluation of correlative models and their corresponding geographic predictions typically employ measures of performance and significance. These measures depend vitally on the careful and appropriate selection of suitable occurrence data sets, with evaluation data that would ideally be fully independent from calibration or training data. Unfortunately, fully independent occurrence data sets rarely exist [16]. Out of necessity, most evaluation efforts are based on partitions of individual samples into calibration and evaluation data sets. Perhaps most commonly, the evaluation data derive from the same study extent as the calibration data, with occurrence data typically selected at random and without respect to geography [16]. However, this approach can yield overly optimistic assessments of model predictive performance due to environmental bias [43]. This is known in the machine learning community as the sample selection bias problem [44].

To address this problem, it is possible to filter occurrence data spatially, thereby reducing the spatial bias present in the data set [16]. Semivariogram analysis, an approach that can be used to provide information about the nature and structure of spatial dependencies, has been recommended for this purpose [43]. While semivariogram analysis provides a way to identify and eliminate correlated, proximal occurrences, this only challenges models to fill gaps in a more densely sampled area. To challenge correlative models to predict accurately across unsampled areas of geographic space, available occurrence data for a species may be separated into quadrants based on whether their coordinates fall above or below the median longitude and median latitude of the species' occurrence data [45]. Models can then be calibrated using one pair of quadrants and evaluated using the other pair, permitting the use of all occurrences for both calibration and evaluation. An advantage of this approach is that the models are challenged to predict into large, unsampled regions, rather than simply interpolating or filling in small gaps in a densely sampled area [46].

Sampling biases in species occurrence data can degrade the performance of correlative models. While some modeling methods are more robust than others to the effects of spatial bias [47], careful consideration of how calibration data is selected can eliminate many sources of bias. In general, so long as focal environmental features are not missing and the range of environmental values that are sampled is similar to the range of values in the overall extent, it has been shown that correlative models can produce reasonable inferences and predictions. Further, in the event that spatial sampling does not fulfill these criteria, it has been shown that it may still be possible to derive reasonable inferences from presence-only data provided with some knowledge of the spatial distribution of sampling effort [34,48].

5.3.4 Presence-Only Data

Due to the paucity of data that characterize where a species is absent, most species data sets consist only of the known species observations (presence data) [49]. Accordingly, the data typically available to a correlative model are comprised of a set of species observations and the remaining, unlabeled geographic points. In the context of machine learning, this is referred to as partially labeled data. Additionally, the number of observations in these data sets is often far outnumbered by the number of unlabeled points—exhibiting a skew or imbalance. In the context of machine learning, this is known as the class imbalance problem [50].

Sample size has been found to be positively related to performance of SDMs [51–53]. Some studies, however, have found that only 50–100 observations of species presence were necessary to achieve acceptable performance [54–56]. As a large random sample of survey sites where the species is mostly absent may not give enough observations of presence, the adequate overall sample size depends on the rarity of the species. This occurs when species prevalence in the sample (i.e., the proportion of localities where the species is present) is allowed to vary with the "natural" prevalence of the species in the study area. Then, the proportion of sites that a species occupies reflects the rarity of the species within the study area [57].

In trying to distinguish environmental suitability based on presence and absence data, some studies have shown that an equal number of presence and absence observations yields an optimal balance between the omission and commission errors in model predictions [58]. These results suggest that perceived relationships between range size and model performance may be an artifact of species prevalence within the study area [59]. However, it has been found in other studies that model performance increases with species rarity [60], independent of manipulated sample prevalence, suggesting that rarer species have more predictable patterns of occurrence relative to environmental variables [58].

The effect of a species' natural prevalence on presence/absence modeling may be different from the effect of prevalence on modeling methods that employ only presence data. In the case of presence-only data, the sample prevalence or skew is a function of the number of background locations [31]. When augmenting presence-only species data via a random sample of background locations, many background points—several orders of magnitude more than the number of presences—may be required to capture and represent the full variation of environmental factors [61,62].

5.3.5 Model Evaluation

The value of a model depends on the accuracy of its outputs [63]. A crucial phase of correlative niche modeling is evaluating the predictive performance and statistical significance of a model. While effective models are always desired, it is not necessarily obvious how to best evaluate a model. Yet, as we discuss below, the manner in which a correlative model is evaluated can seriously influence the perceived quality and utility of its results.

As correlative models discriminate between geographical localities that constitute a species niche and those that do not, quantitative evaluation metrics can be used to measure the accuracy of these models. A variety of evaluation measures exist and have been discussed extensively in the literature [63–67]. In general, they measure either correlation (e.g., kappa coefficient, Pearson's correlation coefficient) or discrimination (e.g., F-measure, precision-recall, ROC, true skill statistic [TSS]). Correlation measures the degree to which the estimation of a predictive model matches the real outcome probability (i.e., the observed proportion of the event). Discrimination measures the ability of a model to distinguish a status (presence/absence).

Table 5.1 provides information on evaluation metrics commonly used in correlative niche modeling. These evaluation metrics are computed based on omission errors (false negatives) and commission errors (false positives). However, special considerations arise because of the questionable notion of what constitutes an "absence" in correlative niche modeling. While omission errors ($Y = \hat{Y} = 0$) are typically

TABLE 5.1   Evaluation Methods Commonly Employed

Method	Type	Reference
Area under the P-R curve (AUPR)	Discrimination	[50]
Area under the ROC curve (AUC)	Discrimination	[46]
F-measure	Discrimination	[71]
Cohen's kappa coefficient (kappa)	Correlation	[72]
Pearson's correlation coefficient (COR)	Correlation	[73]
True skill statistic (TSS)	Discrimination	[65]

genuine indicators of model failure, commission errors ($Y = 0$, $\hat{Y} = 1$) may not be erroneous at all in such applications. Apparent omission errors may arise from georeferencing mistakes or misleading factors. By contrast, apparent commission errors may arise not from mistakes, but from incomplete or unrepresentative sampling of species data. Strategies for addressing these errors have been the subject of extensive work [68–70].

The effect of omission and commission errors on the evaluation of correlative models is important to consider. Results have shown that the area under the ROC curve (AUC), a popular measure of model performance, can be an unreliable guide for model selection when calculated from presence-only data [46,74,75]. Other results have demonstrated theoretical and empirical evidence that kappa has serious limitations that make it similarly unsuitable when calculated from presence-only data [65]. Similar concerns apply to the other measures as well.

In general, measures such as AUC, *F*-measure, kappa, and TSS should be calculated from the presence–absence data that are randomly sampled from the population being modeled [71]. Additionally, using measures in combination may at times be more informative than reporting a solitary value [75].

5.4 Modeling Methods

A wide assortment of correlative models exist that employ different strategies drawn from a variety of related disciplines. Several extensive evaluations have shown that these different methods—some developed solely for application in modeling species niches—can produce markedly different predictions [15,22,76]. In this section, we outline the methods that have found widespread use in species modeling, highlighting their similarities and differences. While the set of models included in our discussion is by no means exhaustive, we believe that it is representative of the variety of correlative models that have been used to investigate species distributions.

Table 5.2 provides summary information for the models discussed. Some important differences exist between correlative models, including: (1) the type of species data that the method requires (i.e., presences only, presences and absences, or presences and background data); (2) the underlying methodological approach (e.g., distance metrics, bounding boxes, statistical modeling, machine learning, etc.); and (3) the form of output (e.g., continuous predictions versus binary or ordinal predictions). As result of these numerous differences, correlative models may be classified and categorized in a variety of ways.

We categorize correlative models according to the type of input data they require. The type of data required by a model is a practical consideration that is of categorical importance. Based on input data requirements, correlative models can be categorized into three groups, as follows:

1. *Presence/absence methods.* These methods function by contrasting the environment at locations where the species has been documented as present with locations where the species has been documented as absent, and therefore require both presence and absence data. Note that, in principle, any presence/absence algorithm can be implemented using background (sampling from the study area as a whole) or pseudoabsence (sampling from areas where the species has not been detected) data.
2. *Presence/background methods.* These methods assess how the environment at locations where the species is known to occur contrasts with the environment across the entire study area (i.e., the "background"). Accordingly, these approaches use presence data along with a sample of environmental data drawn from the study area, potentially including the known species occurrence localities.
3. *Presence-only methods.* These methods rely solely on environmental data where the species is known to occur, without need for reference to any other information drawn from the study area.

5.4.1 Presence/Absence Methods

Presence/absence models discriminate between localities where a species is observed to exist (presence data) and where it is observed not to exist (absence data). Several popular machine learning methods used in other domains can be leveraged to model species distributions with presence/absence data. These

TABLE 5.2   Modeling Methods Commonly Employed

Method	Explanation	Data[a]	Reference
ANN	Artificial neural network	pa	[103,104]
BIOCLIM	Bounding-box/envelope model	p	[97,98]
BIOMAPPER	Factor analysis	pa	[52,105]
BRT	Boosted regression trees	pa	[82]
DOMAIN	Multivariate distance	p	[101]
DT	Decision tree; classification and regression tree	pa	[106,107]
GAM	Regression; generalized additive model	pa	[35,91]
GARP	Rule sets from genetic algorithms	pa	[94,108]
GDM	Generalized dissimilarity modeling; uses community data	pa	[15]
GLM	Regression; generalized linear model	pa	[35,109]
HABITAT	Bounding-box/envelope model	p	[99]
LIVES	Multivariate distance	p	[15,102]
LR	Logistic regression	pa	[110,111]
MARS	Regression; multivariate adaptive regression splines	pa	[93,112]
MAXENT	Maximum entropy	pa	[33,62]
MAXLIKE	Maximum likelihood	pa	[95,96]
RF	Decision tree ensemble	pa	[80,113]
SPECIES	Neural network; uses climate data	pa	[84]
SVM	Support vector machine	p/pa	[114]

[a] p—only presence data used; pa—presence and some form of absence required—e.g., a background sample.

methods include decision trees (DTs), artificial neural networks (ANNs), support vector machines (SVMs), and variants thereof.

Classification trees (CTs) are DTs constructed by repeatedly partitioning species data into pairs of mutually exclusive groups, maximizing the homogeneity of each group. This partitioning procedure is applied to each group separately, and repeated until a tree is "grown." This tree is usually then "pruned" to achieve a desired level of simplicity or complexity [77,78]. The performance of an individual tree can frequently be improved by fitting many trees and combining them to generate predictions. Random forest (RF) is an ensemble DT that grows many CTs. Each CT is fit to a bootstrap sample of the data, but at each node, only a small number of randomly selected variables are available for the binary partitioning of the tree. The trees are fully grown and the predicted class of an observation is calculated by the majority vote from the ensemble for that observation [79,80]. Boosted regression trees (BRTs) build on regression trees fitted previously by focusing iteratively on the observed records that are hardest to predict. At each iteration, a regression tree is fitted on a weighted version of the data set, where the weights are continuously adjusted to take account of observations that are poorly fitted by the preceding models [81,82].

ANNs and SVMs are commonly employed computational models. Inspired by the way the brain works, ANNs "learn" species' responses to environmental predictor variables by repeatedly passing calibration data through a network of artificial "neurons." By adjusting internal structures of the neural network after each iteration, ANNs estimate a response in one part of the network based on inputs (environmental variables or latent factors thereof) at a different point in the network [83]. Spatial evaluation of climate impact on the envelope of species (SPECIES) is a domain-specific method that couples an ANN with a climate-hydrological process to model species distributions [84]. When employed with presence/absence data, two-class SVMs search for an optimal separating hyperplane capable of discriminating between species presences and absences in the environmental space. This is accomplished by nonlinearly mapping the environmental data into a high-dimensional feature space, wherein a linear decision surface is constructed [85,86].

Generalized dissimilarity models (GDMs) model the compositional dissimilarity between pairs of species presence localities as a function of environmental differences between these localities. The approach combines elements of matrix regression and generalized linear modeling in order to model nonlinear responses to the environment that capture ecologically plausible relationships between environmental dissimilarity and ecological distance [87,88]. For predicting species' distributions, an additional kernel regression algorithm [89] may be applied within the transformed or latent environmental space generated by a GDM to estimate the likelihood of occurrence of a given species at each locality [15].

5.4.2 Presence/Background Methods

Presence/background methods discriminate between localities where a species is observed to exist (presence data) and the entire geographic study area (i.e., the "background"). These methods include factor analysis, regression-based techniques, genetic algorithms, and maximum entropy models.

A domain-specific factor analysis technique, known as ecological niche factor analysis (ENFA), calculates a measure of habitat suitability based on the analysis of how the species mean differs from the global mean (marginality) and how the species variance differs from the global variance (environmental tolerance) [49]. A threshold on the suitability value can then be applied to determine the boundaries of the ecological niche [52].

Several regression approaches have been used for modeling species distributions. Generalized linear models (GLMs) generalize linear regression by allowing a linear model to be related to a response variable via a link function. This approach enables the modeling of species responses to a wide range of environmental data types under a single theoretical and computational framework [90]. Generalized additive models (GAMs) are a nonparametric extension of GLMs that use data-defined smoothing functions to fit nonlinear responses. Because of their greater flexibility, GAMs are more capable of modeling complex ecological response shapes than GLMs [91]. Both have found widespread use in correlative niche modeling because of their strong statistical foundation and ability to realistically model ecological relationships [92]. Multivariate adaptive regression splines (MARS) provide an alternative regression-based method for fitting nonlinear responses based on using piecewise linear fits rather than smooth functions, with the advantage of being faster than GAMs [93].

A popular genetic algorithm used for modeling species distributions is known as the genetic algorithm for rule-set production (GARP). GARP produces sets of rules (e.g., adaptations of regression and range specifications) that delineate ecological niches. Like many evolutionary algorithms, GARP works in an iterative process of rule selection, evaluation, testing, and incorporation or rejection. At each iteration, predictive accuracy is evaluated on the test presence data and a set of pseudoabsence points, and the change in predictive accuracy from one iteration to the next is used to evaluate whether or not a particular rule should be incorporated into the model [94].

One of the most widely used methods for modeling species' responses to environmental factors is the maximum entropy approach known as Maxent. Maxent estimates species' distributions by finding the distribution of maximum entropy (i.e., the distribution closest to uniform) subject to the constraint that the expected value of each environmental variable (or its transform and/or interactions) under this estimated distribution matches its empirical average [37]. Recently, Royle et al. [95] advocated a similar model, informally called Maxlike, to remedy issues that they found with Maxent [96].

5.4.3 Presence-Only Methods

Presence-only models rely solely on information at localities where the species is known to exist. As species absence data is relatively uncommon compared to a growing amount of presence data available, there has been increasing focus on developing methods that can be applied to presence-only data (i.e., data lacking information on surveyed locations where a species is absent). Presence-only models include bounding-box, distance-based, and statistical methods.

The first correlative niche model to find widespread use was BIOCLIM, a bounding-box or envelope-style method that characterizes the species niche as the bounding hyper-box that encloses all the records of the species in the climatic space [97,98]. HABITAT defines the distribution as the convex hull of the species observations in the environmental space [99]. In general, HABITAT is more effective than BIOCLIM at adjusting the boundaries of the environmental envelope to the distribution of the species observations in the environmental space [49]. As with BIOCLIM, however, only the outer observations (i.e., those observations that constitute the boundary of the convex hull in environmental space) are used to determine the boundaries of the species distribution.

A variety of distance-based methods exist. One approach ranks potential sites by their Mahalanobis distance to a vector expressing the mean environmental conditions of all the records in the environmental space. A specified distance threshold is then used to define the boundaries of the species distribution, producing an ellipsoidal envelope that explicitly accounts for potential correlations between the environmental variables [100]. DOMAIN and LIVES are distance-based methods that assess new occurrence localities in terms of their environmental similarity to localities of known species presence. DOMAIN uses a point-to-point similarity metric (based on the Gower distance) to assign a value of habitat suitability to each potential site based on its proximity in the environmental space to the closest (most similar) occurrence location [101]. LIVES is based on limiting factor theory, which postulates that the occurrence of a species is solely determined by the factor that most limits its distribution [102].

Note that an important limitation of models fit to presence-only data is that they can only predict the relative likelihood of species presence or habitat suitability at a given location. This is because presence-only data lack information about species prevalence. By contrast, presence/absence data can be used to predict the probability of species presence.

5.5 Deploying Large-Scale Models

A variety of tools are available for generating correlative niche models. However, these tools are becoming increasingly fragmented, implementing different best practices and each only providing access to a small subset of the commonly used methods. Moreover, the large-scale deployment of correlative models demands sophisticated computational resources, technical skills, and access to biological and climatic data. In this section, we outline the tools available for generating correlative models and the efforts underway to unify them into coherent cloud-based platforms.

5.5.1 Modeling in Practice

A variety of approaches to correlative niche modeling exist. These approaches are accessible through a similarly diverse range of platforms, from packages for popular programming languages to stand-alone desktop applications. We highlight some of the focal developments in the practical deployment of correlative niche models.

Presently, the programming language R* is the most widely used modeling platform. Popular R packages include BIOMOD [115], dismo [116], and SDMTools [117]. R provides a free and open-source option for deploying correlative models as it allows users a great deal of flexibility with a variety of relevant packages. However, properly leveraging these packages requires a sophisticated understanding of related programmatic, computational, and statistical concepts. The programming language Python also enables the deployment of free and open correlative models via the package scikit-learn,† which provides functionality for modeling species' geographic distributions [118].

---

* http://www.r-project.org
† http://www.scikit-learn.org

To simplify the model deployment process, stand-alone applications have been developed for certain algorithms that feature graphical user interfaces and more clearly defined workflows. Examples include Maxent [37] and Desktop GARP [94]. While these applications serve to support individual algorithms, openModeller Desktop [119,120] was conceived as a generic desktop framework to provide an open, standardized, and integrated infrastructure for accommodating a flexible, user-friendly, cross-platform environment capable of handling different data formats and multiple algorithms that can be used in correlative niche modeling. Accordingly, openModeller serves an underlying framework that unifies a variety of algorithms into a single coherent desktop application.

5.5.2 Modeling in the Cloud

Despite the availability of modeling software, the large-scale deployment of correlative models demands sophisticated computational resources, technical skills, and access to biological and climatic data. Correlative models require the handling and storage of large spatial data sets that often need intensive formatting, rescaling, projecting from different pieces of data, cropping (clipping) to match certain spatial extents, and reclassifying such that categorical elements are consistent across years. This filtering and processing can be a significant burden and can represent a significant barrier to entry. Further, the required expertise prevents many potential stakeholders (e.g., natural resource managers) from generating their own simulations for use in forming adaptation plans. The use of correlative modeling to inform adaptation plans is also beset with fundamental issues that involve proper use and interpretation of modeling outputs.

"Collaboratories" are emerging as a potential solution to overcome these limitations. As originally coined by William Wulf, prior director of the National Science Foundation's Directorate for Computer and Information Science and Engineering, a "collaboratory" was envisioned as virtual scientific research collaboration: "A 'center without walls,' in which ... researchers can perform their research without regard to geographical location—interacting with colleagues, accessing instrumentation, sharing data and computational resources, accessing information in digital libraries" [121]. Collaboratories are characterized by features that include: (1) a shared interest in a common goal and/or problem; (2) active interaction and contribution by participants; (3) shared information resources; (4) extensive use of technologies; and (5) bridge geographical, temporal, institutional, and disciplinary divides.

To this end, several online, cyberinfrastructure-based collaboratories have been developed that cater to the large-scale application of correlative niche models, including EUBrazil OpenBio* and the Collaboratory for Adaptation to Climate Change.† The EUBrazilOpenBio initiative is a cyberinfrastructure that provides the biodiversity community with a rich set of computational and data resources exploiting existing cloud technologies from the European Union and Brazil. The platform provides ecological practitioners with a workflow through which they may generate, evaluate, and interpret correlative niche models [122]. The Collaboratory for Adaptation to Climate Change relies on cyberinfrastructure, data and knowledge management, simulations, scenario analysis, and visual analytics to provide next-generation biological models for simulating geographic-range change and assessing ecosystem vulnerability due to climate change. The Collaboratory provides access to NatureServe's climate change vulnerability index (CCVI) [123] and hosts the Spatial Portal for Analysis of Climatic Effects on Species (SPACES), an online platform with access to cutting-edge modeling techniques and a multidisciplinary community.

5.6 Future Directions

As ecological data sets continue to grow larger and as the availability of a diverse collection of modeling choices continues to expand, correlative niche models stand poised to contribute more than ever to efforts

---

* http://www.eubrazilopenbio.eu/
† https://adapt.nd.edu/

in ecology and conservation ecology. These trends suggest that the techniques with the most to gain may be those that can effectively leverage the increasingly available assortments of data and algorithms.

With crowdsourced ecological data becoming more readily available, correlative models are being increasingly deployed on the large, longitudinal data sets curated from these "citizen science" efforts. While these efforts can serve to expand the potential for spatial ecology research by producing data sets that are larger than those otherwise available [124,125], they also serve to introduce further concerns about the quality of data gathered. These efforts also risk fragmentation and differing standards, with data potentially gathered in different ways, through different efforts, and at different extents and granularities. Increasingly relevant are techniques that can improve the quality of crowdsourced data and methods that can provide a level of resiliency against the efforts of imperfectly collected data [126,127].

The growth of data is not limited to correlative variables. Promising work is also being applied to the inclusion of information that can supplement correlative data, such as expert- or survey-derived estimates of detectability, of species dispersal or occupancy correlates, or of relative phylogenetic position [126]. These are all powerful forms of information if many species are assessed in a single framework, similar to the practice of using a collection of related species to correct for sample selection bias [128]. The output of resulting integrative models (i.e., models that, unlike ordinary correlative models, manage to combine information from a broader range of data types and uncertainties) can subsequently be used to depict the best possible estimate of the distribution of a species over a specific timeframe [129].

A continuing difficulty in the use of correlative methods to estimate species distributions is the difficulty of reliably estimating species abundance. Though many correlative models provide reliable estimates of the relative abundance rate over the study extent, few methods are able to provide a reliable estimate of the actual rate of abundance over the same extent. Promising lines of research have focused on addressing this deficiency of correlative modeling by introducing new models [95,130], though the effectiveness and validity of such methods are the subject of ongoing debate [131–133].

Acknowledgment

We gratefully acknowledge the support of the National Science Foundation (NSF) Grant OCI-1029584 for this work.

References

1. Joseph F. C. DiMento and Helen Ingram. Science and environmental decision making: The potential role of environmental impact assessment in the pursuit of appropriate information. *Natural Resources Journal*, 45:283, 2005.

2. Jane Elith and John R. Leathwick. Species distribution models: Ecological explanation and prediction across space and time. *Annual Review of Ecology, Evolution, and Systematics*, 40(1):677, 2009.

3. Niklaus E. Zimmermann, Thomas C. Edwards, Catherine H. Graham, Peter B. Pearman, and Jens-Christian Svenning. New trends in species distribution modelling. *Ecography*, 33(6):985–989, 2010.

4. Inés Ibáñez, Elise S. Gornish, Lauren Buckley, Diane M. Debinski, Jessica Hellmann, Brian Helmuth, Janneke HilleRisLambers, Andrew M. Latimer, Abraham J. Miller-Rushing, and Maria Uriarte. Moving forward in global-change ecology: Capitalizing on natural variability. *Ecology and Evolution*, 3(1):170–181, 2013.

5. Sarah E. Gilman, Mark C. Urban, Joshua Tewksbury, George W. Gilchrist, and Robert D. Holt. A framework for community interactions under climate change. *Trends in Ecology & Evolution*, 25(6):325–331, 2010.

6. Jorge Soberon, Jorge Llorente, and Hesiquio Benitez. An international view of national biological surveys. *Annals of the Missouri Botanical Garden*, 83(4):562–573, 1996.

7. Catherine H. Graham, Santiago R. Ron, Juan C. Santos, Christopher J. Schneider, and Craig Moritz. Integrating phylogenetics and environmental niche models to explore speciation mechanisms in dendrobatid frogs. *Evolution*, 58(8):1781–1793, 2004.

8. Jorge Soberón and A. Townsend Peterson. Interpretation of models of fundamental ecological niches and species' distributional areas. *Biodiversity Informatics*, 2:1–10, 2005.

9. N. R. Chalmers. Monitoring and inventorying biodiversity: Collections, data and training. In *Biodiversity, Science and Development: Towards a New Partnership*, eds F. di Castri and T. Younés, p. 171. Wallingford: CAB International, 1996.

10. Leonard Krishtalka and Philip S. Humphrey. Can natural history museums capture the future? *BioScience*, 50(7):611–617, 2000.

11. James L. Edwards. Research and societal benefits of the global biodiversity information facility. *BioScience*, 54(6):485–486, 2004.

12. Camilo Mora, Derek P. Tittensor, Sina Adl, Alastair G. B. Simpson, and Boris Worm. How many species are there on Earth and in the ocean? *PLoS Biology*, 9(8):e1001127, 2011.

13. Robert J. Hijmans, Susan E. Cameron, Juan L. Parra, Peter G. Jones, and Andy Jarvis. Very high resolution interpolated climate surfaces for global land areas. *International Journal of Climatology*, 25:1965–1978, 2005.

14. A. Townsend Peterson. Predicting the geography of species' invasions via ecological niche modeling. *The Quarterly Review of Biology*, 78(4):419–433, 2003.

15. Jane Elith, Catherine H. Graham, Robert P. Anderson, Miroslav Dudík, Simon Ferrier, Antoine Guisan, Robert J. Hijmans, Falk Huettmann, John R. Leathwick, Anthony Lehmann et al. Novel methods improve prediction of species' distributions from occurrence data. *Ecography*, 29(2):129–151, 2006.

16. A. Townsend Peterson, Jorge Soberón, Richard G. Pearson, Robert P. Anderson, Enrique Martínez-Meyer, Miguel Nakamura, and Miguel B. Araújo. *Ecological Niches and Geographic Distributions (MPB-49)*. Princeton, NJ: Princeton University Press, 2011.

17. Steven I. Higgins, Robert B. O'Hara, and Christine Römermann. A niche for biology in species distribution models. *Journal of Biogeography*, 39(12):2091–2095, 2012.

18. A. Townsend Peterson and Jorge Soberón. Species distribution modeling and ecological niche modeling: Getting the concepts right. *Natureza & Conservação*, 10(2):102–107, 2012.

19. Lauren B. Buckley, Mark C. Urban, Michael J. Angilletta, Lisa G. Crozier, Leslie J. Rissler, and Michael W. Sears. Can mechanism in form species' distribution models? *Ecology Letters*, 13(8):1041–1054, 2010.

20. H. Ronald Pulliam. On the relationship between niche and distribution. *Ecology Letters*, 3(4):349–361, 2000.

21. Richard G. Pearson and Terence P. Dawson. Predicting the impacts of climate change on the distribution of species: Are bioclimate envelope models useful? *Global Ecology and Biogeography*, 12(5):361–371, 2003.

22. Richard G. Pearson, Wilfried Thuiller, Miguel B. Araújo, Enrique Martinez-Meyer, Lluís Brotons, Colin McClean, Lera Miles, Pedro Segurado, Terence P. Dawson, and David C. Lees. Model-based uncertainty in species range prediction. *Journal of Biogeography*, 33(10):1704–1711, 2006.

23. John A. Wiens, Diana Stralberg, Dennis Jongsomjit, Christine A. Howell, and Mark A. Snyder. Niches, models, and climate change: Assessing the assumptions and uncertainties. *Proceedings of the National Academy of Sciences*, 106(Supplement 2):19729–19736, 2009.

24. Miguel B. Araújo and A. Townsend Peterson. Uses and misuses of bioclimatic envelope modeling. *Ecology*, 93(7):1527–1539, 2012.

25. Antoine Guisan and Wilfried Thuiller. Predicting species distribution: Offering more than simple habitat models. *Ecology Letters*, 8(9):993–1009, 2005.

26. Mike P. Austin. Species distribution models and ecological theory: A critical assessment and some possible new approaches. *Ecological Modelling*, 200(1):1–19, 2007.

27. E Dooley. Global biodiversity information facility. *Environmental Health Perspectives*, 110(11):A669, 2002.

28. R. J. Hijmans, S. E. Cameron, J. L. Parra, P. G. Jones, and A. Jarvis. *The WorldClim interpolated global terrestrial climate surfaces. Version 1.3*, 2004. Available at http://www.worldclim.org/

29. Andrew J. Davis, Linda S. Jenkinson, John H. Lawton, Bryan Shorrocks, and Simon Wood. Making mistakes when predicting shifts in species range in response to global warming. *Nature*, 391(6669):783–786, 1998.

30. J. L. Lawton. Concluding remarks: A review of some open questions. In *Ecological Consequences of Heterogeneity*, eds M. J. Hutchings, E. A. John, and A. J. A. Stewart, pp. 401–424. Oxford: Blackwell Science, 2000.

31. Janet Franklin. *Mapping Species Distributions: Spatial Inference and Prediction*. Cambridge: Cambridge University Press, 2010.

32. Robert P. Anderson. Harnessing the world's biodiversity data: Promise and peril in ecological niche modeling of species distributions. *Annals of the New York Academy of Sciences*, 1260(1):66–80, 2012.

33. Steven J. Phillips, Miroslav Dudík, and Robert E. Schapire. A maximum entropy approach to species distribution modeling. In *Proceedings of the 21st International Conference on Machine Learning (ICML)*, p. 83. ACM, 2004.

34. Miguel B. Araújo and Antoine Guisan. Five (or so) challenges for species distribution modelling. *Journal of Biogeography*, 33(10):1677–1688, 2006.

35. Christophe F. Randin, Thomas Dirnböck, Stefan Dullinger, Niklaus E. Zimmermann, Massimiliano Zappa, and Antoine Guisan. Are niche based species distribution models transferable in space? *Journal of Biogeography*, 33(10):1689–1703, 2006.

36. Robert P. Anderson and Ali Raza. The effect of the extent of the study region on GIS models of species geographic distributions and estimates of niche evolution: Preliminary tests with montane rodents (genus Nephelomys) in Venezuela. *Journal of Biogeography*, 37(7):1378–1393, 2010.

37. Steven J. Phillips, Robert P. Anderson, and Robert E. Schapire. Maximum entropy modeling of species geographic distributions. *Ecological Modelling*, 190(3):231–259, 2006.

38. Mike P. Austin and Jacqui A. Meyers. Current approaches to modelling the environmental niche of eucalypts: Implication for management of forest biodiversity. *Forest Ecology and Management*, 85(1):95–106, 1996.

39. Jari Oksanen and Peter R. Minchin. Continuum theory revisited: What shape are species responses along ecological gradients? *Ecological Modelling*, 157(2):119–129, 2002.

40. Peter H. Raven, J. Michael Scott, Patricia Heglund, and Michael L. Morrison. *Predicting Species Occurrences: Issues of Accuracy and Scale*. Washington, DC: Island Press, 2002.

41. Antoine Guisan, Catherine H. Graham, Jane Elith, and Falk Huettmann. Sensitivity of predictive species distribution models to change in grain size. *Diversity and Distributions*, 13(3):332–340, 2007.

42. Jesús Aguirre-Gutiérrez, Luísa G. Carvalheiro, Chiara Polce, E. Emiel van Loon, Niels Raes, Menno Reemer, and Jacobus C. Biesmeijer. Fitfor-purpose: Species distribution model performance depends on evaluation criteria—Dutch hoveries as a case study. *PLOS ONE*, 8(5):e63708, 2013.

43. Samuel D. Veloz. Spatially autocorrelated sampling falsely inates measures of accuracy for presence-only niche models. *Journal of Biogeography*, 36(12):2290–2299, 2009.

44. Corinna Cortes, Mehryar Mohri, Michael Riley, and Afshin Rostamizadeh. Sample selection bias correction theory. In Proceedings of the 19th International Conference on *Algorithmic Learning Theory*, pp. 38–53. Heidelberg: Springer, 2008.

45. A. Townsend Peterson, Monica Papeş, and Muir Eaton. Transferability and model evaluation in ecological niche modeling: A comparison of GARP and Maxent. *Ecography*, 30(4):550–560, 2007.

46. A. Townsend Peterson, Monica Papeş, and Jorge Soberón. Rethinking receiver operating characteristic analysis applications in ecological niche modeling. *Ecological Modelling*, 213(1):63–72, 2008.

47. Catherine H. Graham, Jane Elith, Robert J. Hijmans, Antoine Guisan, A. Townsend Peterson, and Bette A. Loiselle. The inuence of spatial errors in species occurrence data used in distribution models. *Journal of Applied Ecology*, 45(1):239–247, 2008.

48. Jennie L. Pearce and Mark S. Boyce. Modelling distribution and abundance with presence-only data. *Journal of Applied Ecology*, 43(3):405–412, 2006.

49. Asaf Tsoar, Omri Allouche, Ofer Steinitz, Dotan Rotem, and Ronen Kadmon. A comparative evaluation of presence-only methods for modelling species distribution. *Diversity and Distributions*, 13(4):397–405, 2007.

50. Reid A. Johnson, Nitesh V. Chawla, and Jessica. J. Hellmann. Species distribution modeling and prediction: A class imbalance problem. In *Proceedings of the Conference on Intelligent Data Understanding (CIDU)*. CIDU, 2012.

51. Graeme S. Cumming. Using between-model comparisons to fine-tune linear models of species ranges. *Journal of Biogeography*, 27(2):441–455, 2000.

52. Alexandre H. Hirzel, J. Hausser, D. Chessel, and N. Perrin. Ecologicalniche factor analysis: How to compute habitat-suitability maps without absence data? *Ecology*, 83(7):2027–2036, 2002.

53. Gordon C. Reese, Kenneth R. Wilson, Jennifer A. Hoeting, and Curtis H. Flather. Factors affecting species distribution predictions: A simulation modeling experiment. *Ecological Applications*, 15(2):554–564, 2005.

54. David R. B. Stockwell and A. Townsend Peterson. Effects of sample size on accuracy of species distribution models. *Ecological Modelling*, 148(1):1–13, 2002.

55. Ronen Kadmon, Oren Farber, and Avinoam Danin. A systematic analysis of factors affecting the performance of climatic envelope models. *Ecological Applications*, 13(3):853–867, 2003.

56. Bette A. Loiselle, Peter M. Jørgensen, Trisha Consiglio, Iván Jiménez, John G. Blake, Lucia G. Lohmann, and Olga Martha Montiel. Predicting species distributions from herbarium collections: Does climate bias in collection sampling inuence model outcomes? *Journal of Biogeography*, 35(1):105–116, 2008.

57. Stéphanie Manel, H. Ceri Williams, and Stephen James Ormerod. Evaluating presence-absence models in ecology: The need to account for prevalence. *Journal of Applied Ecology*, 38(5):921–931, 2001.

58. Janet Franklin, Katherine E. Wejnert, Stacie A. Hathaway, Carlton J. Rochester, and Robert N. Fisher. Effect of species rarity on the accuracy of species distribution models for reptiles and amphibians in southern California. *Diversity and Distributions*, 15(1):167–177, 2009.

59. Jana M. McPherson, Walter Jetz, and David J. Rogers. The effects of species' range sizes on the accuracy of distribution models: Ecological phenomenon or statistical artefact? *Journal of Applied Ecology*, 41(5):811–823, 2004.

60. Pedro Segurado and Miguel B. Araújo. An evaluation of methods for modelling species distributions. *Journal of Biogeography*, 31(10):1555–1568, 2004.

61. Bryan F. J. Manly, Lyman L. McDonald, Dana L. Thomas, Trent L. McDonald, and Wallace P. Erickson. *Resource Selection by Animals: Statistical Design and Analysis for Field Studies*. Netherlands: Springer, 2002.

62. Steven J. Phillips and Miroslav Dudík. Modeling of species distributions with maxent: New extensions and a comprehensive evaluation. *Ecography*, 31(2):161–175, 2008.

63. Giles M. Foody. Impacts of imperfect reference data on the apparent accuracy of species presence–absence models and their predictions. *Global Ecology and Biogeography*, 20(3):498–508, 2011.

64. Alan H. Fielding and John F. Bell. A review of methods for the assessment of prediction errors in conservation presence/absence models. *Environmental Conservation*, 24(01):38–49, 1997.

65. Omri Allouche, Asaf Tsoar, and Ronen Kadmon. Assessing the accuracy of species distribution models: Prevalence, kappa and the true skill statistic (TSS). *Journal of Applied Ecology*, 43(6):1223–1232, 2006.

66. Jorge M. Lobo, Alberto. Jiménez-Valverde, and Raimundo Real. AUC: A misleading measure of the performance of predictive distribution models. *Global Ecology and Biogeography*, 17(2):145–151, 2008.

67. Canran Liu, Matt White, and Graeme Newell. Measuring and comparing the accuracy of species distribution models with presence–absence data. *Ecography*, 34(2):232–243, 2011.

68. Darryl I. MacKenzie, Larissa L. Bailey, and James Nichols. Investigating species co-occurrence patterns when species are detected imperfectly. *Journal of Animal Ecology*, 73(3):546–555, 2004.

69. J. Andrew Royle and William A. Link. Generalized site occupancy models allowing for false positive and false negative errors. *Ecology*, 87(4):835–841, 2006.

70. Trevor J. Heey, David M. Baasch, Andrew J. Tyre, and Erin E. Blankenship. Correction of location errors for presence-only species distribution models. *Methods in Ecology and Evolution*, 5(3):207–214, 2014.

71. Wenkai Li and Qinghua Guo. How to assess the prediction accuracy of species presence–absence models without absence data? *Ecography*, 36(7):788–799, 2013.

72. Guofan Shao and Patrick N. Halpin. Climatic controls of eastern North American coastal tree and shrub distributions. *Journal of Biogeography*, 22(6):1083–1089, 1995.

73. Jane Elith, Michael Kearney, and Steven Phillips. The art of modelling range-shifting species. *Methods in Ecology and Evolution*, 1(4):330–342, 2010.

74. Duncan Golicher, Andrew Ford, Luis Cayuela, and Adrian Newton. Pseudo-absences, pseudo-models and pseudo-niches: Pitfalls of model selection based on the area under the curve. *International Journal of Geographical Information Science*, 26(11):2049–2063, 2012.

75. Alberto Jiménez-Valverde. Insights into the area under the receiver operating characteristic curve (AUC) as a discrimination measure in species distribution modelling. *Global Ecology and Biogeography*, 21(4):498–507, 2012.

76. Richard J. Ladle, Paul Jepson, Miguel B. Araújo, and Robert J. Whittaker. Dangers of crying wolf over risk of extinctions. *Nature*, 428(6985):799–799, 2004.

77. J. Ross Quinlan. Induction of decision trees. *Machine Learning*, 1(1):81–106, 1986.

78. Glenn De'ath and Katharina E. Fabricius. Classification and regression trees: A powerful yet simple technique for ecological data analysis. *Ecology*, 81(11):3178–3192, 2000.

79. Leo Breiman. Random forests. *Machine Learning*, 45(1):5–32, 2001.

80. D. Richard Cutler, Thomas C. Edwards Jr., Karen H. Beard, Adele Cutler, Kyle T. Hess, Jacob Gibson, and Joshua J. Lawler. Random forests for classification in ecology. *Ecology*, 88(11): 2783–2792, 2007.

81. Yoav Freund, Robert E. Schapire. Experiments with a new boosting algorithm. In *Proceedings of the International Conference on Machine Learning (ICML)*, volume 96, pp. 148–156, 1996.

82. Glenn De'Ath. Boosted trees for ecological modeling and prediction. *Ecology*, 88(1):243–251, 2007.

83. Simon Haykin and Richard Lippmann. Neural networks, a comprehensive foundation. *International Journal of Neural Systems*, 5(4):363–364, 1994.

84. Richard G. Pearson, Terence P. Dawson, Pam M. Berry, and P. A. Harrison. SPECIES: A spatial evaluation of climate impact on the envelope of species. *Ecological Modelling*, 154(3):289–300, 2002.

85. Corinna Cortes and Vladimir Vapnik. Support-vector networks. *Machine Learning*, 20(3):273–297, 1995.

86. Qinghua Guo, Maggi Kelly, and Catherine H. Graham. Support vector machines for predicting distribution of Sudden Oak Death in California. *Ecological Modelling*, 182(1):75–90, 2005.

87. Simon Ferrier, Graham Watson, Jennie Pearce, and Michael Drielsma. Extended statistical approaches to modelling spatial pattern in biodiversity in northeast New South Wales. I. species-level modelling. *Biodiversity & Conservation*, 11(12):2275–2307, 2002.

88. Simon Ferrier, Michael Drielsma, Glenn Manion, and Graham Watson. Extended statistical approaches to modelling spatial pattern in biodiversity in northeast New South Wales. II. community-level modelling. *Biodiversity & Conservation*, 11(12):2309–2338, 2002.

89. David G. Lowe. Similarity metric learning for a variable-kernel classifier. *Neural Computation*, 7(1):72–85, 1995.

90. Thomas W. Yee and Monique Mackenzie. Vector generalized additive models in plant ecology. *Ecological Modelling*, 157(2):141–156, 2002.

91. Thomas W. Yee and Neil D. Mitchell. Generalized additive models in plant ecology. *Journal of Vegetation Science*, 2(5):587–602, 1991.

92. Mike P. Austin. Spatial prediction of species distribution: An interface between ecological theory and statistical modelling. *Ecological Modelling*, 157(2):101–118, 2002.

93. J. R. Leathwick, D. Rowe, J. Richardson, Jane Elith, and Trevor J. Hastie. Using multivariate adaptive regression splines to predict the distributions of New Zealand's freshwater diadromous fish. *Freshwater Biology*, 50(12):2034–2052, 2005.

94. David R. B. Stockwell and David Peters. The GARP modelling system: Problems and solutions to automated spatial prediction. *International Journal of Geographical Information Science (IJGIS)*, 13(2):143–158, 1999.

95. J. Andrew Royle, Richard B. Chandler, Charles Yackulic, and James D. Nichols. Likelihood analysis of species occurrence probability from presence-only data for modelling species distributions. *Methods in Ecology and Evolution*, 3(3):545–554, 2012.

96. Cory Merow and John A. Silander. A comparison of Maxlike and Maxent for modelling species distributions. *Methods in Ecology and Evolution*, 5(3):215–225, 2014.

97. J R Busby. BIOCLIM—A bioclimate analysis and prediction system. In: C. R. Margules and M. P. Austin (eds), *Nature conservation: cost effective biological surveys and data analysis*. CSIRO, pp. 64–68, 1991.

98. Trevor H. Booth, Henry A. Nix, John R. Busby, and Michael F. Hutchinson. BIOCLIM: The first species distribution modelling package, its early applications and relevance to most current MAXENT studies. *Diversity and Distributions*, 20(1):1–9, 2014.

99. P. A. Walker and K. D. Cocks. HABITAT: A procedure for modelling a disjoint environmental envelope for a plant or animal species. *Global Ecology and Biogeography Letters*, 1(4):108–118, 1991.

100. Oren Farber and Ronen Kadmon. Assessment of alternative approaches for bioclimatic modeling with special emphasis on the Mahalanobis distance. *Ecological Modelling*, 160(1):115–130, 2003.

101. Guy Carpenter, Andrew N. Gillison, and J. Winter. DOM/AIN: A flexible modelling procedure for mapping potential distributions of plants and animals. *Biodiversity and Conservation*, 2(6):667–680, 1993.

102. Jin Li and David W. Hilbert. LIVES: A new habitat modelling technique for predicting the distribution of species' occurrences using presence only data based on limiting factor theory. *Biodiversity and Conservation*, 17(13):3079–3095, 2008.

103. Heinrich Werner and Michael Obach. New neural network types estimating the accuracy of response for ecological modelling. *Ecological Modelling*, 146(1):289–298, 2001.

104. Stéphanie Manel, J. M. Dias, S. T. Buckton, and S. J. Ormerod. Alternative methods for predicting species distribution: An illustration with Himalayan river birds. *Journal of Applied Ecology*, 36(5):734–747, 1999.

105. Alexandre H. Hirzel, Véronique Helfer, and F. Metral. Assessing habitat-suitability models with a virtual species. *Ecological Modelling*, 145(2):111–121, 2001.

106. David R. B. Stockwell, S. M. Davey, J. R. Davis, and Ian R. Noble. Using induction of decision trees to predict greater glider density. *AI Applications in Natural Resource Management*, 4(4):33–43, 1990.

107. Ana C. Lorena, Luis F. O. Jacintho, Marinez F. Siqueira, Renato De Giovanni, Lúcia G. Lohmann, André C. P. L. F. De Carvalho, and Missae Yamamoto. Comparing machine learning classifiers in potential distribution modelling. *Expert Systems with Applications*, 38(5):5268–5275, 2011.

108. David R. B. Stockwell and Ian R. Noble. Induction of sets of rules from animal distribution data: A robust and informative method of data analysis. *Mathematics and Computers in Simulation*, 33(5):385–390, 1992.

109. Mike P. Austin and R. B. Cunningham. Observational analysis of environmental gradients. *The Proceedings of the Ecological Society of Australia*, 11:109–119, 1981.

110. Janet Franklin. Predictive vegetation mapping: Geographic modelling of biospatial patterns in relation to environmental gradients. *Progress in Physical Geography*, 19(4):474–499, 1995.

111. Kim A. Keating and Steve Cherry. Use and interpretation of logistic regression in habitat-selection studies. *Journal of Wildlife Management*, 68(4):774–789, 2004.

112. Jane Elith and John Leathwick. Predicting species distributions from museum and herbarium records using multiresponse models fitted with multivariate adaptive regression splines. *Diversity and Distributions*, 13(3):265–275, 2007.

113. Jan Peters, Bernard De Baets, Niko E. C. Verhoest, Roeland Samson, Sven Degroeve, Piet De Becker, and Willy Huybrechts. Random forests as a tool for ecohydrological distribution modelling. *Ecological Modelling*, 207(2):304–318, 2007.

114. John M. Drake, Christophe Randin, and Antoine Guisan. Modelling ecological niches with support vector machines. *Journal of Applied Ecology*, 43(3):424–432, 2006.

115. Wilfried Thuiller, Bruno Lafourcade, Robin Engler, and Miguel B. Araújo. BIOMOD–a platform for ensemble forecasting of species distributions. *Ecography*, 32(3):369–373, 2009.

116. Robert J. Hijmans, Steven Phillips, John Leathwick, and Jane Elith. dismo: Species distribution modeling. *R Package Version 1.0–5*, 2012.

117. Jeremy VanDerWal, Lorena Falconi, Stephanie Januchowski, Luke Shoo, and Collin Storlie. SDM-Tools: Species distribution modelling tools: Tools for processing data associated with species distribution modelling exercises. *R Package Version 1.1–221*, 1(5), 2014.

118. Fabian Pedregosa, Gaël Varoquaux, Alexandre Gramfort, Vincent Michel, Bertrand Thirion, Olivier Grisel, Mathieu Blondel, Peter Prettenhofer, Ron Weiss, Vincent Dubourg, et al. Scikit-learn: Machine learning in python. *The Journal of Machine Learning Research*, 12:2825–2830, 2011.

119. Mauro Enrique de Souza Muñoz. openModeller: A framework for biological/environmental modelling. In *Inter-American Workshop on Environmental Data Access, Campinas, SP. Brazil*, 2004.

120. Mauro Enrique de Souza Muñoz, Renato De Giovanni, Marinez Ferreira de Siqueira, Tim Sutton, Peter Brewer, Ricardo Pereira Scachetti, Dora Ann Lange Canhos, and Vanderlei Perez Canhos. Open Modeller: A generic approach to species' potential distribution modelling. *GeoInformatica*, 15(1):111–135, 2011.

121. William A. Wulf. The national collaboratory—A white paper. In *Appendix A in Toward a National Collaboratory, unpublished report of a National Science Foundation Invitational Workshop held at Rockefeller University*, p. 1, 1989.

122. Daniele Lezzi, Roger Rafanell, Erik Torres, Renato De Giovanni, Ignacio Blanquer, and Rosa M. Badia. Programming ecological niche modeling workflows in the cloud. In *Proceedings of the 27th International Conference on Advanced Information Networking and Applications Workshops (WAINA)*, pp. 1223–1228. IEEE, 2013.

123. Bruce Young, Elizabeth Byers, Kelly Gravuer, K. Hall, G. Hammerson, and A. Redder. *Guidelines for Using the NatureServe Climate Change Vulnerability Index*. Arlington, VA: NatureServe, 2010.

124. Janis L. Dickinson, Benjamin Zuckerberg, and David N. Bonter. Citizen science as an ecological research tool: Challenges and benefits. *Annual Review of Ecology, Evolution, and Systematics*, 41:149–172, 2010.

125. Janis L. Dickinson, Jennifer Shirk, David Bonter, Rick Bonney, Rhiannon L. Crain, Jason Martin, Tina B. Phillips, and Karen Purcell. The current state of citizen science as a tool for ecological research and public engagement. *Frontiers in Ecology and the Environment*, 10(6):291–297, 2012.

126. J. Andrew Royle, Marc Kéry, Roland Gautier, and Hans Schmid. Hierarchical spatial models of abundance and occurrence from imperfect survey data. *Ecological Monographs*, 77(3):465–481, 2007.

127. Jorge M. Lobo. More complex distribution models or more representative data? *Biodiversity Informatics*, 5, 2008.

128. Steven J. Phillips, Miroslav Dudík, Jane Elith, Catherine H. Graham, Anthony Lehmann, John Leathwick, and Simon Ferrier. Sample selection bias and presence-only distribution models: Implications for background and pseudo-absence data. *Ecological Applications*, 19(1):181–197, 2009.

129. Walter Jetz, Jana M. McPherson, and Robert P. Guralnick. Integrating biodiversity distribution knowledge: Toward a global map of life. *Trends in Ecology & Evolution*, 27(3):151–159, 2012.

130. J. Andrew Royle, James D. Nichols, and Marc Kéry. Modelling occurrence and abundance of species when detection is imperfect. *Oikos*, 110(2):353–359, 2005.

131. Matthew C. Fitzpatrick, Nicholas J. Gotelli, and Aaron M. Ellison. Max-Ent versus MaxLike: Empirical comparisons with ant species distributions. *Ecosphere*, 4(5):art55, 2013.

132. Trevor J. Hastie and Will Fithian. Inference from presence-only data; the ongoing controversy. *Ecography*, 36(8):864–867, 2013.

133. Steven J. Phillips and Jane Elith. On estimating probability of presence from use-availability or presence-background data. *Ecology*, 94(6):1409–1419, 2013.

6

Using Large-Scale Machine Learning to Improve Our Understanding of the Formation of Tornadoes

Amy McGovern

Corey Potvin

Rodger A. Brown

6.1   Introduction and Motivation 95
6.2   Severe Weather Simulations 97
6.3   Spatiotemporal Relational Random Forests 98
      Spatiotemporally Relational Attributed Data • Training the SRPT
      • Training the SRRF
6.4   Preparing Simulations for Machine Learning 103
      Identifying Storms and Storm Objects • Identifying Strong Vortices
6.5   Empirical Results 106
6.6   Discussion and Future Work 108

6.1 Introduction and Motivation

Meteorologists, like all scientists, are driven by mysteries. One of the greatest mysteries of the atmosphere is how a storm can produce a tornado while a nearby, very similar storm does not. This is much more than an academic curiosity; tornadoes on average produce dozens of fatalities annually in the United States [1], and as demonstrated on April 27, 2011, a single tornado outbreak can still cause hundreds of deaths [2]. This is largely because of the difficulty of predicting the occurrence and tracks of tornadoes with enough lead time and accuracy for those impacted to receive and respond to the warning. Average tornado warning lead times have remained near 17 minutes over the last few decades, 15% of tornadoes still occur without warning, and nearly 75% of warnings are false alarms [3]. This lack of progress in tornado forecasting arises largely from limited understanding of the internal storm processes leading up to tornadogenesis. We seek to fundamentally transform understanding and prediction of tornadogenesis through the development and application of spatiotemporal data mining techniques to simulations of tornadic and nontornadic thunderstorms.

The vast majority of significant tornadoes are spawned by supercell thunderstorms but it should be noted that most supercells do not produce tornadoes [8]. Supercells are especially long-lived thunderstorms (often lasting several hours) that can form when strong vertical wind shear and moderate-to-strong thermodynamic instability are present in the atmosphere. Supercells are characterized by a deep, rotating updraft known as a mesocyclone (Figure 6.1a). This updraft is sustained by warm, converging inflow, the disruption of which (e.g., by rain-cooled air) often initiates storm demise. Upon the onset of significant

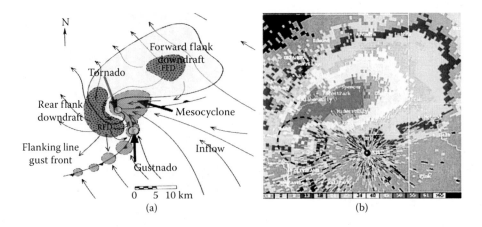

FIGURE 6.1 These figures are best viewed in color. (a) Structure of a classic supercell adapted from [4–6]. (b) Radar reflectivity observations of the 3 May 1999 tornadic supercell near Oklahoma City, OK (from [7]). The hook echo reflectivity signature (dBZ) at 4 km is indicated by a dashed circle.

precipitation, regions of descending air develop in the forward and rear flanks of the supercell. When significant tornadogenesis occurs, it is nearly always preceded by the development of the rear-flank downdraft (RFD). The presence of an RFD is often signified by a hook echo signature in the radar reflectivity field, which provides a measure of precipitation intensity (Figure 6.1b). Low-level mesocyclones (LLMs), typically defined as mesocyclones whose base extends to within 1 or 2 km of the ground, are also closely associated with tornadogenesis. The presence of an LLM signifies a substantially higher probability of tornadogenesis [8] and, conversely, significant tornadoes are usually associated with an LLM.

There are two primary pathways by which better understanding of tornadogenesis would improve tornado warnings. First, better understanding of the physical processes and scales associated with tornadogenesis would help guide and prioritize efforts to improve numerical weather prediction (NWP) models. NWP models play an increasingly important role in severe weather forecasting as observational, modeling, and computational capabilities advance. This transition to the "Warn-on-Forecast" paradigm [3,9] is critical for improving the tornado forecasting process, which is currently overly dependent on visual and radar cues that are often ambiguous or arise once a tornado is already in progress. Second, discoveries of tornado precursors that can be accurately represented in real-time storm analyses could be used to develop automated tornado prediction algorithms. Automated algorithms will be increasingly needed as the volume of observational and model data available to forecasters increases beyond their ability to process subjectively.

Present lack of understanding of tornadogenesis arises largely from the strong nonlinearity and complexity of the relevant processes and the importance of scales down to O(10 m, 1 s). At present, such small scales can only be observed by fast-scanning mobile radars positioned within a few kilometers of developing tornadoes (e.g., [10]). Due to the unpredictability of storms in general and tornadoes in particular, collecting such observations is difficult and dangerous, making high-quality data sets rare. Furthermore, no tornado observations have yet been collected within 10 m of the ground, thought to be a critical layer for tornadogenesis [11].

The lack of observations of tornadogenesis motivates the use of high-resolution supercell simulations as a complementary research tool. Idealized tornadic supercell simulations have been used since the 1980s to investigate tornadoes (e.g., [12,13]). The advent of mobile radars and sophisticated storm-scale data assimilation techniques have recently enabled the blending of observations and simulations into highly descriptive analyses of real-world tornadoes [14]. Case studies of both simulations and data assimilation analyses have yielded some critical insights into the genesis, evolution, and decay of tornadoes.

An important downside, however, of the time-intensive case studies traditionally performed by tornado researchers is the limited number of storms that can be examined in a reasonable timeframe. This reduces the generality of conclusions drawn about tornadogenesis. Data mining techniques, on the other hand, can be used to analyze numerous cases simultaneously, allowing more rigorous hypothesis testing. Even more critically, by introducing an element of automation into the hypothesis development and testing process, the data mining approach can discover important relationships whose relevance a human researcher may not have ever suspected. This approach is particularly suited to investigations of highly complex phenomena that generate huge data sets and may therefore prove invaluable to unlocking the secrets of tornadogenesis.

Simulations provide a valuable picture of the atmosphere for studying tornadogenesis. Mobile radars can measure the intensity of the precipitation near a tornadic storm but they cannot measure the full state of the atmosphere because they are not designed to measure variables such as temperature and pressure. Simulations can produce a complete picture of the atmosphere, which is necessary to achieve an improved fundamental understanding of the processes involved in tornadogenesis. However, simulations produce a large volume of data, necessitating that our machine learning and data mining methods be capable of learning with large data. This chapter describes our current approach to performing large-scale data mining on the supercell simulations.

6.2 Severe Weather Simulations

The supercell simulations used to train the SRRF were generated using the compressible mode of the Bryan Cloud Model 1 (CM1) [15]. The CM1 is a three-dimensional, time-dependent, nonhydrostatic numerical model designed primarily for research on deep precipitating convection (i.e., thunderstorms). The horizontal and vertical grid domain lengths were set to approximately 200 and 20 km, respectively. To reduce computational cost, both horizontal and vertical grid stretching were used. Vertical grid spacing increased from 40 m within the lowest model layer to 500 m above 18.2 km. Horizontal grid spacing was set to 100 m within the middle 125 km of the domain, and increased to 400 m at the lateral boundaries. A total of $1536 \times 1536 \times 99 \approx 234$ million model grid points were used. Cloud microphysics were parameterized using a sophisticated three-moment scheme developed at the National Severe Storms Laboratory [16]. Typical of idealized storm simulations, a horizontally uniform analytical base state was used, and storms were initiated using a thermal bubble. Additional simulation details are listed in Table 6.1.

In order to produce a variety of tornadic and nontornadic supercells, each simulation used a different base state. The first simulation was initialized using a vertical sounding of the atmosphere collected near a tornadic supercell that impacted Geary, Oklahoma, on 29 May 2004 [17]. The base states used in the remaining simulations were produced by perturbing the original sounding. The sounding perturbations were computed by scaling typical operational model sounding errors [18] to 25% (20 simulations), 50% (30 simulations), or 100% (20 simulations) of their original magnitudes, resulting in a total of 71 simulations. For this chapter, we focused on the 50% simulations, of which seven were discarded due to model boundary artifacts.

TABLE 6.1 CM1 Configuration for Supercell Simulations

Large/small time steps	0.5 s / 0.083 s
Lateral boundary conditions	open-radiative
Turbulence parameterization	1.5-order TKE closure
Coriolis effect	Off
Surface drag	Off
Radiation parameterization	Off
Cumulus parameterization	Off
Surface layer physics	Off
Boundary layer parameterization	Off
Explicit numerical diffusion	Off

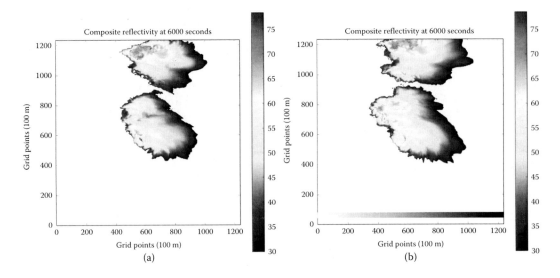

FIGURE 6.2 (a) Control sounding and (b) perturbation 29 simulations. Composite reflectivity for these two simulations at 6000 seconds. In both cases, one of the storms is moving out of the top boundary of the domain and a new storm is in the center.

The simulations were produced using XSEDE resources,* specifically at the National Institute for Computational Science (NICS) using Darter [19] and Nautilus.† Each simulation used 4096 cores and ran for 6 hours, producing 2.5 hours of simulated storm time. Each simulation produced approximately 5 TB of data.

Figure 6.2 shows the simulated radar reflectivity from two of the simulations at 6000 seconds into the simulation. The redder areas are regions with more intense precipitation and the more blue regions have less precipitation. Any values below 30 dBZ are not shown. The left panel shows the control sounding, and the right panel shows one of the perturbations. It is clear that there is significant variation in the storms from the perturbations.

6.3 Spatiotemporal Relational Random Forests

Meteorologists study storms using high-level features such as updrafts (regions where the air flows upward) and downdrafts (regions where the air descends). To facilitate learning a physically plausible model, we use statistical relational learning (SRL) [20–22] to study the simulations. SRL methods learn about objects (e.g., the high-level features that meteorologists already use to study storms) and the relationships between these objects. SRL has proven to be very successful in a wide variety of applications (e.g., [23–26]).

The relationships between the high-level features, or objects, are crucial to fully understanding the causes of tornadogenesis. The particular goal of our project is to identify critical spatiotemporal relationships, that is, the relationships that change as a function of space, time, or space and time together. We have previously developed the spatiotemporal relational probability tree (SRPT) and its related spatiotemporal relational random forest (SRRF) techniques [27–31]. We chose to use the SRRF in this work since we have previously shown that it can be successfully applied to severe weather investigations.

We describe the SRRF algorithm in sufficient detail here for the reader to understand the results. Interested readers are referred to [31] for further implementation details.

---

* https://www.xsede.org
† https://www.nics.tennessee.edu

6.3.1 Spatiotemporally Relational Attributed Data

A traditional Random Forest [32] is an ensemble of decision trees learned using ID3 or C4.5 [33,34]. SRPTs and SRRFs are trained using a similar method to the Random Forest but the two approaches use very different data. A traditional decision tree uses data that can be described by attribute–value pairs. For example, the storm data would be reduced to a series of attributes such as the maximum temperature within a storm, the duration of the storm, the maximum updraft strength, and so on. Although such an approach can work in many situations, it does not take advantage of the objects and relationships already in the data and which we know are critical to understanding the data.

The SRRF learns from spatiotemporal relational attributed data, specifically from an enhanced version of the relational attributed graph representation developed by [35]. For this work, our relational data contain only *objects*, which are defined to be high-level concepts that meteorologists already use to describe the data. The SRRF can learn from data that also contain predefined *relationships* between the objects but we focus on the objects here and enable the relationships to be discovered automatically by the SRRF. In previous work, we enabled objects to have spatiotemporally varying *fields* of scalar and vector data associated with them. We call these *fielded objects*, following the convention in geographic information systems [36,37].

Table 6.2 lists the objects we extracted from each simulation and for each storm. Each object was temporally varying and had an associated start and end time. Because the SRRF can autonomously identify spatial, temporal, and spatiotemporal relationships, we did not identify predefined relationships between the objects.

The SRRF can learn from *static* attributes, meaning they don't change during the lifetime of the object, or *dynamic* ones, which can change throughout the lifetime of the object. Dynamic attributes can either be scalar (such as a measurement of the maximum wind speed inside an updraft) or fielded, where there

TABLE 6.2 Objects Extracted for the SRRF and a Description of Why We Chose Those Objects

Name	Description
Updraft	A concentrated region of ascending air. The updraft forms the core of the supercell and is responsible for stretching near-ground rotation to tornadic magnitude.
Downdraft	A concentrated region of descending air. The rear-flank downdraft has been shown to play a critical role in generating near-ground rotation.
Reflectivity	Measure of the precipitation intensity (units of dBZ). We use reflectivity objects to help identify and track storms.
Cold pool	A mass of cold air at the surface produced by a storm downdraft. The strength and location of the cold pool strongly influence whether tornadogenesis occurs.
Convergence (Divergence)	A measure of the contraction (spreading out) of the wind field. Due to conservation of mass, strong low-level convergence tends to intensify the storm updraft.
Vertical vorticity	A measure of local rotation about the vertical dimension. Once substantial vertical vorticity occurs near the ground, it may be stretched to tornadic magnitude by the low-level updraft. Positive (negative) vorticity corresponds to counterclockwise (clockwise) rotation.
Tilting	Tilting of horizontal vorticity into the vertical dimension. This mechanism is often important as air descends toward the ground or ascends as it enters the updraft. Positive (negative) tilting corresponds to increasing (decreasing) vorticity.
Stretching	Intensification of vorticity as air accelerates within an updraft. Positive (negative) stretching corresponds to increasing (decreasing) vorticity.
Mesocyclone	A deep column of rotating air associated with a supercell updraft. The existence of a mesocyclone identifies a storm as a supercell. We define a positive (negative) mesocyclone as one that rotates counterclockwise (clockwise) when viewed from above.
Low-level mesocyclone	A mesocyclone with base 1–2 km above the ground. LLMs are significantly more likely to be associated with tornadoes than higher based mesocyclones.
Tornado	A strong low-level rotation defined herein to be associated with an LLM.

TABLE 6.3 Attributes Extracted for Each Object in Our Data

Name	Type	Description
Hydrometeor buoyancy	3D scalar field	Vertical drag force exerted by precipitation particles. Negative hydrometeor buoyancy contributes to the intensity of downdrafts.
Thermal buoyancy	3D scalar field	Vertical force arising from horizontal gradients in air density. Positive thermal buoyancy contributes to the intensity of the main updraft.
Total buoyancy	3D scalar field	Sum of hydrometeor and thermal buoyancy.
Reflectivity	3D scalar field	Measure of precipitation intensity.
Potential temperature perturbations	3D scalar field	Potential temperature is the temperature an air parcel would have if brought adiabatically down to a reference level, typically 1000 mb. Large negative potential temperature perturbations (deviations from the storm environment) in the cold pool can promote baroclinic (buoyancy-gradient-induced) vorticity generation but also decelerate parcels rising within the main updraft.
Wind	3D vector field	A full field of the wind vectors inside an object.
Vorticity	3D vector field	A full field of the vorticity (measure of instantaneous spin) vectors inside an object.

is a gridded series of measurements. The fields can be either two or three dimensional and can comprise either scalar values or vectors. For this work, we extracted only temporally and spatially varying attributes. Table 6.3 lists the attributes extracted within the 3D boundaries of each object.

Each simulation and each storm within a simulation provided a different set of objects and attributes. The SRPT and SRRF are classifiers, which means they learn from labeled data. We describe the details of how we labeled each storm below. The storms were broken into three categories: tornadic, tornado-failure, and nontornadic. We trained the SRPT and SRRF on the tornadic and tornado-failure storms.

6.3.2 Training the SRPT

SRPTs are probability estimation trees [38] that learn with the spatiotemporally varying relational data described above. SRPTs differ from existing tree-based relational learning approaches such as TILDE [39,40] through their ability to handle the discovery of multidimensional spatial, temporal, and spatiotemporal relationships and their ability to handle spatially and temporally varying data. SRPTs also differ from the relational probability tree (RPT) [35] and the temporal extensions to the RPT [41,42] in their ability to handle spatially and spatiotemporally varying relational data. This is critical for applications to severe weather.

We first discuss how to grow an SRPT from the spatiotemporal relational attributed data and then we discuss how to train the SRRF. We follow the standard approach for training a random forest and do not introduce pruning mechanisms into the individual model training. Algorithm 6.1 describes the approach for growing an individual SRPT.

Algorithm 6.1: Train-SRPT: Algorithm for growing an individual SRPT

> **Input**: Data = Spatiotemporal relational attributed graphs used for training data
> **Input**: numSamples = Number of questions to sample when growing the tree
> **Input**: maxDepth = Maximum depth of tree
> **Input**: currentDepth = Current tree depth
> **Input**: p = p-value used to stop tree growth
> **Output**: A single SRPT

```
if maxDepth is reached or Data is too small then
    │ return a leaf node using Data;
end
split ← Find-Best-Split(Data,numSamples,p);
if split ≠ Ø then
    │ tree ← Ø;
    │ for all possible values v in split do
    │   │ add a child to tree by calling Train-SRPT recursively and using only Data where split = v;
    │ end
    │ Return tree
else
    │ return a leaf node using Data;
end
```

Each SRPT is grown using the standard greedy approach also used by ID3 and C4.5 [33,34]. The primary difference between these approaches and that outlined in Algorithm 6.1 is in the way the best split is identified. In a standard ID3 or C4.5 decision tree, the number of possible splits is finite and each is examined. In a Random Forest [32], the number of attributes is limited significantly in order to introduce diversity into the forest. The trees are then grown using the standard algorithm but only examining the limited set of attributes available at each node. Due to the continuous nature of the spatiotemporal relational attributed data, for the SRPT, the number of possible questions or ways to split the tree is essentially infinite and the best split is identified using a random sampling of the possible splits. Algorithm 6.2 describes how the best split is identified in an SRPT.

Algorithm 6.2: Find-Best-Split: Finds the best split for the data

---

Input: Data = Spatiotemporal relational attributed graphs used for training data
Input: numSamples = Number of questions to sample
Input: p = p-value used to stop tree growth
Output: A split if one exists that satisfies the criteria or Ø otherwise
```
best ← Ø;
for i = 1 to numSamples do
    │ split ← generate a random split by choosing a question type randomly and filling in the
    │ values by randomly sampling from the training data;
    │ eval ← evaluate the quality of split using chi-squared;
    │ if eval is statistically significant (using p) and eval is best evaluation so far then
    │   │ best ← split;
    │ end
end
Return best;
```

The splits for the SRPT provide the key differences between the SRPT and other decision tree algorithms. Decision trees must split the data in a way that provides meaningful information. In a typical decision tree, these splits take the form of yes/no questions such as "Is it raining?" In the SRPT, the splits are derived from a set of spatial, temporal, and spatiotemporal questions. The full list of questions is given in [31]. We describe the types of questions that are possible, using examples from the tornado domain. Since there are essentially an infinite number of possible values for each split, the specific values are chosen by sampling from the training data.

Nonspatial, nontemporal questions: Does the storm contain a downdraft? Does the storm contain an updraft with a maximum speed of at least 30 m s^{-1}?

Temporal questions: Does the storm contain a downdraft that lasts at least 5 minutes? Does the reflectivity region have a reflectivity maximum that exceeds 60 dBZ for at least 5 minutes? Is the partial derivative of the maximum updraft strength (computed using finite differences) at least 0.02 m s^{-2} over 5 minutes?

Spatial and spatiotemporal questions for fielded attributes: Does the updraft have a reflectivity value of at least 60 dBZ? Does an updraft have a maximum magnitude of the gradient of its reflectivity field of at least x? Does the downdraft have a potential temperature value (mean/maximum/minimum/standard deviation) of at least x at any or all time steps?

Spatial and spatiotemporal questions for wind vectors: Does the updraft have a maximum divergence or minimum convergence of at least x? Does the downdraft have a maximum/minimum stretching or shearing of x? Does the updraft have a vorticity (computed from the wind vector field) of at least x? Is the average direction of the wind vector in a cold pool object to the northwest?

Two-dimensional shape questions: Does the shapelet [43,44] of the updraft's composite match the template (trained from the data)?

Three-dimensional shape questions: Does the shape of the updraft match the template (trained from the data)? Does the shape of the space between the updraft and the downdraft match this template? Does the shape change over time in a way that matches two templates? Does the shape stay the same for 5 minutes?

Combination questions: Are there at least five downdraft objects? Is there a temporal relationship between the items matching question 1 and question 2? The temporal relationships (from [45]) are before, meets, overlaps, equals, starts, finishes, and during. Are the centroids of the objects matching question 1 within Euclidean distance d of the objects matching question 2?

Since severe weather features are unlikely to correspond to canonical shapes such as a circle or a sphere, we have developed two approaches that can distinguish arbitrary shapes. For two-dimensional data, we use shapelets as developed by [43,44]. Shapelets are pieces of a time series that can be used to distinguish different time series. Shapelets can be applied to either univariate time series data, such as the maximum updraft within a field at each time step, or to two dimensional shapes. For the latter, we convert the outline of a two dimensional shape into what [46] calls a time series.

Three-dimensional shapes are more difficult to distinguish as they cannot be easily reduced to a single time series. Instead, we use shape distributions, which are statistical distributions that characterize a 3D shape [47]. These can be formed by sampling from random points on the surface of the shape and calculating a simple statistic, such as the distance between the two points. We use this idea to distinguish shapes from one another and to distinguish spatial and spatiotemporal relationships by sampling from the space between objects. The distributions are distinguished using Kolmogorov–Smirnov.

In standard decision trees, the best split is chosen by measuring which split provides the largest information gain. For the SRPT, we use the chi-squared statistic, which we found to be a more robust solution than information gain. To handle smaller numbers of examples, we also used Fisher's exact test when there were fewer than 20 examples at that point in the tree. Chi-squared was used for 20 or more examples.

6.3.3 Training the SRRF

The SRRF is an ensemble of SRPTs, trained using a very similar approach to that of Random Forests [32]. Random Forests are a powerful machine learning method that make use of the fact that an ensemble of weak learners can become a strong learner through diverse models. One way that Random Forests provide diversity to the underlying models is to train each model on a different set of data. These data are constructed using bootstrap resampling on the training data. The other way that Random Forests ensure diversity is through sampling the attributes available for each split (our approach for this was described above).

Algorithm 6.3 describes the procedure for training a SRRF given our spatiotemporal relational attributed data. Although this can be computationally intensive on large data sets such as those we use here, it is easily implemented in parallel since the training of each tree is independent from one another. We implemented it using a single thread for each tree, which significantly improved the running time of the algorithm. Further improvements could also multithread the Train-SRPT algorithm but that was not necessary for our data, despite the largeness thereof.

Algorithm 6.3: Train-SRRF: Trains a full spatiotemporal relational random forest

---

Input: Data = Spatiotemporal relational attributed graphs used for training data
Input: numSamples = Number of questions to sample
Input: p = p-value used to stop tree growth
Input: numTrees = Number of trees in the forest
Output: A SRRF
forest ← Ø;
for i = 1 to numTrees **do**
  newData ← create a new data set using bootstrap resampling from Data;
  tree ← call Train-SRPT on newData;
  add the tree to the forest;
end
Return forest;

6.4 Preparing Simulations for Machine Learning

Each simulation produces over 5 TB of raw data that we must process prior to inputting to our machine learning techniques. This section explains the details of how we identified and tracked distinct storms within each simulation, how we extracted each type of object, and how we labeled each storm.

6.4.1 Identifying Storms and Storm Objects

Preparing the simulations for machine learning and data mining requires an automated approach to processing the data. The first step is to automatically identify and track the distinct storms from within each simulation. As Figure 6.2 shows, there can be multiple distinct storms within a given simulation. In order to identify and track individual storm cells, we need a definition of what a storm cell is. We define a storm using the updraft since that is the central feature of supercell thunderstorms.

Most objects are extracted from the simulation data using thresholds to define their boundaries. The thresholds and heights over which the thresholds are applied are given in Table 6.4. Because the simulations can have small areas of noisy data that appear to cross a threshold but are not really useful, objects are also subjected to minimum area and volume constraints. Objects must also persist for at least two consecutive time steps (60 seconds) in order to be saved for machine learning.

With the threshold approach, objects can be extracted from each time step in parallel, facilitating large-scale machine learning. Before sending the final data to the SRRF, the objects must be identified and

TABLE 6.4 Definitions for A Objects Extracted Using Thresholds on the Simulation Data

Name	Definition	Heights	Min area	Min volume
Updraft	$w \geq 15\text{m s}^{-1}$	0–8 km	5 km²	10 km³
Downdraft	$w \leq -6\text{m s}^{-1}$	0–8 km	2 km²	4 km³
Intense reflectivity	$dBZ \geq 60$	0–8 km	5 km²	20 km³
Radar reflectivity	$dBZ \geq 40$	0–8 km	5 km²	20 km³
Cold pool	$\theta' \leq -2°\text{K}$	0–500 m	2 km²	0.5 km³
Positive vertical vorticity	$\zeta \geq 0.05\text{s}^{-1}$	0–8 km	1 km²	1 km³
Negative vertical vorticity	$\zeta \leq -0.05\text{s}^{-1}$	0–8 km	1 km²	1 km³
Convergence	$\nabla \geq 0.05\text{s}^{-1}$	0–500 m	1 km²	0 km³
Positive tilting	$T \geq 0.03\text{s}^{-2}$	0–3 km	1 km²	0 km³
Negative tilting	$T \leq -0.03\text{s}^{-2}$	0–3 km	1 km²	0 km³
Positive stretching	$S \geq 0.03\text{s}^{-1}$	0–5 km	1 km²	0 km³
Negative stretching	$S \leq -0.03\text{s}^{-1}$	0–5 km	1 km²	0 km³

tracked across time, with each object being tracked uniquely. Since the data are available at 30-second intervals, approaches that have been developed for less frequent real-time weather data [48] are not needed. Instead, we track objects from one time to another by looking for the nearest object in the previous time step. Objects do not move significantly within 30 seconds and so this approach has proven very reliable.

Figure 6.3 shows the updraft and downdraft objects identified at 6000 seconds in one simulation. In order to show the three-dimensional structure of the objects, Figure 6.3 shows the composites across the vertical and the two horizontal dimensions. The updraft regions are the central feature of each storm and each storm can only have one updraft. Other objects are associated with the nearest updraft and all objects within a specified radius (20 km for this work) are assumed to belong to the same storm. Figure 6.3 shows three identifiable storms that are present at the time.

6.4.2 Identifying Strong Vortices

In order to label storms as tornadic, tornado-failures, or nontornadic, we identify mesocyclones and tornadoes in the simulation data using an updated version of the vortex detection and characterization (VDAC) technique [49]. The VDAC algorithm variationally produces a least-squares fit of the local wind field to a low-order model of a vortex. The vortex model parameters retrieved by this process include location, radius of maximum wind, and maximum tangential wind. Vortex detections are made based on thresholds of the retrieved radius and maximum tangential wind. Each retrieval operates on a single height and time, making it necessary to impose continuity criteria to identify 3D or 4D vortex objects. The 3D mesocyclone and tornado objects at each time are constructed from 2D objects at each vertical level. The location and boundary of each 2D object are defined by the location and radius of the corresponding VDAC detection. VDAC detections have a sign indicating which way the vortex is spinning and detections are only matched with other detections of the same sign.

Mesocyclone objects must have at least one VDAC detection within the 2–4 km height interval, and at least four detections total from 0 to 8 km. Any detections below 2 km must have a radius of more than 750 m. Tornado objects require a VDAC detection at height 100 m with vortex radius of less than 500 m as well as one or more additional detections over the 0.5–3.0 km height interval. Any tornado detections in between are added to the tornado object. Detections at successively higher levels are associated with the same object if they are within 60° from the vertical of the next-highest detection. Both mesocyclone and tornado objects extend from their lowest altitude VDAC detection to their highest.

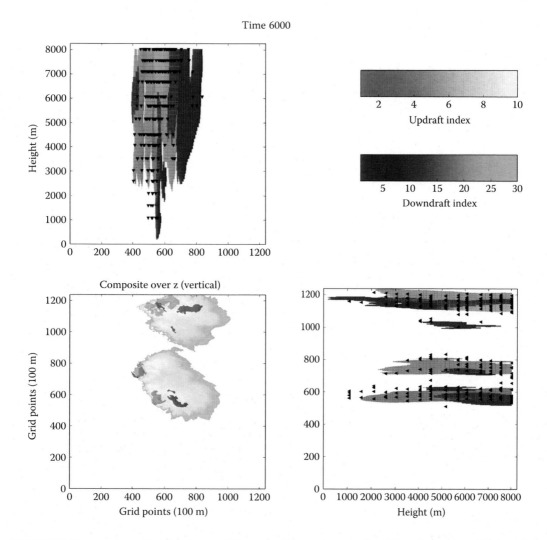

FIGURE 6.3 (See color insert.) Visualization of the three-dimensional updraft and downdraft objects. The lower left panel shows the composite over the vertical and the top and right panels show the composites across the north–south and east–west directions, respectively. Updraft objects are shown in shades of red, and downdraft objects are shown in shades of blue. The colors correspond to the object number, not to values within the updraft or downdraft themselves. The vertical composite also includes reflectivity. The triangles in the horizontal composites indicate locations of strong vortices (described in Section 6.4.2).

Figure 6.4 shows the mesocyclone objects identified in the same simulation and the same time step as Figure 6.3. The figure also shows all of the VDAC detections in the vertical and horizontal composites. The tornado objects do not appear until later in this simulation.

The mesocyclone and tornado objects are used to label each storm for the SRRF for training. A tornadic storm is defined as having a LLM and colocated tornado object at the same time (for at least 60 seconds). A tornado-failure storm has an LLM (lowest height of 2 km or less) but no tornado. A nontornadic storm has neither an LLM nor a tornado object. The SRRF learns from the 30 minutes prior to the start of the tornado or the time of tornado-failure (defined as the start of the LLM). This focuses the SRRF on the environment directly associated with tornadogenesis. The training data consisted of 34 tornadic storms and 21 tornado-failure storms.

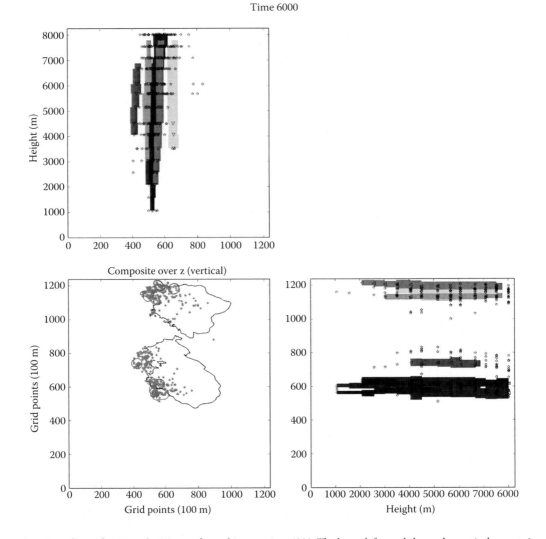

FIGURE 6.4 (See color insert.) Mesocyclone objects at time 6000. The lower left panel shows the vertical composite with the VDAC detections shown using either a star or a triangle. Detections shown using a triangle were used in the mesocyclone objects. Positive mesocyclones are shown in red and negative in blue. There were no tornado objects at this time step.

6.5 Empirical Results

We first examine the performance of the SRRF as a function of the main parameters to the algorithm. The main parameters are the number of trees in the forest and the number of samples of questions asked when building each tree. In previous work, we have demonstrated that the performance of the tree increases initially as both of these parameters are increased but it quickly reaches an asymptote [30,31]. The results shown in Figure 6.5 confirm this. Panel (a) shows the performance as a function of the number of trees and panel (b) shows the performance as a function of the number of samples. For the graphs, we measured performance using the area under the ROC curve. A score of 1 means the SRRF scored perfectly, and a score of 0.5 means it was performing as a random algorithm would. The figure shows the mean AUC and the 95% confidence interval of the AUCs as a function of the two main parameters. These results are

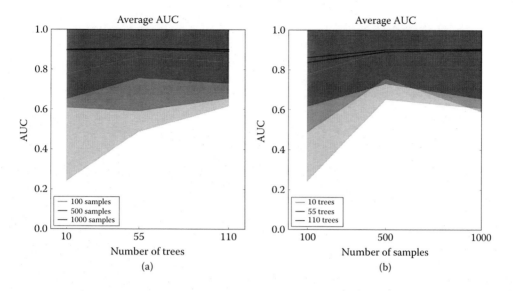

FIGURE 6.5   Average AUC with the 95% confidence interval in shading for (a) as a function of the number of trees in the forest and (b) as a function of the number of samples used to grow the trees.

TABLE 6.5   TSS for the SRRFs as a Function of the Two Main Parameters. The 95% Confidence Interval Is Shown in Parentheses and the Average Is Shown in Bold

Number of Trees	Number of Samples		
	100	500	1000
10	**0.31** (−0.45,0.88)	**0.59** (0.23,1.00)	**0.70** (0.18,1.00)
55	**0.44** (−0.31,1.00)	**0.62** (0.12,1.00)	**0.67** (0.13,1.00)
110	**0.47** (−0.03,1.00)	**0.56** (0.07,1.00)	**0.59** (0.00,0.86)

averaged over 30 runs. As the number of trees increases, the confidence interval is tighter and the mean AUC increases. This is also true as the number of samples increases.

We also evaluated the results using the true skill statistic (TSS) [50] as calculated on the contingency tables for each forest. This statistic is very similar to the area under the curve except that it varies from −1 to 1 with 0 being the performance of a random classifier. A TSS value of 1 is perfect. Table 6.5 shows the TSS with the confidence intervals for each parameter setting. As with the AUC results shown graphically in Figure 6.5, the confidence interval bounds grow tighter as either the number of trees increases or the number of samples increases.

Given these results, we chose to use 110 trees and 1000 question samples for the remaining experiments. Here, we examine individual forests for physical insights into the causes of the tornadogenesis in the simulations. The first step is to examine the variable importance results. The SRRF implements variable importance in the same manner as Random Forests [32]. Table 6.6 shows the variable importance for a forest with 1000 samples and 110 trees. Mesocyclone wind attributes are found to best discriminate between tornado and tornado-failure cases in the forest.

Our next step in examining what is learned from the forests is to see what types of questions are asked in the trees and what objects or relations are being referenced in those questions. Table 6.7 shows the top questions and object types used for each question. The mesocyclone wind attributes that were shown by the variable importance values to be most critical to discriminating between tornado and tornado-failure cases are here revealed as vorticity, deformation, and divergence. Questions involving these attributes comprise nearly 60% of all the tree branches of the forest.

TABLE 6.6   Top Five Variables as Measured by Variable Importance

Object	Attribute	Score
Positive mesocyclone	Wind	10.9
Negative mesocyclone	Wind	5.0
Updraft	Wind-shape	3.5
Downdraft	Wind-shape	1.4
Updraft	Wind	0.8

TABLE 6.7   Top Eight Question Types and the Percentages of Tree Branches Containing Them

Question Type	Object	Percentage (%)
Vector field vorticity	Positive mesocylone	24.2
Vector field vorticity	Negative mesocyclone	12.7
Vector field deformation	Positive mesocyclone	10.4
Vector field divergence	Positive mesocyclone	6.5
Vector field deformation	Negative mesocyclone	5.1
Array temporal duration	Updraft	5.1
Array shapelet	Updraft	4.8
Vector field vorticity	Updraft	3.4

Our final step in examining the SRRF output was to assess the roles of the aforementioned questions within their trees. Only one question category was found to consistently and frequently discriminate between tornado and tornado-failure cases. An answer of "yes" to the question of whether the horizontal vorticity vector within a positive mesocyclone points within 22.5° of southwest led to a 100% tornado probability in 15 out of 17 trees. In 14 of those 15 tornado classifications, the question was asked of a model level 700–1500 m above the surface. This result is consistent with the known importance of tilting and stretching of low-level vorticity near the mesocyclone, and therefore helps validate our methodology. The significance of the southwestward orientation of the horizontal vorticity vector in the tornadic cases is intriguing and requires further investigation. For now, we hypothesize that vorticity generated within the buoyancy gradient along the forward flank of the storm, which in nature often extends northeastward from the LLM, contributes strongly to tornadogenesis in our simulations.

Figure 6.6 shows a sample tree from the forest. In this tree, the wind and vorticity vector average directions are examined and found to be critical in distinguishing tornadic storms from tornado-failure storms.

6.6 Discussion and Future Work

While we have worked with large-scale data before, including severe weather data, the data we used here were on a much larger scale than anything we have used before (such as the results presented in [30,31]). Our lower resolution simulations were less than 1 TB each and other severe weather data such as aircraft turbulence still comprised significantly less data than our simulations used here, which were over 5 TB each. While applying the SRRF to the new simulations, we learned several valuable lessons about large-scale machine learning applications. Many of the lessons are computational in nature but they also include statistical and algorithmic lessons.

First, creating realistic large data sets for machine learning requires significant computational effort that should not be underestimated. Simulations of complex events such as severe weather require extensive

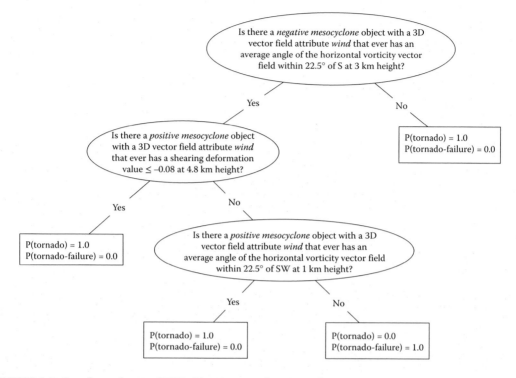

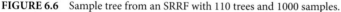

FIGURE 6.6 Sample tree from an SRRF with 110 trees and 1000 samples.

high-performance computing experience and a significant involvement from domain scientists to ensure that the data are realistic. We have been working with high-resolution simulations for over a decade and the results presented here are our first simulations that can resolve tornadoes. This was made possible through a combination of improvements in supercomputing power and improvements in simulations.

Second, data handling techniques for large-scale data must be very efficient. It is important to properly store and process the data so that the automated post-processing techniques do not spend the majority of their time in disk input/output.

Third, the algorithms for learning from large data sets must be designed to learn from large data from the beginning. The SRRF had been applied to other large data sets but these simulations were significantly larger. As discussed above, each simulation was approximately 5 TB of raw data. The reduced meta-data for learning was approximately 1 TB per simulation. Although the time complexity of the algorithm remained steady as the data increased, the space complexity increased significantly and required a supercomputer with large memory.

Fourth, because generating and processing large data sets requires significant computational effort, statistics may need to be adjusted to handle smaller data sets than expected. Our original SRPT algorithm relied on chi-squared to score potential tree splits but chi-squared is most reliable with more than 20 examples. With so few training examples available and 20% being used for a test set, we adjusted the tree to use Fisher's exact test for smaller contingency tables and to use chi-squared only when there were more than 20 examples.

Finally, successfully applying machine learning to a domain such as meteorology cannot have the final result of a black-box method. The output must be interpretable and able to guide new physical intuitions into the causes of the event being studied. When the meteorologists inspected the SRRF output, it became clear that objective post-processing techniques need to be developed in order to quickly and accurately identify those questions that most consistently and frequently distinguish between tornadogenesis and tornado-failure cases.

We emphasize that the results presented herein are very preliminary. We will implement many methodological improvements before drawing firm conclusions about how storm processes differ preceding tornadogenesis versus tornado failure. Object definitions will be refined to ensure features are properly captured, and new objects and attributes related to tornadogenesis will be added. Since the processes contributing to tornadogenesis evolve leading up to the beginning of the tornado, some of our future experiments will restrict the data input to the machine learning technique to, say, 15–30 or 0–15 minutes (rather than 0–30 minutes) prior to tornado formation or failure. Physical inferences from the SRRFs will be validated against nature using high-quality analyses of real supercells observed by mobile radars. Given the demonstrated value of machine learning to gaining insights into large data sets in other domains, we anticipate our efforts will culminate in new understanding of tornadogenesis, and ultimately improved tornado warnings. In future work, we will also apply our techniques to assimilated data from observed tornadic storms.

Acknowledgments

This material is based on work supported by the National Science Foundation under Grant No. IIS .0746816. Any opinions, findings, and conclusions or recommendations expressed in this material are those of the author(s) and do not necessarily reflect the views of the National Science Foundation.

The data presented here were produced using computational resources supported by the University of Tennessee and Oak National Laboratory's Joint Institute for Computational Sciences (http://www.jics.utk.edu). Any opinions, findings, and conclusions or recommendations expressed in this material are those of the author(s) and do not necessarily reflect the views of the University of Tennessee, Oak Ridge National Laboratory, or the Joint Institute for Computational Sciences.

References

1. HE Brooks and CA Doswell, III. Deaths in the 3 May 1999 Oklahoma City tornado from a historical perspective. *Weather and Forecasting*, 17:354–361, 2002.
2. KR Knupp, TA Murphy, TA Coleman, RA Wade, SA Mullins, CJ Schultz, EV Schultz, L Carey, A Sherrer, EW McCaul Jr., B Carcione, S Latimer, A Kula, K Laws, PT Marsh, and K Klocko. Meteorological overview of the devastating 27 April 2011 tornado outbreak. *Bulletin of the American Meteorological Society*, 95:1041–1062, 2014.
3. DJ Stensrud, LJ Wicker, M Xue, and KA Brewster. Progress and challenges with Warn-on-Forecast. *Atmospheric Research*, 123:2–16, 2013.
4. LR Lemon and CA Doswell, III. Severe thunderstorm evolution and mesocyclone structure as related to tornadogenesis. *Monthly Weather Review*, 107:1184–1197, 1979.
5. RP Davies-Jones. Tornado dynamics. In E Kessler, ed., *Thunderstorm Morphology and Dynamics*, Chapter 10, pp. 197–236. Norman, OK: University of Oklahoma Press, 1986.
6. HB Bluestein. *Synoptic-Dynamic Meteorology in Midlatitudes: Observations and Theory of Weather Systems*, Volume 2. New York, NY: Oxford University Press, 1993.
7. DW Burgess, MA Magsig, J Wurman, DC Dowell, and Y Richardson. Radar observations of the 3 May 1999 Oklahoma City tornado. *Weather and Forecasting*, 17:456–471, 2002.
8. RJ Trapp, GJ Stumpf, and KL Manross. A reassessment of the percentage of tornadic mesocyclones. *Weather and Forecasting*, 20(4):680–687, 2005.
9. DJ Stensrud, M Xue, LJ Wicker, KE Kelleher, MP Foster, JT Schaefer, RS Schneider, SG Benjamin, SS Weygandt, JT Ferree, and JP Tuell. Convective-scale warn-on-Forecast system: A vision for 2020. *Bulletin of the American Meteorological Society*, 90:1487–1499, 2009.
10. AL Pazmany, JB Mead, HB Bluestein, JC Snyder, and JB Houser. A mobile rapid-scanning polarimetric (RaXPol) Doppler radar system. *Journal of Atmospheric and Oceanic Technology*, 30:1398–1413, 2013.

11. J Wurman, K Kosiba, and P Robinson. In situ, Doppler radar, and video observations of the interior structure of a tornado and the wind–damage relationship. *Bulletin of the American Meteorological Society*, 94:835–846, 2013.

12. JB Klemp and R Rotunno. A study of the tornadic region within a supercell thunderstorm. *Journal of the Atmospheric Sciences*, 40(2):359–377, 1983.

13. LJ Wicker and RB Wilhelmson. Simulation and analysis of tornado development and decay within a three-dimensional supercell thunderstorm. *Journal of the Atmospheric Sciences*, 52(15):2675–2703, 1995.

14. J Marquis, Y Richardson, PM Markowski, D Dowell, and J Wurman. Tornado maintenance investigated with high resolution dual-Doppler and EnKF analysis. *Monthly Weather Review*, 140:3–27, 2012.

15. GH Bryan and JM Fritsch. A benchmark simulation for moist nonhydrostatic numerical models. *Monthly Weather Review*, 130:2917–2928, 2002.

16. ER Mansell, CL Ziegler, and EC Bruning. Simulated electrification of a small thunderstorm with two-moment bulk microphysics. *Journal of the Atmospheric Sciences*, 67:171–194, 2010.

17. HB Bluestein, MM French, RL Tanamachi, S Frasier, K Hardwick, F Junyent, and AL Pazmany. Close-range observations of tornadoes in supercells made with a dual-polarization, X-band, mobile Doppler radar. *Monthly Weather Review*, 135:1522–1543, 2007.

18. R Cintineo and DJ Stensrud. On the predictability of supercell thunderstorm evolution. *Journal of the Atmospheric Sciences*, 70:1993–2011, 2013.

19. MR Fahey, R Budiardja, L Crosby, and S McNally. *Deploying Darter—A Cray XC30 System*, pp. 430–439. Lecture Notes in Computer Science, LNCS 8488. Switzerland, Springer International Publishing, 2014.

20. D Jensen and L Getoor. IJCAI 2003 workshop on learning statistical models from relational data. http://scholarworks.umass.edu/cgi/viewcontent.cgi?article=1163&context=cs_faculty_pubs, 2003.

21. A Fern, L Getoor, and B Milch. SRL2006: Open problems in statistical relational learning. http://www.cs.umd.edu/projects/srl2006/, 2006.

22. L Getoor and B Taskar, editors. *Introduction to Statistical Relational Learning*. Adaptive Computation and Machine Learning series. Cambridge, MA: MIT Press, 2007.

23. J Neville, Ö Şimşek, D Jensen, J Komoroske, K Palmer, and H Goldberg. Using relational knowledge discovery to prevent securities fraud. In *Proceedings of the 11th ACM SIGKDD International Conference on Knowledge Discovery and Data Mining*, pp. 449–458, 2005.

24. A Fast, L Friedland, M Maier, B Taylor, D Jensen, H Goldberg, and K Komoroske. Relational data pre-processing techniques for improved securities fraud detection. In *Proceedings of the 13th International Conference on Knowledge Discovery and Data Mining*, pp. 941–949, 2007.

25. J Neville and D Jensen. Relational dependency networks. *Journal of Machine Learning Research*, 8:653–692, 2007.

26. S Raghavan, RJ Mooney, and H Ku. Learning to "read between the lines" using Bayesian logic programs. In *Proceedings of the 50th Annual Meeting of the Association for Computational Linguistics (ACL-2012)*, pp. 349–358, July 2012.

27. A McGovern, N Hiers, M Collier, DJ Gagne II, and RA Brown. Spatiotemporal relational probability trees. In *Proceedings of the 2008 IEEE International Conference on Data Mining*, pp. 935–940, Pisa, Italy, 2008.

28. A McGovern, T Supinie, DJ Gagne II, N Troutman, M Collier, RA Brown, J Basara, and J Williams. Understanding severe weather processes through spatiotemporal relational random forests. In *Proceedings of the 2010 NASA Conference on Intelligent Data Understanding*, pp. 213–227, 2010.

29. A McGovern, DJ Gagne II, N Troutman, RA Brown, J Basara, and J Williams. Using spatiotemporal relational random forests to improve our understanding of severe weather processes. *Statistical Analysis and Data Mining*, 4(4):407–429, 2011.

30. A McGovern, N Troutman, RA Brown, JK Williams, and J Abernethy. Enhanced spatiotemporal relational probability trees and forests. *Data Mining and Knowledge Discovery*, 26(2):398–433, 2013.

31. A McGovern, DJ Gagne II, JK Williams, RA Brown, and JB Basara. Enhancing understanding and improving prediction of severe weather through spatiotemporal relational learning. *Machine Learning*, 95(1):27–50, 2014.

32. L Breiman. Random forests. *Machine Learning*, 45(1):5–32, 2001.

33. JR Quinlan. Induction of decision trees. *Machine Learning*, 1:81–106, 1986.

34. JR Quinlan. *C4.5: Programs for Machine Learning*. San Francisco, CA: Morgan Kaufmann, 1993.

35. J Neville, D Jensen, L Friedland, and M Hay. Learning relational probability trees. In *Proceedings of the Ninth ACM SIGKDD International Conference on Knowledge Discovery and Data Mining*, pp. 625–630, 2003.

36. M Goodchild, M Yuan, and T Cova. Towards a general theory of geographic representation in GIS. *International Journal of Geographical Information Science*, 21(3):239–260, January 2007.

37. T Cova and M Goodchild. Extending geographical representation to include fields of spatial objects. *International Journal of Geographical Information Science*, 16(6):509–532, 2002.

38. FJ Provost and P Domingos. Tree induction for probability-based ranking. *Machine Learning*, 52:199–215, 2003.

39. H Blockeel and L De Raedt. Top-down induction of first order logical decision trees. *Artificial Intelligence*, 101(1–2):285–297, 1998.

40. J Ramon, T Francis, and H Blockeel. Learning a go heuristic with TILDE. In *Revised Papers from the Second International Conference on Computers and Games*, pp. 151–169, London: Springer-Verlag, 2002.

41. U Sharan and J Neville. Exploiting time-varying relationships in statistical relational models. In *Proceedings of the 1st SNA-KDD Workshop, 13th ACM SIGKDD Conference on Knowledge Discovery and Data Mining*, 2007.

42. U Sharan and J Neville. Temporal-relational classifiers for prediction in evolving domains. In *Proceedings of the IEEE International Conference on Data Mining*, 2008.

43. L Ye and E Keogh. Time series shapelets: A new primitive for data mining. In *Proceedings of the ACM SIGKDD International Conference on Knowledge Discovery and Data Mining*, pp. 947–956, 2009.

44. A Mueen, E Keogh, and N Young. Logical-shapelets: An expressive primitive for time series classification. In *Proceedings of ACM SIGKDD*, pp. 1154–1162, 2011.

45. JF Allen. Time and time again: The many ways to represent time. *International Journal of Intelligent Systems*, 6(4):341–355, 1991.

46. E Keogh, L Wei, X Xi, S-H Lee, and M Vlachos. LB Keogh supports exact indexing of shapes under rotation invariance with arbitrary representations and distance measures. In *Proceedings of the 32nd international conference on Very large data bases, VLDB '06*, pp. 882–893. VLDB Endowment, 2006.

47. R Osada, T Funkhouser, B Chazelle, and D Dobkin. Shape distributions. *ACM Transactions on Graphics*, 21(4):807–832, 2002.

48. JT Johnson, PL MacKeen, A Witt, ED Mitchell, GJ Stumpf, MD Eilts, and KW Thomas. The storm cell identification and tracking algorithm: An enhanced WSR-88D algorithm. *Weather and Forecasting*, 13(2):263–276, 1998.

49. CK Potvin. A variational method for detecting and characterizing convective vortices in Cartesian wind fields. *Monthly Weather Review*, 141:3102–3115, 2013.

50. DS Wilks. *Statistical Methods in the Atmospheric Sciences*. New York, NY: Academic Press, 2006.

7

Deep Learning for Very High-Resolution Imagery Classification

Sangram Ganguly

Saikat Basu

Ramakrishna
Nemani

Supratik
Mukhopadhyay

Andrew Michaelis

Petr Votava

Cristina Milesi

Uttam Kumar

7.1   Proposed Approach..115
       Unsupervised Segmentation • Feature Extraction • Classification
       Using DBN • Conditional Random Field • Online Update of the
       Training Database
7.2   The NEX HPC Architecture 121
7.3   The AWS Computing Infrastructure...................................122
7.4   Results and Comparative Studies......................................124
7.5   Related Work..127
7.6   Summary 128

Very high-resolution (VHR) land cover classification maps are needed to increase the accuracy of current land ecosystem and climate model outputs. Limited studies are in place that demonstrate the state-of-the-art in deriving VHR land cover products [1–4]. Additionally, most methods heavily rely on commercial softwares that are difficult to scale given the area of study (e.g., continents to globe). Complexities in present methods relate to (1) scalability of the algorithm, (2) large image data processing (compute and memory intensive), (3) computational cost, (4) massively parallel architecture, and (5) machine learning automation. VHR satellite data sets are of the order of terabytes and features extracted from these data sets are of the order of petabytes. This chapter demonstrates the use of a scalable machine learning algorithm using airborne imagery data acquired by the National Agriculture Imagery Program (NAIP) for the Continental United States (CONUS) at an optimal spatial resolution of 1 m [5]. These data come as image tiles (a total of quarter million image scenes with ~60 million pixels) that are multispectral in nature (red, green, blue, and near-infrared [NIR] spectral channels) and have a total size of ~60 terabytes for an individual acquisition over CONUS. Features extracted from the entire data set would amount to ~8–10 petabytes. In the proposed approach, a novel semiautomated machine learning algorithm rooted on the principles of "deep learning" is implemented to delineate the percentage of canopy tree cover. In order to perform image analytics in such a granular system, it is mandatory to devise an intelligent archiving and query system for image retrieval, file structuring, metadata processing, and filtering of all available image scenes. This chapter showcases an end-to-end architecture for designing the learning algorithm, namely deep belief network (DBN) (stacked restricted Boltzmann machines [RBMs] as an unsupervised classifier) followed by a backpropagation neural network (BPNN) for image classification, a statistical region merging (SRM)-based segmentation algorithm to perform unsupervised segmentation, and a structured prediction framework using conditional random field (CRF) that integrates the results of the classification module and the segmentation module to create the final classification labels. In order to scale this process across quarter million NAIP tiles that cover the entire CONUS, we provide two architectures, one using the National Aeronautics and Space Administration (NASA) high-performance computing (HPC) infrastructure [6,7] and the other using the Amazon Web Services (AWS) cloud compute platform [8]. The HPC framework describes the granular parallelism architecture that can be designed to implement the process across multiple cores with low-to-medium memory requirements in a distributed manner. The AWS framework showcases use-case scenarios of deploying multiple AWS services like the Simple Storage Service (S3) for data storage [9], Simple Queuing Service (SQS) [10] for coordinating the worker nodes and the compute-optimized Elastic Cloud Compute (EC2) [11] along with spot instances for implementing the machine learning algorithm.

Deep learning has gained popularity over the last decade due to its ability to learn data representations in an unsupervised manner and generalized to unseen data samples using hierarchical representations. The most recent and best-known deep learning model is the DBN [12]. Over the last decade, numerous breakthroughs have been made in the field of deep learning; a notable one being [13], where a locally connected sparse autoencoder was used to classify objects in the ImageNet data set [14] producing state-of-the-art results. In [15], DBNs have been used for modeling acoustic signals and have been shown to outperform traditional approaches using Gaussian mixture models for automatic speech recognition (ASR). Another closely related approach, which has gained much popularity over the last decade, is the convolutional neural network (CNN) [16]. This has been shown to outperform DBN in classical object recognition tasks like Mixed National Institute of Standards and Technology database (MNIST) [17] and Canadian Institute for Advanced Research (CIFAR) [18].

A related and equally hard problem is image classification from remotely sensed images from satellite and airborne platforms. It comprises of terabytes of data and exhibits variations due to conditions in data acquisition, preprocessing, and filtering. Traditional supervised learning methods like random forests [19] do not work well for such a large-scale learning problem. This is primarily because of the fact that these algorithms require labeled data and are unable to generalize when the amount of labeled data is very low as compared to the total size of the data set. A novel classification algorithm for detecting roads in aerial imagery using deep neural networks was proposed in [1], however, the problem of classifying multiple land cover types is a much harder problem considering the significantly higher intraclass variability in land cover types such as trees, grasslands, barren lands, water bodies, etc. relative to that of roads.

In this chapter, we focus on leveraging the power of DBNs for the representation and classification of aerial imagery data acquired at very high spatial resolutions of the order of 1 m. Aerial image representation and classification is a challenging problem that lies at the intersection of machine learning, computer vision, and remote sensing. To address this large-scale classification problem, an automated probabilistic framework is proposed for the segmentation and classification of 1-m VHR NAIP data to derive large-scale estimates of tree cover from the image tiles. The results from the classification and segmentation algorithms are integrated using a discriminative undirected probabilistic graphical model [20] based on CRF that performs structured prediction and helps in capturing the higher order contextual dependencies between neighboring pixels.

This chapter is organized as follows: Section 7.1.1 provides a detailed description of the unsupervised segmentation phase using the SRM algorithm. Sections 7.1.2 and 7.1.3 enumerate the key components of the feature extraction step and the learning phase based on DBN. The details of the CRF algorithm and the online update procedure for the training data are illustrated in Sections 7.1.4 and 7.1.5, respectively. An overview of the NASA Earth Exchange high-performance computing (NEX HPC) architecture is provided in Section 7.2 and details of the AWS infrastructure are provided in Section 7.3. Section 7.4 contains results and comparative studies with National Land Cover Database (NLCD), and light detection and ranging (LiDAR). We finally conclude this chapter with related work in Section 7.5 and a comprehensive summarization in Section 7.6.

7.1 Proposed Approach

7.1.1 Unsupervised Segmentation

An image segment can be considered to be any region having pixels with uniform values for the various spectral bands. The aim of the segmentation is to find regions with uniform spectral band values representing a particular land cover class. Segmentation is performed using the SRM algorithm [21]. A generalized SRM algorithm is used that incorporates values from all the four multispectral bands from the NAIP data. The SRM algorithm initially considers each pixel as a region and merges them to form larger regions based on a merging criterion. The merging criterion used is as follows: Given the differences in red, green, blue, and NIR channel values of neighboring pixels that correspond to ΔR, ΔG, ΔB, and ΔNIR, respectively, merge two regions if ($\Delta R < \tau$ & $\Delta G < \tau$ & $\Delta B < \tau$ & $\Delta NIR < \tau$), where τ is a threshold that determines the coarseness of segmentation. The merging criterion can be formalized as a merging predicate that is evaluated as "true" if two regions are merged and "false" otherwise. The generalized version of the merging predicate (adopted from [21]) can be formally written as follows:

$$P(\nabla, \nabla') = \begin{cases} \text{true,} & \text{if } \forall c \in \{R, G, B, NIR\} \left| \overline{\nabla'}_c - \nabla_c \right| \leq \sqrt{b^2(\nabla) + b^2(\nabla')} \\ \text{false,} & \text{otherwise} \end{cases} \quad (7.1)$$

where $\overline{R}_c$ and $\overline{R'}_c$ denote the mean value of the color channel c for regions R and R', respectively. b is a function defined as follows:

$$b(\nabla) = g \sqrt{\left(\frac{1}{2Q|\nabla|}\right) \ln\left(\frac{|\nabla_{|\nabla|}|}{\delta}\right)} \quad (7.2)$$

where g is the number of possible values for each color channel (256 in this case). $|\nabla|$ denotes the cardinality of a segment, that is, the number of pixels within the boundaries of an image region ∇. $\nabla_{|\nabla|}$ represents the set of all regions that have the same cardinality as ∇. δ is a parameter that is inversely proportional to the image size. Q is the quantization parameter that controls the coarseness of the segmentation. A careful analysis of Equations 7.1 and 7.2 shows that a higher value of Q results in a lower threshold thereby reducing the probability of two segments getting merged into a bigger segment, henceforth resulting in a finer segmentation. On the other hand, a lower value of Q results in a higher threshold and a coarser segmentation.

The algorithm calculates the differences between neighboring pixels and sorts the pairs using radix sort. If the merging criterion is met, then it merges corresponding segments into one. Algorithm 7.1 provides the details of the SRM algorithm. A low threshold (or a higher Q value) is used in order to perform oversegmentation. In essence, each class (e.g., forest, grass, etc.) might be divided into multiple segments, but one segment would ideally not contain more than one class. This is achieved by selecting a quantization level of 2^{15}. This is useful for eliminating the possibility of interclass overlap within a segment.

In the case of an oversegmented image, areas within large homogeneous patches of vegetated pixels are also split into multiple segments owing to slight variability in spectral characteristics, for instance, shadows arising amid tree/nontree regions and grassy areas with occasional dry brown patches. This ensures that the chances of interclass overlap are averted. SRM is more efficient compared to other segmentation algorithms like k-means clustering [22]. The lists of merging tests can be sorted using radix sort with color difference as the keys and hence has a time complexity of $O[|I|\log(g)]$ which is linear in $|I|$. Here, $|I|$ is the cardinality or size of the input image. SRM segments a 512×512 image in about 1 second on an Intel Pentium 4 2.4G processor and hence is better suited for the current application since the size of the data set is of the order of terabytes. However, SRM has high memory requirements—around 3 gigabytes per 6000×7000 image. This is mitigated by splitting the input image into 256×256 windows as illustrated in Section 7.2.

Algorithm 7.1: Statistical region merging algorithm

---

1. Compute the set of $2\,|I|$ couples of adjacent pixels in a 4-connected neighborhood in image I.
2. Sort the pixel pairs in this list using radix sort.
3. For each pixel pair (∇, ∇'), compute the merging predicate $P(\nabla, \nabla')$ as defined in Equation 7.1.
4. If the merging predicate returns true, merge the two pixels into a single unified region.
5. Continue the merging of segments recursively until none of the remaining segments satisfy the merging predicate.

7.1.2 Feature Extraction

Prior to the classification process, the feature extraction phase computes 150 features from the input imagery. Some of the key features used for classification are mean, standard deviation, variance, second moment, direct cosine transforms, correlation, covariance, autocorrelation, energy, entropy, homogeneity, contrast, maximum probability, and sum of variance of the hue, saturation, intensity, and NIR channels as well as those of the color co-occurrence matrices. These features were shown to be useful descriptors for classification of satellite imagery in previous studies [23–25]. The red and NIR bands already provide useful features for delineating forests and nonforests based on chlorophyll absorptance and reflectance, however, derived features such as vegetation indices from spectral band combinations are more representative of vegetation greenness—these include the enhanced vegetation index (EVI) [26], normalized difference vegetation index (NDVI) [27,28], and atmospherically resistant vegetation index (ARVI) [29].

These indices are expressed as follows:

$$EVI = G \times \frac{NIR - Red}{NIR + c_{red} \times Red - c_{blue} \times Blue + L} \tag{7.3}$$

Here, the coefficients G, c_{red}, c_{blue}, and L are chosen to be 2.5, 6, 7.5, and 1 following those adopted in the moderate resolution imaging spectroradiometer (MODIS) EVI algorithm [30].

$$NDVI = \frac{NIR - Red}{NIR + Red} \tag{7.4}$$

$$ARVI = \frac{NIR - (2 \times Red - Blue)}{NIR + (2 \times Red + Blue)} \tag{7.5}$$

The performance of the deep learning algorithm and its convergence to the optima depends to a large extent on the selected features. Some features contribute more than others toward optimal classification. The 150 features extracted are narrowed down to 24 using a feature-ranking algorithm that uses a statistical t-test [31]. Examples of computed image features in the proposed framework are shown in Figure 7.1.

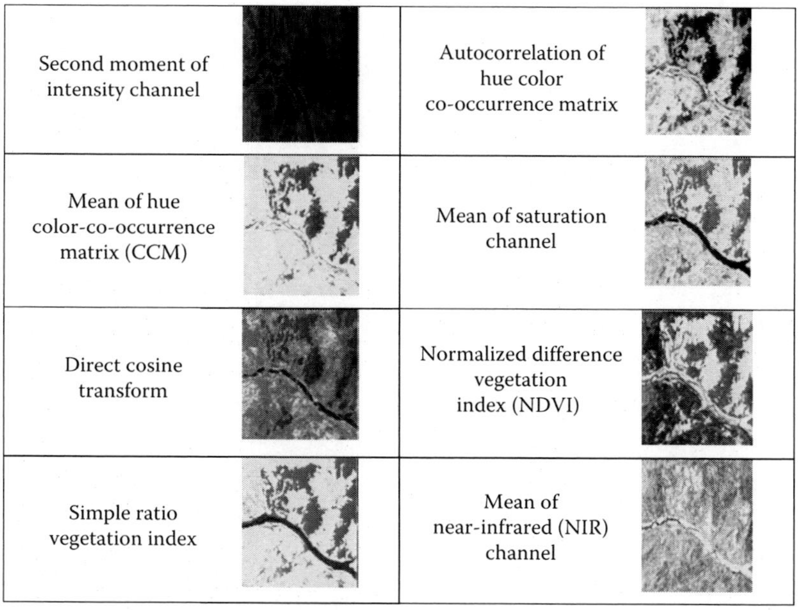

FIGURE 7.1 Some example features computed from a sample National Agricultural Imagery Program image tile in California.

7.1.3 Classification Using DBN

DBN consists of multiple layers of stochastic, latent variables trained using an unsupervised learning algorithm followed by a supervised learning phase using feedforward BPNN. In the unsupervised pretraining stage, each layer is trained using an RBM. Once trained, the weights of the DBN are used to initialize the corresponding weights of a neural network [32]. A neural network initialized in this manner converges much faster than an otherwise uninitialized one. Unsupervised pretraining is an important step in solving a classification problem with terabytes of data and high variability. A DBN is a graphical model [33] where neurons of the hidden layer are conditionally independent of each other, given a particular configuration of the visible layer and vice versa. A DBN can be trained layer-wise by iteratively maximizing the conditional probability of the input vectors or visible vectors, given the hidden vectors and a particular set of layer weights. As shown in [12], this layer-wise training can help in improving the variational lower bound on the probability of the input training data, which in essence leads to an improvement of the overall generative model. Algorithm 7.2 details this greedy layer-wise training procedure for DBNs. We first provide a formal introduction to the RBM. The RBM can be denoted by the energy function

$$E(v, h) = -\sum_i a_i v_i - \sum_j b_j h_j - \sum_i \sum_j h_j w_{i,j} v_i \tag{7.6}$$

where each RBM consists of a matrix of layer weights $W = (w_{ij})$ between the hidden units h_j and the visible units v_i. a_i and b_j are the bias weights for the visible units and the hidden units, respectively. The RBM takes the structure of a bipartite graph and hence it only has interlayer connections between the hidden or visible layer neurons but no intralayer connections within the hidden or visible layers. So, the visible unit activations are mutually independent given a particular set of hidden unit activations and vice versa [34]. Hence, by setting either h or v constant, the conditional distribution of the other can be computed as follows:

$$P(h_j = 1|v) = \sigma\left(b_j + \sum_{i=1}^{m} w_{i,j}v_i\right) \tag{7.7}$$

$$P(v_i = 1|h) = \sigma\left(a_i + \sum_{j=1}^{n} w_{i,j}h_j\right) \tag{7.8}$$

where σ denotes the log sigmoid function:

$$f(x) = \frac{1}{1 + e^{-x}} \tag{7.9}$$

The training algorithm maximizes the expected log probability assigned to the training data set V. So if the training data set V consists of the visible vectors v, then the objective function is as follows:

$$\underset{W}{\text{argmax}}\, E\left[\sum_{v \in V} \log P(v)\right] \tag{7.10}$$

Algorithm 7.3 details the contrastive divergence algorithm for training RBMs.

Algorithm 7.2: Greedy layer-wise training algorithm for DBNs

---

1. The first layer of the DBN is modeled as an RBM and its input neurons are connected to the visible vectors v.
2. Using this RBM, an input data representation can be obtained for the second layer as the conditional distribution $p(h = 1|v)$ using the activation function given in Equation 7.1.
3. Train the next layer of the DBN as an RBM with the vector h as the input data vector.
4. Repeat steps 2 and 3 for all the RBMs in the DBN model.
5. Perform supervised fine-tuning of the parameters using a supervised learning algorithm (a feedforward BPNN in this case).

Once trained, the DBN is used to initialize the weights and biases of a feedforward BPNN. The neural network gives an estimate of the posterior probabilities of the class labels, given the input vectors, which is the feature vector in this case. As illustrated in [35], the outputs of a neural network trained by minimizing the sum of squares error function approximate the conditional averages of the target data

$$y_k(x) = t_k|x = \int t_k p(t_k|x)dt_k \tag{7.11}$$

Here, t_k are the set of target values that represent the class membership of the input vector x_k. For a binary classification problem, in order to map the outputs of the neural network to the posterior probabilities of the labeling, we use a single output y and a target coding that sets $t^n = 1$ if x^n is from class C_1 and $t^n = 0$ if x^n is from class C_2. The target distribution would then be given as

$$p(t_k|x) = \delta(t - 1)P(C_1|x) + \delta(t)P(C_2|x) \tag{7.12}$$

Here, δ denotes the Dirac delta function which has the properties $\delta(x) = 0$ if $x \neq 0$ and $\int_{-\infty}^{\infty} \delta(x)dx = 1$.

Combining these equations, we get $y(x) = P(C_1|x)$.

So, the network output $y(x)$ represents the posterior probability of the input vector x having the class membership C_1 and the probability of the class membership C_2 is given by $P(C_2|x) = 1 - y(x)$. This argument can easily be extended to multiple class labels for a generalized multiclass classification problem.

Algorithm 7.3: The contrastive divergence algorithm for training restricted boltzmann machines

---

1. For a training sample v, compute the probabilities of the hidden layer neurons and sample a hidden activation vector h from this probability distribution.
2. Compute the *positive gradient* as the outer product of v and h.
3. From h, reconstruct a vector sampling v' of the visible units, then resample the hidden activations h' from this using Gibbs sampling.
4. Compute the *negative gradient* as the outer product of v' and h'.
5. The update rule for the weights can be defined as the difference between the *positive gradient* and the *negative gradient*, multiplied by the learning rate η as follows:

$$\Delta w_{i,j} = \eta(vh^T - v'h'^T)$$

The biases can be also be updated using the same update rule.

7.1.4 Conditional Random Field

A CRF has been used in the pattern recognition literature for performing structured prediction [36]. In structured prediction, the labeling of a pixel depends not only on the label assigned to that particular pixel but also on the values assumed by "neighboring" pixels. The word "neighboring" here can mean either a 4-connected or 8-connected neighborhood or some custom metric defining the notion of neighborhood. The concept of neighborhood is useful in encoding contextual information. The final labeling of a pixel as a vegetated pixel depends not only on whether that pixel is classified as a tree but also on the classification results of neighboring pixels. For example, if a pixel has been classified as a tree pixel by the classifier and all the neighboring pixels have been classified as nontree pixels, then it is safe to assume with a high probability that the result of the classifier is due to random classification noise. A CRF is a type of discriminative undirected probabilistic graphical model that encodes contextual information using an undirected graph [20]. The probability distributions are defined using a random variable X over a set of observations and another random variable Y over corresponding label sequences. Y is indexed by the vertices of an undirected graph $G = (V, E)$ such that $Y = (Y_v)_{v \in V}$. The tuple (X, Y) is known as a CRF if the random variable Y conditioned on X exhibits the Markov property with respect to the graphical model, that is, $p(Y_v|X, Y_w, w \neq v) = p(Y_v|X, Y_w, w \sim v)$, where $w \sim v$ means that w and v are neighbors in G. Following the conventions defined in [37], the random variable X is defined over a lattice $V = \{1, 2, \ldots, N\}$ and a neighborhood system N. A CRF defines a set of random variables X_C conditionally dependent on each other as a clique c. So, a probability distribution associated with any random variable X_i of a clique is conditionally dependent on the distributions of all other random variables in the clique. The posterior probability distribution $\Pr(x|D)$ over a set of label assignments x to a set of data D in a CRF is given by a Gibb's distribution where the Gibb's energy function can be written as

$$E(x) = -\log \Pr(x|D) - \log Z = \sum_{c \in C} \psi_c(x_c) \tag{7.13}$$

Here, Z is a normalizing constant known as partition function, C is the set of all cliques and $\psi_c(x_c)$ is the potential function that defines the Gibb's energy over the clique c.

Thus, estimating the maximum a posteriori (MAP) probability of a label assigned to a random variable X involves estimating the label assignment that minimizes the Gibbs energy. Denoting this optimum label assignment as x^{opt}, we have

$$x^{opt} = \text{argmax}_{x \in L} \Pr(x|D) = \text{argmin}_{x \in L} E(x) \tag{7.14}$$

Here, $L = \{L_1, L_2, \dots, L_n\}$ is the set of all labels that can be assigned, given the input data D, which is {tree, nontree} in this case.

The Gibbs energy function defined in [37] takes the form

$$E(x) = \sum_{i \in V} \psi_i(x_i) + \sum_{i \in V, j \in N_i} \psi_{ij}(x_i, x_j) + \sum_{c \in S} \psi_c(x_c) \tag{7.15}$$

where $\psi_i(x_i)$ is the unary potential, $\psi_{ij}(x_i, x_j)$ is the pairwise potential, and $\psi_c(x_c)$ is the function associated with robust higher order region based and quality sensitive consistency potentials defined over a segment S. We use this same energy function, but define the unary, and pairwise potentials differently in order to capture the semantics of the aerial image classification problem. The unary potential defined in [37] takes the form

$$\psi_i(x_i) = \theta_T \psi_T(x_i) + \theta_{col} \psi_{col}(x_i) + \theta_l \psi_l(x_i) \tag{7.16}$$

where $\theta_T \psi_T(x_i)$ denotes the potential obtained from [38], $\theta_{col} \psi_{col}(x_i)$ the potential from color and $\theta_l \psi_l(x_i)$ the potential from location. However, we use an updated form of the unary potential function as follows:

$$\psi_i(x_i) = \theta_N \psi_N(x_i) + \theta_{band} \psi_{band}(x_i) \tag{7.17}$$

Here, we get rid of the location term $\theta_l \psi_l(x_i)$ defined in Equation 7.16 in order to achieve shift-invariance. The term $\theta_N \psi_N(x_i)$ denotes the potential due to the output produced by the neural network classifier described in Section 7.3. This is similar to the shape-texture term defined in [38]. So, we can define the potential $\psi_N(x_i)$ as

$$\psi_N(x_i) = -\log P(C_i|x) \tag{7.18}$$

Here, $P(C_i|x)$ denotes the normalized distribution generated by the classifier. Plugging this into Equation 7.11, we have

$$\psi_N(x_i) = -\log y_i \tag{7.19}$$

where y_i denotes the output distribution from the classifier.

We also update the $\theta_{col} \psi_{col}(x_i)$ term defined in [37] with the $\theta_{band} \psi_{band}(x_i)$ term because images in the NAIP data set consist of four bands—red, green, blue, and NIR. This is just for notational clarity and is a generalization of the $\psi_{col}(x_i)$ potential. Similarly, the pairwise term $\psi_{ij}(x_i, x_j)$ is updated to encode the band information as

$$\psi_{ij}(x_i, x_j) = \begin{cases} 0 & \text{if } x_i = x_j \\ \theta_P + \theta_v \exp(-\theta_\beta B_i - B_j^2), & \text{otherwise} \end{cases} \tag{7.20}$$

Here, B_i and B_j are the band vectors for pixels i and j, respectively. The model parameters $\theta_N, \theta_{band}, \theta_P, \theta_v$, and θ_β are learned from the training data.

The term $\psi_c(x_c)$ denotes the *region consistency potential* as defined in [37] and is given by

$$\psi_c(x_c) = \begin{cases} 0 & \text{if} x_i = l_k \\ \theta_P \, |c|^{\theta_\alpha}, & \text{otherwise} \end{cases} \qquad (7.21)$$

Here, $|c|$ is the number of pixels in the segment. $\theta_P \, |c|^{\theta_\alpha}$ denotes the cost associated with labeling that do not confirm with the labeling of the other pixels in the segment. This term ensures that the labels assigned to the pixels belonging to the same segment are consistent with one another, that is, pixels belonging to the same segment are likely to belong to the same object/class. This is useful in obtaining object segmentations with fine boundaries and particularly helpful for accurate delineation of tree-cover areas in aerial images, where a single pixel denotes an area of 1 m^2. In the unsupervised segmentation stage using SRM algorithm, an oversegmentation of the image is performed and hence, the likeliness of interclass overlaps is minimized in the framework. So, we can safely get rid of the quality term in the potential function.

7.1.5 Online Update of the Training Database

Once the final results are obtained, the training database is updated online with incorrectly labeled examples using expert knowledge on the fly. This is done as follows:

After the generation of tree-cover maps from a certain number of NAIP tiles (100, here), 10 (10% in general) maps are chosen at random and a reference to the NAIP tiles corresponding to these maps are saved to a database. An automated image-rendering tool (developed as part of the framework), then allows experts to relabel misclassified image patches. These relabeled patches are then saved to the training database with the correct labeling. This improves the quality of results produced by the classifier in subsequent iterations. 100 consecutively processed NAIP tiles cover a relatively small geographic area. Hence, a random 10% of the tiles represent a uniform selection of tiles from every spatial window and choosing every 100 images ensures a uniform selection from the entire mapped region. Every time the training data set is updated, automated online training is done. The online update phase helps in reducing the false positive rate (FPR) and at the same time, significantly increases the true positive rate (TPR). Algorithm 7.4 provides details of the online update algorithm.

Algorithm 7.4: Algorithm for online update of the training data set

---

1. For every 100 images from the output data set O, select 10 images at random.
2. For each selected image, use expert knowledge to label 100 randomly selected pixels in the corresponding tiles from the NAIP input data set.
3. Compare these pixel labels with those generated by the deep learning framework.
4. For each misclassified pixel take the corresponding input feature vectors and append it to the training data set.
5. Using the new training data set retrain the classifier for future iterations.
6. Repeat steps 1–5 for the entire test data set.

7.2 The NEX HPC Architecture

The abovementioned modules have been deployed as standalones on the NEX supercomputing cluster [39]. NEX is a collaborative computing platform that brings together state-of-the-art computing facility with large volumes (hundreds of terabytes) of satellite and climate data from NASA along with multiple climate and ecosystem models. NEX facilitates the execution of various research projects related to the

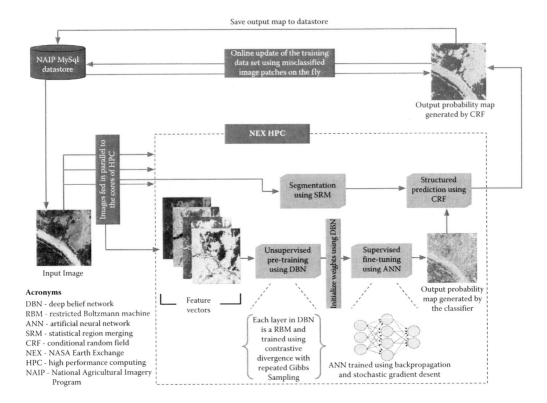

FIGURE 7.2   High-performance computing architecture.

Earth Sciences and provides an end-to-end computing infrastructure along with data acquisition, analysis, model executions, and result sharing. The deployment of the modules was done through QSub routines and the message passing interface (MPI). The data were accessed through a MySQL database. The NAIP tiles were processed in parallel in the cores of the NEX HPC platform. Each node in the cluster having Harpertown CPUs consists of 8 gigabytes of memory and eight cores with 3 GHz processors per node. So, in order to process eight tiles in parallel, one tile per core, the memory requirement per core has to be kept lower than 1 gigabyte. However, the problem arises with the use of the SRM algorithm illustrated in Section 7.1. Despite being fast, the algorithm has to store all the indices of the image gradients in memory while sorting them using radix sort as it makes decisions about region boundaries using global scene-level image descriptors. This has space complexity of the order of $O(n^2)$, which indicates all image gradients in an $n \times n$ image, which is of the order of ~3 gigabytes for a typical NAIP tile. So, in order to address this memory-performance tradeoff, each image is being split into $\lambda \times \lambda$ windows and then fed in a pipeline to each core in the HPC node. λ was chosen to be 256 for the experiments, because higher values led to a higher memory requirement while lower values resulted in a substantial increase of processing time. The current architecture takes a maximum of approximately 4 hours to process each NAIP tile. Figure 7.2 shows the details of the architecture.

7.3 The AWS Computing Infrastructure

An alternative implementation of the classification framework was done on the AWS cloud compute platform. AWS system components like S3, EC2, and SQS were used to provide a cost-effective yet efficient computing architecture that is fault tolerant and adaptive to processing requirements for the various classification modules being deployed in the framework.

Figure 7.3 depicts how one can construct a low-cost processing system using some of the many services that AWS has to offer. Currently, the effort to architect and deploy such a batch processing system is amazingly easy and can be done very quickly as compared to building out an "in house" compute cluster and "machine learning" software stack. Our demonstration AWS setup is semiresilient, cheap relative to the alternatives, reasonably secure, and effective at solving our problem of classifying a relatively large set of aerial imagery.

We first configure a base set of AWS services to build the processing pipeline and initialize our setup. Using the AWS Identity and Access Management (IAM) facility and keeping in mind the *principle of minimal privilege* [40], we created *identities* to enable services to access each other at the appropriate interface as well as identities for humans to allow them to initiate the processes described in the article. Next, we stage our aerial imagery data in the *persistent* AWS S3 input bucket. Then, we construct a virtual private cloud (VPC), which is a network-based logically isolated section of our *transient* EC2 compute nodes; note, compute nodes are only active or instantiated, when there is work to be done. We then setup and activate our SQS message queue with the appropriate parameters, such as queue item visibility timeouts, and so on [10].

Once the base level services are setup and in a ready state, we must build a single Amazon Machine Image (AMI) template that can perform the image classification. Within our template AMI, we use a python script to interface with the SQS message queue, which is a basic loop *consumer* that pulls a queue item containing

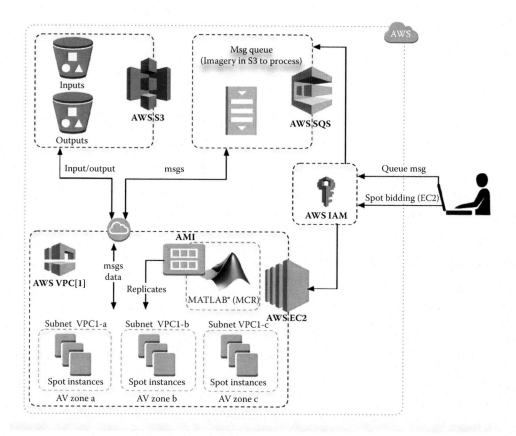

FIGURE 7.3 (See color insert.) The AWS computing infrastructure. AMI, Amazon Machine Image; AV, availability; AWS, Amazon Web Services; EC2, Elastic Cloud Compute; IAM, AWS identity and access management; MCR, MATLAB® compiler runtime; S3, Simple Storage Service; SQS, Simple Queuing Service; VPC, virtual private cloud.

only the location of the input image in S3 that is to be processed and some algorithm parameters. This python script stages the current image to be processed from S3, instantiates an existing MATLAB®-based code that performs the classification described in this chapter, and then writes the resultant data to our *persistent* S3 output bucket. The classification package runs on the MATLAB® compiler runtime (MCR), a default installed package in the AMI. The MATLAB® package that preforms the classification described in this chapter only processes one full image at a time. As our AMI boots, the python *consumer* is automatically started via a simple system V init script and if the *consumer* "starves" for more than several minutes it will terminate the instance it is running in.

Once we build our simple template AMI described above and then test it on small image set, it is saved and then ready for use. The AMI can now be concurrently instantiated as many times as needed, giving us a consistent replicate compute environment that "leans" on the message queue to coordinate the processing.

The pipeline is now constructed and ready for our human user to send small messages from their personal computers to SQS and then begin placing "bids" in the AWS spot market. As the message queue fills, spot instance bids are fulfilled, machines are instantiated, and the processing of the data set can commence. Note that the processing, bidding, and bid fulfillment all occur asynchronously. The rate of input *consumption* depends on the ability to fulfill the bids we place in the spot market (and any other limits imposed by AWS). This brings us to the importance of having a bid placement strategy to determine the optimal price evaluation and selection strategy.

We use our VPC and its associated subnets as logical units for our bidding and simplistic pricing strategy. Each subnet within the VPC is assigned to an availability (AV) zone. AWS provides an API call for retrieving price histories for each AV zone, which is the smallest unit of price variability for EC2 computing. Using the expected runtime of one unit of processing, our *urgency* factor, the pricing history for each AV, and our overall processing budget goals, we place meaningful bids on a per instance basis to each zone within the spot market in an effort to reduce our overall costs of the project. One must take care when submitting a price to the bidding system, if it is too low, the machines may terminate too soon, since a spot instance can be taken away at any time and given to a user that places a higher bid or pays the on-demand price. If the bid is too high we pay more than necessary, but the machines are less likely to terminate. The previous sentence reemphasizes the importance of the *transient* and *persistent* components of our pipeline as the expensive computing is *transient* (and procured in the spot market) and the other less expensive components are *persistent* (procured in a standard fashion), without this fact, it may not make sense to construct such a system. Note that using the message queue (SQS) we can achieve a better level of bidding and price granularity as opposed to using something like Hadoop, which often requires one to bid in larger chunks to be meaningful.

This brief introduction gives the reader an overview on how one may create a simple yet effective batch processing systems on AWS, usually within a matter of days to hours, and use the spot market to reduce processing costs. Note that in our system we are able to scale out our almost desktop like machine learning stack to hundreds of machines without expensive software modification or sophisticated development, without any large hardware outlays and time-consuming procurements, and at a reduced cost in our cloud processing environment.

7.4 Results and Comparative Studies

A pilot study was performed to compare the feature extraction-based DBN framework with various traditional deep learning algorithms, namely Stacked Autoencoder, DBN, and CNN that learns features from the raw pixel values of the multispectral data. For this experiment, a total of 405,000 image patches each measuring 28 × 28 pixels covering six broad land cover classes were extracted from a randomly selected subset of NAIP tiles for the state of California. In order to maintain the high variance inherent in the entire NAIP data set, image patches were sampled from a multitude of scenes (a total of 1500 image tiles) covering different landscapes like rural areas, urban areas, densely forested, mountainous terrain, small to large water bodies, agricultural areas, and so on.

The various land cover classes include barren land, trees, grassland, roads, buildings, and water bodies. 324,000 images were chosen as the training data set and the rest were chosen as the testing data set. Table 7.1 reports the results from the classification framework. Figure 7.4 shows the output probability maps generated by the feature extraction-based DBN classifier and the final probability map generated by the CRF. Figure 7.5 shows a sample NAIP tile extracted from the Blocksburg area in California and the corresponding tree-cover map generated by the framework. To generate this binary tree-cover map, the output probability map from the CRF is filtered with a threshold τ to eliminate pixels with output probability less than the threshold. The threshold τ is set as 0.5. So, for the CRF output probability for a pixel x being $\Pr(x)$, the final output map value for pixel x

$$O(x) = \begin{cases} 1, & \text{if } \Pr(x) \geq 0.5 \\ 0, & \text{otherwise} \end{cases}. \tag{7.22}$$

Figure 7.6 shows the receiver operating characteristic (ROC) curve generated by varying the sample size of the training data set from 1000 samples per class to 2500 per class in steps of 100 for a particular tile in the NAIP data set from the Blocksburg area in California. The flat response toward the end of the curve indicates that increasing the training sample size has minimal effect beyond a point, which is around 2200 training samples for this exercise. This shows the robustness of the algorithm and the fact that a minimal amount of training samples is sufficient for training the classifier.

In a separate experimental setup, the tree-cover maps generated by our algorithm were validated against a high-resolution airborne LiDAR data footprint. The data were obtained in the Chester area in California, using the NASA Goddard's LiDAR, hyperspectral and thermal (G-LiHT) airborne imager [41]. NASA's Cessna 206 was used for acquiring the G-LiHT data. The Cessna was fitted with the VQ-480 (Riegl USA, Orlando, FL, USA) airborne laser scanning (ALS) instrument and was flown at an altitude of 335 m. The data acquired had a swath width of 387 m and a field of view of 60°. The sampling density was 6 pulses/m².

TABLE 7.1 Comparative Results with Various Traditional Deep Learning Algorithms

Classification Accuracy of Stacked Autoencoder (%)	Classification Accuracy of Deep Belief Network (%)	Classification Accuracy of Convolutional Neural Network (%)	Classification Accuracy of DBN with Feature Extraction Framework (%)
78.43	76.51	79.063	**93.916**

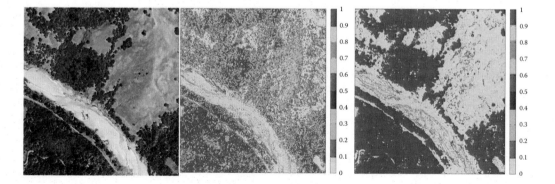

FIGURE 7.4 (See color insert.) The output probability maps. A sample NAIP tile from the Blocksburg region in California (left), the output probability map from the DBN-based classifier (center) and the output from the CRF algorithm (right).

FIGURE 7.5 The output tree-cover map from a sample NAIP tile from the Blocksburg area in northwestern California (7610 m × 6000 m). The green pixels denote tree-cover areas while the white pixels denote nontree areas.

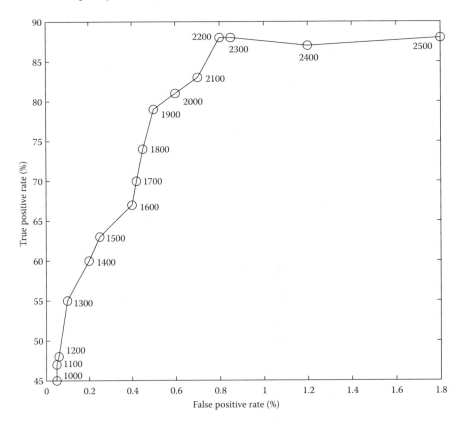

FIGURE 7.6 ROC curve generated by varying the sample size of the training data. The region of study was the same as that of Figure 7.4, an area in the Blocksburg region. The numbers alongside the circles represent the number of training samples used.

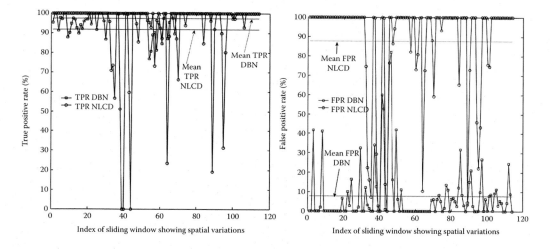

FIGURE 7.7 True positive rate (TPR) (left) and false positive rate (FPR) (right) of the DBN and NLCD with LiDAR as ground truth for the Chester area in California. The sliding window size was 100 × 100. NLCD, National Land Cover Database; LiDAR, light detection and ranging.

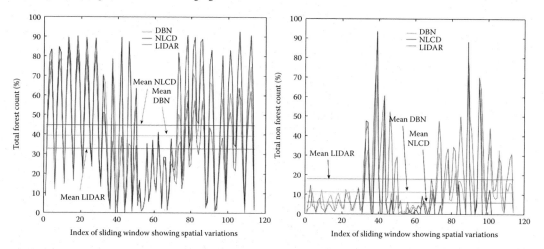

FIGURE 7.8 Percentage of tree-cover (left) and nontree areas (right) obtained using DBN, NLCD, and LiDAR for the same area as Figure 7.6. A 100 × 100 sliding window was used to obtain the percentage of tree-cover pixels in NAIP, NLCD, and LiDAR.

The spatial resolution of the final LiDAR data was 1 m. Tree-cover maps were also obtained from the NLCD [42] algorithm and Figure 7.7 shows the TPR and FPR of the feature extraction-based DBN framework and the NLCD algorithm with the results from the LiDAR data considered as the ground truth. It can be seen that the DBN-based framework produces higher TPR as compared to the NLCD results and has significantly lower FPR. Figure 7.8 shows the percentage of tree-cover areas and nontree areas for the same area as Figure 7.7.

7.5 Related Work

Present classification algorithms used for MODIS [43] or Landsat-based land cover maps like NLCD produce accuracies of 75% and 78%, respectively. The MODIS algorithm works on 500-m resolution imagery, and the NLCD works at 30-m resolution. The accuracy assessment is performed on a per-pixel basis and the relatively lower resolution of the data set makes it difficult to analyze the performance of these algorithms for 1-m imagery. A method based on object detection using a Bayes framework and a subsequent clustering of the objects into a hierarchical model using latent Dirichlet allocation was proposed in [44]. However, their approach detected object groups at a higher level of abstraction like parking lots. However, detecting the objects like cars or trees in itself was not addressed in their work. A deep convolutional hierarchical framework was proposed recently by Romero et al. [45]. However, they reported results on the AVIRIS Indiana's Indian Pines test site. The spatial resolution of the data set is limited to 20 m and it is difficult to evaluate the performance of their algorithm for object recognition tasks at a higher resolution.

7.6 Summary

The probabilistic framework based on feature extraction and DBNs has proved to be a useful tool for analyzing 1-m NAIP imagery for large-scale tree-cover mapping. The algorithm scales seamlessly to millions of scenes and can handle high variations, as is often the case for aerial imagery. The use of handcrafted features extracted from the hue, saturation, intensity, and NIR channels provides a useful framework for classifying NAIP imagery. The integration of the structured prediction framework based on CRF helped in increasing the TPRs while at the same time reducing the FPR by incorporating classifier outputs from the neighboring pixels located within the same neighborhood system. The NIR channel in the NAIP data set was also useful in segregating chlorophyll from nonchlorophyll regions and proved to be a very useful discriminative feature for addressing the tree/nontree classification of the 1-m NAIP data set. Comparative studies with various traditional deep learning algorithms have proved the effectiveness of the feature extraction step in creating better representations for the classification of very high-resolution imagery data. The NEX HPC infrastructure and the AWS cloud compute platform have shown to be useful for scaling up machine learning for the classification of terabytes of very high-resolution imagery data.

References

1. Mnih, V. and Hinton, G. (2010). Learning to detect roads in high-resolution aerial images. In *Proceedings of the 11th European Conference on Computer Vision (ECCV)*, vol. 10(1), pp. 5–15.

2. Pouliot, D.A., King, D.J., Bell, F.W., and Pitt, D.G. (2002). Automated tree crown detection and delineation in high-resolution digital camera imagery of coniferous forest regeneration. *Remote Sens. Environ.* 82(2–3), 322–334.

3. Komura, R., Kubo, M., and Muramoto, K.-I. (2004). Delineation of tree crown in high resolution satellite image using circle expression and watershed algorithm. In *Geoscience and Remote Sensing Symposium, 2004. IGARSS '04. Proceedings. 2004 IEEE International*, vol. 3, pp. 1577–1580.

4. Leckie, D.G., Gougeon, F.A., Walsworth, N., and Paradine, D. (2003). Stand delineation and composition estimation using semi-automated individual tree crown analysis. *Remote Sens. Environ.* 85(3), 355–369 Available at: http://www.sciencedirect.com/science/article/pii/S0034425703000130.

5. http://www.fsa.usda.gov/FSA/apfoapp?area=home&subject=prog&topic=nai.

6. http://www.nas.nasa.gov/hecc/resources/pleiades.html.

7. https://nex.nasa.gov/nex/.

8. http://aws.amazon.com/.

9. http://aws.amazon.com/s3/.

10. http://aws.amazon.com/sqs/.

11. http://aws.amazon.com/ec2/.

12. Hinton, G.E. and Osindero, S. (2006). A fast learning algorithm for deep belief nets. *Neural Comput.* 18, 2006.

13. Le, Q.V., Ranzato, M., Monga, R., Devin, M., Corrado, G., Chen, K., Dean, J., and Ng, A.Y. (2012). Building high-level features using large scale unsupervised learning. In *Proceedings of the 29th International Conference Machine Learning*.

14. Deng, J., Dong, W., Socher, R., jia Li, L., Li, K., and Fei-fei, L. (2009). Imagenet: A large-scale hierarchical image database. In *Proceedings of the IEEE International Conference on Computer Vision and Pattern Recognition*, pp. 248–255.

15. Mohamed, A.-R., Dahl, G.E., and Hinton, G.E. (2012). Acoustic modeling using deep belief networks. *IEEE Trans. Audio Speech Language Process.* 20(1), 14–22.

16. Lecun, Y., Bottou, L., Bengio, Y., and Haffner, P. (1998). Gradient-based learning applied to document recognition. In *Proceedings of the IEEE*, pp. 2278–2324.

17. MNIST. http://yann.lecun.com/exdb/ mnist/.

18. Krizhevsky, A. (2009). Learning multiple layers of features from tiny images. Technical report.

19. Breiman, L. (2001). Random forests. *Mach. Learn.* 45(1), 5–32, ISSN 0885–6125.

20. Lafferty, J., McCallum, A., Pereira, F. (2001). Conditional random fields: Probabilistic models for segmenting and labeling sequence data. In *Proceedings of the 18th International Conference on Machine Learning*. Morgan Kaufmann, San Francisco, CA. pp. 282–289.

21. Nock, R. and Nielsen, F. (2004). Statistical region merging. *IEEE Trans. Pattern Anal. Mach. Intell.* 26(11), 1452–1458.

22. Chen, T., Chen, Y., and Chien, S. (2008). Fast image segmentation based on K-Means clustering with histograms in HSV color space, In *Proceedings of IEEE 10th Workshop on Multimedia Signal Processing*.

23. Haralick, R.M., Shanmugam, K., and Dinstein, I. (1973). Textural features of image classification, *IEEE Trans Syst Man Cybernet.* SMC-3, pp. 610–621.

24. Soh, L. and Tsatsoulis, C. (1999). Texture analysis of SAR sea ice imagery using gray level co-occurrence matrices, *IEEE Trans. Geosci. Remote Sens.* 37(2), 780–795.

25. Clausi, D.A. (2002). An analysis of co-occurrence texture statistics as a function of grey level quantization, *Can. J. Remote Sensing.* 28(1), 45–62.

26. Huete, A., Didan, K., Miura, T., Rodriguez, E.P., Gao, X., Ferreira, L.G. (2002). Overview of the radiometric and biophysical performance of the MODIS vegetation indices, *Remote Sens. Environ.* 83(1–2), 195–213.

27. Rouse, J.W., Haas, R.H., Schell, J.A., and Deering, D.W. (1973). Monitoring vegetation systems in the Great Plains with ERTS, In *Third ERTS Symposium, NASA SP-351 I*, 309–317.

28. Tucker, C.J. (1979). Red and photographic infrared linear combinations for monitoring vegetation, *Remote Sens. Environ.* 8(2), 127–150.

29. Kaufman, Y.J. and Tanre, D. (1992). Atmospherically resistant vegetation index (ARVI) for EOS-MODIS, *IEEE Trans. Geosci. Remote Sens.* 30(2), 261, 270.

30. WWW1: http://vip.arizona.edu/documents/MODIS/MODIS_VI_UsersGuide_01_2012.pdf

31. Guyon, I. and Elisseeff, A. (2003). An introduction to variable and feature selection, *J. Mach. Learn Res.* 3, 1157–1182.

32. Bengio, Y. (2009). Learning deep architectures for AI. *Found. Trends Mach. Learn.* 2(1), 1–127, ISSN 1935–8237.

33. Koller, D. and Friedman, N. (2009). *Probabilistic Graphical Models: Principles and Techniques—Adaptive Computation and Machine Learning.* Cambridge, MA: MIT Press. ISBN 0262013193, 9780262013192.

34. Carreira-Perpinan, M.A. and Hinton, G.E. (2005). On contrastive divergence learning. In *Proceedings of the tenth International Workshop on Artificial Intelligence and Statistics*, pp. 59–66.

35. Bishop, C.M. (1995). *Neural Networks for Pattern Recognition.* Oxford: Clarendon Press.

36. Bakir, G., Taskar, B., Hofmann, T., Schölkopf, B., Smola, A., and Vishwanathan, S.V.N. (2007). *Predicting Structured Data.* Cambridge, MA: MIT Press. ISBN 0262026171.

37. Kohli, P., Ladický, L., Torr, P.HS. (2008). Robust higher order potentials for enforcing label consistency, In *Computer Vision and Pattern Recognition, 2008. CVPR 2008. IEEE Conference on*, pp.1, 8, 23–28 June 2008.

38. Shotton, J., Winn, J., Rother, C., and Criminisi, A. (2006). TextonBoost: Joint appearance, shape and context modeling for multi-class object recognition and segmentation. In *European conference on computer vision*. pp. 1–15.

39. NASA Earth Exchange (NEX). https://nex.nasa.gov/nex/

40. Saltzer, J.H. and Schroeder, M.D. (1975). The protection of information in computer systems, *Proc. IEEE*. 63, pp. 1278–1308.

41. Cook, B.D., Corp, L.A., Nelson, R.F., Middleton, E.M., Morton, D.C., McCorkel, J.T., Masek, J.G., Ranson, K.J., Ly, V., and Montesano, P.M. (2013). NASA Goddards LiDAR, hyperspectral and thermal (G-LiHT) airborne imager. *Remote Sens.* 5(8), 4045–4066.

42. Wickham, J.D., Stehman, S.V., Gass, L., Dewitz, J., Fry, J.A., and Wade, T.G. (2013). Accuracy assessment of NLCD 2006 land cover and impervious surface. *Remote Sens Environ.* 130, 294–304.

43. Friedl, M.A., Sulla-Menashe, D., Tan, B., Schneider, A., Ramankutty, N., Sibley, A., and Huang, X. (2009). Modis collection 5 global land cover: Algorithm refinements and characterization of new datasets. *Remote Sens. Environ.* 114, 168–182.

44. Vaduva, C., Gavat, I., and Datcu, M. (2012). Deep learning in very high resolution remote sensing image information mining communication concept. In *Signal Processing Conference (EUSIPCO), 2012 Proceedings of the 20th European*, pp. 2506–2510.

45. Romero, A., Gatta, C., and Camps-Valls, G. (2014). Unsupervised deep feature extraction of hyperspectral images. IEEE Trans. Geosci. *Remote Sens.* 54(3), 1349–1362.

8

Unmixing Algorithms: A Review of Techniques for Spectral Detection and Classification of Land Cover from Mixed Pixels on NASA Earth Exchange

Uttam Kumar

Cristina Milesi

S. Kumar Raja

Ramakrishna Nemani

Sangram Ganguly

Weile Wang

Petr Votava

Andrew Michaelis

Saikat Basu

8.1   Introduction.. 132
8.2   Unmixing Algorithm... 134
8.3   Unconstrained Unmixing Algorithms.............................. 134
      Unconstrained Least Squares • Matched Filtering • Mixed Tuned
      Matched Filtering • Constrained Energy Minimization • Sparse
      Unmixing via Variable Splitting and Augmented Lagrangian •
      SUnSAL Total Variation
8.4   Partially and Fully Constrained Unmixing Algorithms........... 140
      Sum-to-One Constrained Least Squares • Normalized Sum-to-One
      Constrained Least Squares • Nonnegative Constrained Least Squares •
      Normalized Nonnegative Constrained Least Squares • Fully
      Constrained Least Squares • Modified Fully Constrained Least Squares
8.5   Data 143
      Computer Simulations • Landsat Data • Endmember Generation
8.6   Methodology.. 147
8.7   Experimental Results.. 149
      Computer-Simulated Data • Landsat Data—An Agricultural Setup •
      Landsat Data—An Urban Setup
8.8   Discussion 162
8.9   Conclusion.. 167

8.1 Introduction

Land cover (LC) signifies the physical and biological cover present over the surface of the land, including water, vegetation, bare soil, and/or artificial structures and is a way of portraying the surface of the Earth. Changes in the LC induced by humans or caused due to certain environmental phenomena play a major role in global as well as regional-scale patterns, which in turn influence weather and climate. Hence, understanding LC at the regional to global levels is essential to evolve appropriate management strategies to mitigate the impacts of these changes. The LC information in conjunction with other primary and ancillary data is indispensable in addressing a variety of remote sensing (RS) applications such as geological research [1,2], wetland mapping [3–5], crop estimation [6,7], vegetation classification [8], forest classification [9], urban studies [10], feature extraction [11], LC change detection [12], Earth system modeling [13,14], global change research [15], etc.

The LC patterns can be captured through multiresolution spaceborne RS data that facilitate observations across a larger extent of Earth's surface compared to the ground-based observations [16]. However, the accuracy of LC mapping for larger spatial extent is limited by spatial and temporal characteristics of the currently available RS data. In this context, assignment of a single LC category to a pixel is appropriate if the spatial resolution of the imagery is comparable to the size of the object of interest [17]. Since many landscape features occur at spatial scales much finer than the resolution of the primary satellites used for continental or global LC mapping, observed data are a mixture of spectral signatures of individual objects resulting in mixed pixels. Regardless of the spectral resolution of the RS data, the spectral signals collected by most satellite sensors are undoubtedly mixtures of different class signatures within the sensor's instantaneous field of view (IFOV). This scale–resolution mismatch is one of the greatest challenges in modeling LC changes, where the spatial resolution of detail is less than what is required and the pixels consist of a mixture of multiple LC classes. Even at the spatial resolution of Landsat, which is of the order of tens of meters, each observed pixel is a mixed spectrum. These mixed spectra are combinations of the characteristics of the spectrally homogeneous components at the surface within the pixel [18]. It also means that the observed reflected electromagnetic spectrum from a pixel has not interacted with an area composed of a single homogeneous material. This necessitates identifying appropriate techniques for LC mapping at subpixel level and are thus of great interest. Quantitative interpretation of these mixed pixel spectra requires an understanding of the models and spatial scales of mixing of the spectrally homogeneous components in the area of interest. The solution to mixed pixel problem typically centers on spectral unmixing techniques [19] for estimating the proportion of each class within individual pixels. The main objective is to find out the proportion of each category in a given pixel, or in other words, unmix the pixel to identify the categories present.

During the last two decades, numerous methods have been proposed ranging from modeling the component mixtures to solving the linear combinations to obtain abundances through geometrical, statistical, and sparse regression-based approaches [20,21]. Commonly used approaches to mixed pixel classification include linear spectral unmixing [22,23], supervised fuzzy-c means classification [24], artificial neural networks (ANNs) [25], Gaussian mixture discriminant analysis [26], linear regression and regression tree [27], and spatial correlation-based unmixing [28], which use the linear mixture model (LMM) to estimate the abundance fractions of spectral signatures lying within a pixel. The LMM assumes that the reflectance spectrum of a mixture is systematic combination of the component reflectance spectra in the mixture (called endmembers). The combination of these endmembers is linear if the component of interest in a pixel appears in spatially segregated patterns, that is, no interaction between materials is assumed, and a pixel is treated as a linear combination of signatures with relative concentrations, where these concentrations or fractions correspond to the area occupied by that LC type. However, if the components are in intimate association, the electromagnetic spectrum typically interacts with more than one component as it is multiple scattered, and the mixing systematics between the different components are highly nonlinear [29,30].

There have been a few studies comparing the unmixing techniques but most of them have focused on the theoretical foundation of the algorithms. Ichoku and Karnieli [31] reviewed linear, probabilistic, geometric-optical, stochastic geometric, and fuzzy models concluding that there was some difference in the number and nature of components that could be resolved with the different models. Keshava [32] surveyed spectral unmixing algorithms characteristics through hierarchical taxonomies that revealed the commonalities and differences according to their philosophical assumptions such as interpretation of data, randomness, and optimization criterion. Pu et al. [10] found that ANN-based unmixing outperformed unconstrained least squares (UCLS) unmixing of Advanced Spaceborne Thermal Emission and Reflection Radiometer (ASTER) data when compared to a hard classified map obtained from linear discriminant analysis. Parente and Plaza [33] carried out a survey on the geometric and statistical approaches for unmixing. Zandifar et al. [34] compared a projected gradient nonnegative matrix factorization (NMF) independent component analysis (PG-NMFICA) algorithm with minimum spectral dispersion-minimum spatial dispersion NMF and vertex component analysis. However they pointed that unmixing of real data was not very successful with these algorithms. Bioucas-Dias et al. [20] reviewed signal-subspace, geometrical, statistical, sparsity-based, and spatial–contextual unmixing algorithms and their characteristics. To this end, studies assessing the performance of the classical, state-of-the-art linear unmixing algorithms using a unified framework through quantitative approaches and intercomparative assessment are still lacking, that can be used as a reference to perform continental to global LC mapping. Most of the existing literature describing new methods of unmixing compare their result only to a few standard unmixing techniques to establish their advantages.

In this chapter, analyses of the theory of different mixture models along with a quantitative and inter-comparative assessment of the popular unmixing algorithms is performed based on the evaluation of the derived LC fractional estimates in a comprehensive and rigorous manner on the NASA Earth Exchange (NEX) computing platform [15]. The objective is to perform a quantitative study of six unconstrained and eight partially and fully constrained unmixing algorithms. The unconstrained algorithms are (1) UCLS, (2) matched filtering (MF), (3) mixed-tuned matched filtering (MTMF), (4) constrained energy minimization (CEM), (5) sparse unmixing via variable splitting and augmented Lagrangian (SUnSAL), and (6) SUnSAL total variation (TV). The partially and fully constrained algorithms considered are (1) sum-to-one constrained least squares (SCLS), (2) normalized sum-to-one constrained least squares (NSCLS), (3) nonnegative constrained least squares (NCLS), (4) normalized nonnegative constrained least squares (NNCLS), (5) fully constrained least squares (FCLS), (6) modified fully constrained least squares (MFCLS), (7) constrained SUnSAL (CSUnSAL), and (8) constrained SUnSAL TV (CSUnSAL TV). These algorithms have different genesis and are based on different underlying assumptions. In the first set of experiments, the fraction estimates of different LC types obtained from the unmixing algorithms on computer-simulated noise-free data and noisy data of different signal-to-noise (SNR) ratio were evaluated. In separate tests, Gaussian noise (a random variable with 0 mean and fixed variance) was added to the data to judge the robustness of the algorithms in unmixing. The error in the fractional estimate was examined as the noise power (variance) was set to 2, 4, 8, 16, 32, 64, 128, and 256. In the second and third set of experiments with the real-world data, 11 Landsat scenes of an agricultural environment (near Fresno, California) and a Landsat scene of an urban setup (San Francisco, California) were used to evaluate the algorithms. These data were analyzed by deriving vegetation, substrate, and dark objects endmember fractions with respect to the ground measurements for an agricultural area and comparing with the high-resolution World View-2 (WV-2) unmixed image for an urban area, respectively. The idea was to identify the most robust technique in performance on different real landscapes. This study provides practical observations regarding the utility and type of technique needed to analyze different data sets pertaining to various landscapes. The results were evaluated and the sources of errors were analyzed using descriptive statistics—Pearson product–moment correlation coefficient (cc), root mean square error (RMSE), probability of success (p_s), boxplot, and bivariate distribution function (BDF).

This chapter is organized as follows: Section 8.2 formulates linear spectral unmixing problem, Sections 8.3 and 8.4 lay down the theoretical foundation of unconstrained and partially and fully constrained unmixing algorithms while Section 8.5 details the data used in this study and the endmember generation procedure, followed by methodology in Section 8.6. Section 8.7 presents the experimental results and their evaluation from the computer-simulated data (Section 8.7.1) followed by assessment of the algorithms on Landsat data of an agricultural scenario (Section 8.7.2) and of an urban setup (Section 8.7.3). Section 8.8 discusses the evaluation results along with the merits and demerits of the algorithms with concluding remarks in Section 8.9.

8.2 Unmixing Algorithm

At the outset, if there are M spectral bands and N classes, associated with each pixel is an M-dimensional vector $\mathbf{y}$ whose components are the gray values corresponding to the M bands. Let $\mathbf{E} = [e_1, \dots e_{n-1}, e_n, e_{n+1} \dots, e_N]$ be a $M \times N$ matrix, where $\{e_n\}$ is a column vector representing the spectral signature (endmember) of the nth target material. For a given pixel, the abundance or fraction of the nth target material present in the pixel is denoted by α_n, and these values are the components of the N-dimensional abundance vector $\boldsymbol{\alpha}$.

Assuming LMM [35], the spectral response of a pixel in any given spectral band is a linear combination of all of the endmembers present in the pixel at the respective spectral band. For each pixel, the observation vector $\mathbf{y}$ is related to $\mathbf{E}$ by a linear model written as

$$\mathbf{y} = \mathbf{E}\boldsymbol{\alpha} + \boldsymbol{\eta} \tag{8.1}$$

where $\boldsymbol{\eta}$ accounts for the measurement noise. We further assume that the components of the noise vector $\boldsymbol{\eta}$ are zero-mean random variables that are independent and identically distributed (i.i.d). Therefore, covariance matrix of the noise vector is $\sigma^2 \mathbf{I}$, where σ^2 is the variance, and $\mathbf{I}$ is a $M \times M$ identity matrix.

8.3 Unconstrained Unmixing Algorithms

In the unmixing jargon, nonnegativity and sum-to-one constraints are termed abundance nonnegativity constraint (ANC) and abundance sum-to-one constraint (ASC) (details about these constraints are discussed in Section 8.4). Sometimes, when these constraints are not imposed on the solution, the individual fractions can be negative, and/or abundance fractions sum to less than or greater than one, since the algorithm does not account for every class in a pixel. The algorithms producing negative fraction estimates or non–sum-to-unity abundance in a pixel are called unconstrained algorithms. Below, we discuss the six unconstrained unmixing algorithms.

8.3.1 Unconstrained Least Squares

The conventional approach [36] to extract the abundance values is to minimize $\|\mathbf{y} - \mathbf{E}\boldsymbol{\alpha}\|$

$$\hat{\boldsymbol{\alpha}}_{\text{UCLS}} = (\mathbf{E}^T \mathbf{E})^{-1} \mathbf{E}^T \mathbf{y} \tag{8.2}$$

which is termed as UCLS estimate of the abundance. The value of α_n is the abundance of the nth class in an abundance map. UCLS with full additivity is a nonstatistical, nonparametric algorithm that optimizes a squared-error criterion but does not enforce the nonnegativity and unity conditions. It does not incorporate noise in its signal model and assumes that the endmembers are deterministic [32].

8.3.2 Matched Filtering

MF was intended for target signal detection processing [37–39] that involves calculation of a linear operator, which seeks to balance target detection and background suppression. MF eliminates the requirement of knowing all the endmembers by maximizing the response of a known endmember and suppressing the response of the unknown background, thus matching the known signature [37,40–44].

The MF score is calculated for each pixel by matching minimum noise fraction (MNF) transformed input data to a spectrally pure endmember target spectra while suppressing the background. Spectra that closely match the training spectrum will have a score near one while background noise will have a score near zero [45]. This is achieved through a set of orthogonal matched filter vectors (MFVs) (target spectrum in MNF space) constructed to estimate the unmixing coefficients via a dot product with the observed pixel spectra:

$$\hat{\alpha} = \mathbf{MFV}.\mathbf{y} \tag{8.3}$$

Each filter vector is chosen to maximize the SNR ratio and is orthogonal to all the endmember spectra except the one that it represents. Using the calculus of variations, Bowles et al. [46] suggested a filter vector given by

$$\mathbf{MFV} = (\mathbf{AE})^{-1}\mathbf{A} \tag{8.4}$$

$$\text{where} \quad \mathbf{A} = \mathbf{E}^T - \left(\frac{\mathbf{J}}{\mathbf{M}}\mathbf{E}\right)^T$$

and J is an N by M matrix with each element as 1 [47].

This filter vector is projected onto the inverse covariance of the MNF transformed data and normalized to the magnitude of the target spectra [45]. The length of the MF vector equates to target abundance estimations that range from 0% to 100% [48], which provide means of estimating relative degree of match to the reference spectrum. Since the detection scheme is linear, a half-filled target pixel is expected to score 0.5 in a normalized MF method [49]. The MF vector linearly divides spectral space into equal MF scores. A high MF score with low infeasibility value is assigned to a correctly classified pixel. Discrimination among spectrally similar infrequent targets is problematic for the normal MF.

MF rapidly detects desired classes of interest, having wide applications in mineral mapping. Yet, it generally suffers from false positives. Other techniques like MTMF, which is discussed next, are used to reduce the false positive cases by considering noise variance and by estimating the probability of MF estimation error in each pixel [48].

8.3.3 Mixed Tuned Matched Filtering

MTMF resolves the selectivity problem inherent in the existing classical MF technique between a target class and the background [37,40,42,49]. It combines the strength of MF (no requirement to know all the endmembers) with physical constraints imposed by mixing theory (the signature at any given pixel is a linear combination of the individual components contained in that pixel).

The algorithm consists of three phases: (1) MNF transformation of the data, (2) MF calculation for abundance estimation and background suppression, and (3) mixture tuning (MT) calculation for the identification and rejection of false positives [45,49]. The known target signature defines a corner of a mixing simplex and the background covariance (eigenvectors and eigenvalues) gives the shape, orientation, and size of the background spectral scatter. Since MTMF works with data that have been whitened and decorrelated with the MNF transform, the noise distribution about a given point is a unit variance isotropic Gaussian distribution [49]. Now, the algorithm assesses the probability of MF estimation error for each pixel by calculating infeasibility values (a measure of goodness of match or the likelihood of each pixel being a mixture of the known class and the background materials) in the following three stages: (1) determination of the target vector component of the pixel, (2) interpolation of variance eigenvalues respective to the target vector component, and (3) calculation of the standardized separation between a pixel and its ideal target vector component [48]. The MTMF algorithm [49] is summarized below:

1. **MNF transformation of the data:** Computation of noise covariance, noise-whitening transformation of data, mean subtraction from noise-whitened data, and data decorrelation by projecting it to its eigenvectors.
2. **Abundance estimation through matched filter:** Projection of target spectrum to the MNF space, estimation of background covariance matrix followed by inversion, calculation of normalized MF projection vector, and projection of the MNF data to the MF vector to obtain target proportion.
3. **Measurement of feasibility:** Interpolation of mixture distributions, calculation of mixture infeasibilities, and segregation of true detections from false positives.

8.3.4 Constrained Energy Minimization

CEM is suitable for target detection as an unconstrained algorithm [50] but constrains a desired target signature by using a specific gain. It only utilizes knowledge of the target class signature and designs a finite impulse response (FIR) filter to pass through the desired target class while minimizing its output energy resulting from the undesired classes [21,42,51–55]. The following CEM filter is derived based on [21].

Let $\mathbf{X} = (\mathbf{y}_1, \mathbf{y}_2, \dots, \mathbf{y}_{p-1}, \mathbf{y}_p, \mathbf{y}_{p+1}, \dots, \mathbf{y}_P)$ be a data matrix representing P pixels in an image, where $\mathbf{y}_p = (y_{p1}, y_{p2}, ..., y_{PM})^T$ for $1 \leq p \leq P$ is an M-dimensional pixel vector. $\mathbf{E} = [e_1, \dots e_{n-1}, e_n, e_{n+1}, \dots, e_N]$ is the spectral signature of N targets $(t_1, \dots, t_{n-1}, t_n, \dots, t_N)$ present in the image. We intend to design a constrained FIR linear filter with a M-dimensional weight vector $\mathbf{wv} = (wv_1, wv_2, \dots, wv_M)^T$ specified by a set of M filter coefficients that minimize the filter output energy subject to the following constraints:

$$\mathbf{E}^T \mathbf{wv} = \mathbf{c}, \text{ where } \mathbf{e}_n^T \mathbf{wv} = \sum_{m=1}^{M} e_{nm} wv_m = c_n \text{ for } 1 \leq n \leq N \tag{8.5}$$

where $\mathbf{c} = (c_1, \dots c_{n-1}, c_n, c_{n+1}, \dots, c_N)^T$ is a constraint vector. If O_p represents the output of the designed FIR filter resulting from input $\mathbf{y}_p$, then

$$O_p = \sum_{m=1}^{M} wv_m y_{pm} = \mathbf{wv}^T \mathbf{y}_p = \mathbf{y}_p^T \mathbf{wv} \tag{8.6}$$

From Equations 8.5 and 8.6, we obtain a linearly constrained minimum variance-based target class detector, which minimizes the average energy of the total filter outputs

$$\frac{1}{P} \left[\sum_{p=1}^{P} O_p^2 \right] = \frac{1}{P} \left[\sum_{p=1}^{P} (\mathbf{y}_p^T \mathbf{wv})^T \mathbf{y}_p^T \mathbf{wv} \right] = \mathbf{wv}^T \left(\frac{1}{P} \left[\sum_{p=1}^{P} \mathbf{y}_p \mathbf{y}_p^T \right] \right) \mathbf{wv} = \mathbf{wv}^T \mathbf{S\_mat}_{M \times M} \mathbf{wv} \tag{8.7}$$

where $\mathbf{S\_mat}_{M \times M} = \frac{1}{P} \left[\sum_{p=1}^{P} \mathbf{y}_p \mathbf{y}_p^T \right]$ is a sample spectral matrix of the image. So, $\mathbf{S\_mat}_{M \times M} = \frac{1}{P} [\mathbf{X} \mathbf{X}^T]$. Combining Equation 8.7 with Equation 8.5 results in constrained optimization problem of the form

$$\min_{wv} \mathbf{wv}^T \mathbf{S\_mat}_{MM} \mathbf{wv} \text{ subject to } \mathbf{E}^T \mathbf{wv} = \mathbf{c}. \tag{8.8}$$

The solution to Equation 8.8 is achieved by Chang et al. [56,57]. For details, the readers are referred to [21, p. 54]. CEM works for a single target signature $\mathbf{t}_n$, $(\mathbf{E} = \mathbf{t}_n)$, the constraint vector $\mathbf{c}$ in Equation 8.5 can be substituted by $\mathbf{t}_n^T \mathbf{wv} = \sum_{m=1}^{M} \mathbf{t}_{nm} wv_m = 1$ and becomes a constraint unity scalar. Then, Equation 8.8 reduces to $\min_{wv} \left\{ \mathbf{wv}^T \mathbf{S\_mat}_{M \times M} \mathbf{wv} \right\}$ subject to $\mathbf{t}_n^T \mathbf{wv} = 1$ with the optimal solution $\mathbf{wv}^{\mathrm{CEM}}$ given by

$$\mathbf{wv}^{\mathrm{CEM}} = \frac{\mathbf{S\_mat}_{M \times M}^{-1} \mathbf{t}_n}{\mathbf{t}_n^T \mathbf{S\_mat}_{M \times M}^{-1} \mathbf{t}_n}.$$

Substituting the optimal weight $\mathbf{wv}^{\mathrm{CEM}}$ for $\mathbf{wv}$ in Equation 8.6 gives CEM filter with a detector, $\mathrm{detector}_{\mathrm{CEM}}(\mathbf{y})$ given by Harsanyi [58] in Chang [21]

$$\mathrm{detector}_{\mathrm{CEM}}(\mathbf{y}) = (\mathbf{wv}^{\mathrm{CEM}})^T \mathbf{y} = (\mathbf{t}_n^T \mathbf{S\_mat}_{M \times M}^{-1} \mathbf{t}_n)^{-1} (\mathbf{S\_mat}_{M \times M}^{-1} \mathbf{t}_n)^T \mathbf{y}. \tag{8.9}$$

8.3.5 Sparse Unmixing via Variable Splitting and Augmented Lagrangian

Sparse regression [59] is a new direction for unmixing which is related to both statistical and the geometrical frameworks. A mixed pixel with sparse linear mixtures of spectral signatures from a library (available *a priori*) is fitted. Estimating or generating the endmembers from the data is not required here. The method depends on searching an optimal subset of signatures that can model each mixed pixel in the data. The degree of coherence (isometry) between the data matrix and sparseness of the endmember signals [60] decides the algorithm's ability to obtain sparse solutions for an underdetermined system. When the data matrix has low coherence and the signals are sparse, it is an ideal situation for sparse unmixing.

Endmember search is conducted in a large library, say $\mathbf{E} \in \mathbb{R}^{M \times N}$, where M is the number of spectral bands and N is the number of spectral class (or target material) signatures in the library, $M < N$ and $\boldsymbol{\alpha} \in \mathbb{R}^N$. It is possible that only a few of the signature contained in $\mathbf{E}$ would involve in the mixed pixel spectrum. Therefore, $\boldsymbol{\alpha}$ will contain many values of zero, so $\boldsymbol{\alpha}$ is a sparse vector. The sparse regression problem [60] is expressed as

$$\min_{\boldsymbol{\alpha}} \|\boldsymbol{\alpha}\|_0 \text{ subject to } \|\mathbf{y} - \mathbf{E}\boldsymbol{\alpha}\|_2 \leq \delta, \tag{8.10}$$

$$\boldsymbol{\alpha} \geq 0, \ \mathbf{1}^T \boldsymbol{\alpha} = 1$$

where $\|\boldsymbol{\alpha}\|_0$ denotes the number of nonzero components of $\boldsymbol{\alpha}$, and $\delta \geq 0$ is the noise and modeling error tolerance. Note that δ in Equation 8.10 is the error tolerance and $\boldsymbol{\eta}$ in Equation 8.1 is a $M \times 1$ vector collecting errors affecting the measurements at each spectral band. $\boldsymbol{\alpha} \geq 0$ and $\mathbf{1}^T \boldsymbol{\alpha} = 1$ refer to ANC and ASC, respectively. A set of sparsest signals belonging to the $(N-1)$ probability simplex satisfying error tolerance inequality defines the solution of Equation 8.10.

When the fractional abundances from sparse regression follow ANC and ASC, the problem is referred to as constrained sparse regression (CSR) given by Equation 8.11, which is used to find sparse linear mixtures of spectra selected from large libraries [59]

$$\min_{\alpha}(1/2) \|\mathbf{E}\boldsymbol{\alpha} - \mathbf{y}\|_2^2 + \lambda \|\boldsymbol{\alpha}\|_1 \tag{8.11}$$

$$\text{subject to } \boldsymbol{\alpha} \geq 0, \quad \mathbf{1}^T \boldsymbol{\alpha} = 1$$

where $\|\boldsymbol{\alpha}\|_2$ and $\|\boldsymbol{\alpha}\|_1$ are the l_2 and l_1 norms and $\lambda \geq 0$ is a weighing factor between the l_2 and l_1 terms. SUnSAL solves the problem stated by Equation 8.11 based on the alternating direction method of multipliers (ADMM) [59,61,62]. ADMM can be derived as a variable splitting procedure followed by the adoption of an augmented Lagrangian method to solve the constrained problem. The algorithm is briefly stated below. For more detailed derivation, the readers are encouraged to refer Bioucas-Dias and Figueiredo [59]. Assume that matrix $\mathbf{E}$ (library of spectral signatures) is known and corresponds to under-determined systems ($N > M$) rather than obtained from an endmember extraction algorithm (where $N < M$). Consider arbitrary $\mu > 0, \mathbf{u}_0, \mathbf{d}_0 \in \mathbb{R}^{aff\_dim}$ (where *aff_dim* is an affine dimension), and $\{\boldsymbol{\alpha}_i \in \mathbb{R}^N, \mathbf{u}_i, \mathbf{d}_i \in \mathbb{R}^{aff\_dim}, \text{where } i = 0, 1, \dots\}$.

Step 1: Let $i = 0$, select $\mu > 0$, $\mathbf{u}_0, \mathbf{d}_0$.
Step 2: Continue step 3 to step 8 until specified condition is achieved.
Step 3: Compute $\mathbf{w} = \mathbf{E}^T \mathbf{y} + \mu(\mathbf{u}_i + \mathbf{d}_i)$.
Step 4: Compute $\boldsymbol{\alpha}_{i+1} = \mathbf{B}^{-1}\mathbf{w} - \mathbf{C}(1^T \mathbf{B}^{-1}\mathbf{w} - 1)$.
Step 5: Compute $\boldsymbol{\nu}_i = \boldsymbol{\alpha}_{i+1} - \mathbf{d}_i$.
Step 6: Compute $\mathbf{u}_{i+1} = \max\{0, \text{soft}\left(\boldsymbol{\nu}_i, {\lambda}/{\mu}\right)\}$.
Step 7: Compute $\mathbf{d}_{i+1} = \mathbf{d}_i - (\boldsymbol{\alpha}_{i+1} - \mathbf{u}_{i+1})$.
Step 8: Increment i by 1.
Step 9: Exit.

where $\mathbf{B} \equiv \mathbf{E}^T\mathbf{E} + \mu\mathbf{I}$, $\mathbf{C} \equiv \mathbf{B}^{-1}1(1^T\mathbf{B}^{-1}1)^{-1}$, and λ is a parameter controlling the relative weight. (*Note*: the symbol $\equiv$ means "is defined as" or "equivalence"). Soft-threshold function is discussed in Chen et al. [63]. SUnSAL with ANC and ASC imposed has been referred to as CSUnSAL in this chapter.

8.3.6 SUnSAL Total Variation

While sparse unmixing techniques characterize mixed pixel problems using spectral libraries, they do not deal with the neighboring pixels and tend to ignore the spatial context. SUnSAL TV [64] take into account spatial information (the relationship between each pixel vector and its neighbors) on the sparse unmixing formulation by means of the TV regularizer [65] assuming that it is very likely that two neighboring pixels have similar fractional abundances for the same endmember. The TV regularizer acts as *a priori* information and unmixing is achieved by a large nonsmooth convex optimization problem.

Let $\boldsymbol{\alpha} \in \mathbb{R}^{E \times P}$ (E = matrix containing endmembers and P = number of observed M-dimensional pixel vector) represent a fractional abundance matrix. Below, the SUnSAL TV algorithm is briefly discussed based on the derivation by Iordache et al. [64]. Assume $\|\boldsymbol{\alpha}\|_F \equiv \sqrt{\text{trace}\{\boldsymbol{\alpha\alpha}\}^T}$ be the Frobenius norm of $\boldsymbol{\alpha}$, let $\|\boldsymbol{\alpha}\|_{1,1} \equiv \sum_{p=1}^{P} \|\boldsymbol{\alpha}_p\|_1$ ($\boldsymbol{\alpha}_p$ denote the pth column of $\boldsymbol{\alpha}$), let $\lambda \geq 0$ and $\lambda_{TV} \geq 0$ be the regularization parameters. Then the sparse unmixing optimization problem is

$$\min_{\boldsymbol{\alpha}} \frac{1}{2} \|\mathbf{E}\boldsymbol{\alpha} - \mathbf{y}\|_F^2 + \lambda \|\boldsymbol{\alpha}\|_{1,1} + \lambda_{TV} TV(\boldsymbol{\alpha}) \text{ subject to } \boldsymbol{\alpha} \geq 0 \qquad (8.12)$$

where $TV(\boldsymbol{\alpha}) \equiv \sum_{\{p,n\} \in \varepsilon} \|\boldsymbol{\alpha}_p - \boldsymbol{\alpha}_n\|_1$, p is the pth observed M-dimensional pixel vector, n is the nth class, and ε is a set of horizontal and vertical neighbors of the pixel in consideration in the image. Also note that ASC has not been imposed. Equation 8.12 can also be written as

$$\min_{\boldsymbol{\alpha}} \frac{1}{2} \|\mathbf{E}\boldsymbol{\alpha} - \mathbf{y}\|_F^2 + \lambda \|\boldsymbol{\alpha}\|_{1,1} + \lambda_{TV} \|\mathbf{H}\boldsymbol{\alpha}\|_{1,1} + \iota_{R+}(\boldsymbol{\alpha}) \qquad (8.13)$$

where $\iota_{R+}(\boldsymbol{\alpha}) = \sum_{p=1}^{P} \iota_{R+}(\boldsymbol{\alpha}_p)$ is the indicator function, $\mathbf{H}\boldsymbol{\alpha}$ accounts for the horizontal and vertical neighboring pixel's differences ($\mathbf{H_h}\boldsymbol{\alpha} = [d_p, \dots, d_P]$, where $p = 1$ to P, $d_p = \alpha_p - \alpha_{ph}$ with p and ph denoting a pixel and its horizontal neighbor and $\mathbf{H_v\alpha} = [v_p, \ldots, v_P])$ where $p = 1$ to P, $v_p = \alpha_p - \alpha_{pv}$ with p and *pv* denoting a pixel and its vertical neighbor. So, $\mathbf{H\alpha} \equiv \begin{bmatrix} \mathbf{H_h\alpha} \\ \mathbf{H_v\alpha} \end{bmatrix}$. Equation 8.13 can be written in a constrained form as

$$\min_{\mathbf{U,V_1,V_2,V_3,V_4,V_5}} \frac{1}{2} \|\mathbf{V_1} - \mathbf{y}\|_F^2 + \lambda \|\mathbf{V_2}\|_{1,1} + \lambda_{TV} \|\mathbf{V_4}\|_{1,1} + \imath_{R+}(\mathbf{V_5}) \tag{8.14}$$

where $\mathbf{V_1} = \mathbf{EU}$, $\mathbf{V_2, V_3, V_5} = \mathbf{U}$, and $\mathbf{V_4} = \mathbf{HV_3}$ and Equation 8.14 can be expressed in a smaller form as

$$\min_{\mathbf{U,V}} g(\mathbf{V}) \text{ when } \mathbf{GU + BV} = 0 \tag{8.15}$$

where $V = (\mathbf{V_1, V_2, V_3, V_4, V_5})$, $g(\mathbf{V}) \equiv \frac{1}{2} \|\mathbf{V_1} - \mathbf{y}\|_F^2 + \lambda \|\mathbf{V_2}\|_{1,1} + \lambda_{TV} \|\mathbf{V_4}\|_{1,1} + \imath_{R+}(\mathbf{V_5})$,

$$\mathbf{G} = \begin{bmatrix} \mathbf{E} \\ \mathbf{I} \\ \mathbf{I} \\ 0 \\ \mathbf{I} \end{bmatrix}, \text{ and } \mathbf{B} = \begin{bmatrix} -\mathbf{I} & 0 & 0 & 0 & 0 \\ 0 & -\mathbf{I} & 0 & 0 & 0 \\ 0 & 0 & -\mathbf{I} & 0 & 0 \\ 0 & 0 & \mathbf{H} & -\mathbf{I} & 0 \\ 0 & 0 & 0 & 0 & -\mathbf{I} \end{bmatrix}.$$

Equation 8.15 is solved using ADMM algorithm. The overall steps in SUnSAL TV are summarized below:

Step 1: Let $i = 0$, select $\mu \geq 0$, $\mathbf{U}^{(0)}, \mathbf{V}_1^{(0)}, \mathbf{V}_2^{(0)}, \mathbf{V}_3^{(0)}, \mathbf{V}_4^{(0)}, \mathbf{V}_5^{(0)}, \mathbf{D}_1^{(0)}, \mathbf{D}_2^{(0)}, \mathbf{D}_3^{(0)}, \mathbf{D}_4^{(0)}, \mathbf{D}_5^{(0)}$.
Step 2: Continue step 3 to step 6 until specified condition is achieved.
Step 3: Compute $\mathbf{U}^{(i+1)} = \arg \min_U \mathcal{L}(\mathbf{U}, \mathbf{V}_1^{(i)}, \mathbf{V}_2^{(i)}, \mathbf{V}_3^{(i)}, \mathbf{V}_4^{(i)}, \mathbf{V}_5^{(i)}, \mathbf{D}_1^{(i)}, \mathbf{D}_2^{(i)}, \mathbf{D}_3^{(i)}, \mathbf{D}_4^{(i)}, \mathbf{D}_5^{(i)})$.
Step 4: For j in 1:5

$$\mathbf{V}_j^{(i+1)} = \arg \min_{V_j} \mathcal{L}(\mathbf{U}^{(i)}, \mathbf{V}_1^{(i)}, \ldots, \mathbf{V}_j^{(i)}, \ldots, \mathbf{V}_5^{(i)}).$$

Step 5: Update Lagrangian multipliers as follows:

$$\mathbf{D}_1^{(i+1)} = \mathbf{D}_1^{(i)} - \mathbf{EU}^{(i+1)} + \mathbf{V}_1^{(i+1)}$$

$$\mathbf{D}_2^{(i+1)} = \mathbf{D}_2^{(i)} - \mathbf{U}^{(i+1)} + \mathbf{V}_2^{(i+1)}$$

$$\mathbf{D}_3^{(i+1)} = \mathbf{D}_3^{(i)} - \mathbf{U}^{(i+1)} + \mathbf{V}_3^{(i+1)}$$

$$\mathbf{D}_4^{(i+1)} = \mathbf{D}_4^{(i)} - \mathbf{HV}_3^{(i+1)} + \mathbf{V}_4^{(i+1)}$$

$$\mathbf{D}_5^{(i+1)} = \mathbf{D}_5^{(i)} - \mathbf{U}^{(i+1)} + \mathbf{V}_5^{(i+1)}.$$

Step 6: Increment i by 1.
Step 7: Exit.

where $\mathcal{L}(\mathbf{U, V, D}) \equiv g(\mathbf{U, V}) + \frac{\mu}{2} \|\mathbf{GU + BV - D}\|_F^2$ is the augmented Lagrangian, μ is a positive constant, $\mathbf{D}/\mu$ is a Lagrangian multiplier with the constraint $\mathbf{GU + BV} = 0$. For detailed derivation and optimizations, the readers are requested to refer to the Appendix of [64]. SUnSAL TV with ANC and ASC imposed has been referred to as CSUnSAL TV among the constrained unmixing algorithms, discussed in Section 8.4 of this chapter.

8.4 Partially and Fully Constrained Unmixing Algorithms

If no constraints are imposed on abundance, the estimated abundance fractions may deviate with a wide range. To avoid such conditions, generally, two constraints are imposed on the model described in Equation 8.1: the ANC given as Equation 8.16 and the ASC expressed as Equation 8.17

$$\alpha_n \geq 0 \; \forall \, n : \; 1 \leq n \leq N \tag{8.16}$$

and

$$\sum_{n=1}^{N} \alpha_n = 1 \tag{8.17}$$

This allows proportion of each pixel to be partitioned between classes. Below we discuss constrained unmixing algorithms.

8.4.1 Sum-to-One Constrained Least Squares

Imposing the unity constraint on the abundance values (i.e., $\alpha_1 + \ldots + \alpha_{n-1} + \alpha_n + \alpha_{n+1} + \ldots + \alpha_N = 1$) as in Equation 8.17 while minimizing $\|\mathbf{y} - \mathbf{E}\alpha\|$, gives the SCLS estimate of the abundance as

$$\hat{\alpha}_{\text{SCLS}} = \mathbf{E}^T\mathbf{E}^{-1} \left(\mathbf{E}^T\mathbf{y} - \frac{\lambda}{2}\mathbf{1} \right) \tag{8.18}$$

where

$$\lambda = \frac{2(\mathbf{1}^T(\mathbf{E}^T\mathbf{E})^{-1}\,\mathbf{E}^T\mathbf{y} - 1)}{\mathbf{1}^T(\mathbf{E}^T\mathbf{E})^{-1}\mathbf{1}}. \tag{8.19}$$

The SCLS solution may have negative abundance values but they add to unity.

8.4.2 Normalized Sum-to-One Constrained Least Squares

Here, the negative values are considered to be 0 and the abundance fractions of the remaining target signature are normalized to unity resulting in NSCLS solution. The abundance values range from 0 to 1.

8.4.3 Nonnegative Constrained Least Squares

NCLS utilizes the nonnegative least squares (NNLS) algorithm to minimize $\|\mathbf{E}\alpha - \mathbf{y}\|$ subject to $\alpha \geq 0$. The N-vectors x and z specify the working dimension, and index sets $\wp$ and Z are initialized and adjusted during execution. Variables indexed in Z are initialized to zero and variables indexed in $\wp$ can take any nonzero values. In case the variable in $\wp$ takes negative value, the algorithm either assumes a positive value for the variable or assigns it to zero and moves its index from $\wp$ to Z. Once the algorithm terminates, α is the solution vector and x is the dual vector. Algorithm NNLS $(M, N, \mathbf{y}, \mathbf{E}, \alpha, x, z, \wp, Z)$ is briefly summarized below [66].

Step 1:   Set $\wp = $ NULL, $Z = \{1, 2, \ldots, N\}$, and $\alpha = 0$.
Step 2:   Compute $x = \mathbf{E}^T(\mathbf{y} - \mathbf{E}\alpha)$.
Step 3:   If $Z = \Phi$ or $x_j \leq 0 \forall j \epsilon Z$, stop ($\alpha$ is the solution).
Step 4:   Find an index $k \in Z$ such that $x_k = \max\{x_j : j \in Z\}$.
Step 5:   Move the index k from Z to $\wp$.

Step 6:   Let $\mathbf{E}_{\wp}$ denote the $M \times N$ matrix defined by

$$\text{Column } j \text{ of } \mathbf{E}_{\wp} = \begin{cases} \text{column } j \text{ of } \mathbf{E} \text{ if } j \in \wp \\ 0 \text{ if } j \in Z \end{cases}$$

compute N-dimensional vector $\mathbf{z}$ as least squares solution of $\left\| \mathbf{E}_{\wp}\mathbf{z} - \mathbf{y} \right\|$. Note that only the components z_j, $j \in \wp$ are determined by this problem. Define $z_j = 0 \forall j \in Z$.

Step 7:   If $z_j > 0 \forall j \in \wp$, then $\boldsymbol{\alpha} = \mathbf{z}$. Go to step 2.

Step 8:   Find an index $q \in \wp$ such that

$$\frac{\alpha_q}{\alpha_q - z_q} = \min \left\{ \frac{\alpha_j}{\alpha_j - z_j} : z_j \leq 0, j \in \wp \right\}.$$

Step 9:   Let $\beta = \frac{\alpha_q}{\alpha_q - z_q}$.

Step 10:   Let $\boldsymbol{\alpha} = \boldsymbol{\alpha} + \beta(\mathbf{z} - \boldsymbol{\alpha})$.

Step 11:   Move all indices $j \in \wp$ from set $\wp$ to set Z for which $\alpha_j = 0$. Go to Step 6.

Step 12:   Exit.

Upon execution, $\boldsymbol{\alpha}$ satisfies two conditions: $\alpha_j > 0$ for $j \in \wp$ and $\alpha_j = 0$ for $j \in Z$ and is a solution for the least squares problem $\mathbf{E}_{\wp}\boldsymbol{\alpha} \cong \mathbf{y}$. The dual vector $\mathbf{x}$ satisfies the following conditions: $x_j = 0$ for $j \in \wp$, $x_j \leq 0$ for $j \in Z$ and $\mathbf{x} = \mathbf{E}^T(\mathbf{y} - \mathbf{E}\boldsymbol{\alpha})$. The above settings together constitute the Kuhn–Tucker conditions characterizing a solution vector $\boldsymbol{\alpha}$ for NNLS. This algorithm finitely converges, the discussion of which is beyond the scope of this chapter, and readers are encouraged to refer Lawson and Hanson [66, p. 162], for details.

8.4.4 Normalized Nonnegative Constrained Least Squares

The NNCLS solution is obtained by dividing each class abundance (α_n) of the NCLS solution by the sum of all the abundances ($\alpha_1 + \ldots + \alpha_{n-1} + \alpha_n + \alpha_{n+1} + \ldots + \alpha_N$) in that pixel resulting in abundance values adding to unity.

8.4.5 Fully Constrained Least Squares

Most of the constrained algorithms discussed above impose ANC and ASC sequentially (if at all both the constraints are forced on the solution). As discussed earlier, for a SCLS solution, the target signatures with negative abundances can be discarded and the abundance fractions of the remaining target signatures can be normalized to unity, resulting in an NSCLS solution. The NCLS results can be normalized to unity to obtain NNCLS solution. However, Chang [21] claims that neither NSCLS nor NNCLS yields optimal solutions since the two constraints ANC and ASC are carried out sequentially and not simultaneously. Heinz et al. [50] produced a nearly optimal solution and recently Chang [21] proposed a FCLS solution that produces accurate abundance fraction of target signatures similar to Haskell and Hanson [67] and extends the NNLS algorithm in Lawson and Hanson [66] discussed above by including ASC. ASC is included in the signature matrix $\mathbf{E}$ by a new signature matrix (**SME**) defined by

$$\mathbf{SME} = \begin{bmatrix} \theta\mathbf{E} \\ \mathbf{1}^T \end{bmatrix} \tag{8.20}$$

with $\mathbf{1} = (\underbrace{11111 \dots 1}_{N})^T$, and

$$\mathbf{s} = \begin{bmatrix} \theta\mathbf{y} \\ 1 \end{bmatrix}. \tag{8.21}$$

θ in Equations 8.20 and 8.21 regulate ASC. Using these two equations, FCLS algorithm can be derived directly from the NNLS algorithm by replacing signature matrix $\mathbf{E}$ with $\mathbf{SME}$ and the pixel vector $\mathbf{y}$ with $\mathbf{s}$ [21].

8.4.6 Modified Fully Constrained Least Squares

ANC is a major problem in solving constrained linear unmixing problems as it forbids use of the Lagrange multiplier. Chang [21] proposed the replacement of $\alpha_n \geq 0 \; \forall n : 1 \leq n \leq N$ with absolute ASC (AASC), $\sum_{n=1}^{N} |\alpha_n| = 1$. AASC allows usage of the Lagrange multiplier along with exclusion of negative abundance fractions leading to optimal constrained least squares solution satisfying both ASC and AASC with all nonnegative fractions. This method is called MFCLS, expressed as

$$\min_{\alpha \in \Delta}\{(\mathbf{y} - \mathbf{E}\alpha)(\mathbf{y} - \mathbf{E}\alpha)^T\} \tag{8.22}$$

subject to $\Delta = \left\{ \alpha \middle| \sum_{n=1}^{N} \alpha_n = 1 \text{ and } \sum_{n=1}^{N} |\alpha_n| = 1 \right\}$.

It turns out that the solution to Equation 8.22 is

$$\hat{\alpha}_{\text{MFCLS}} = \hat{\alpha}_{\text{UCLS}} - (\mathbf{E}^T\mathbf{E})^{-1}[\lambda_1 \mathbf{1} + \lambda_2 \text{ sign } (\alpha)] \tag{8.23}$$

where $\hat{\alpha}_{\text{UCLS}} = (\mathbf{E}^T\mathbf{E})^{-1}\mathbf{E}^T\mathbf{y}$ which is the unconstrained solution as in Equation 8.2. ASC and AASC are now used to compute λ_1 and λ_2 by replacing α with $\hat{\alpha}_{\text{UCLS}}$ with the following constraints: $\sum_{n=1}^{N} \alpha_n = \mathbf{1}^T\alpha = 1$, and $\sum_{n=1}^{N} |\alpha_n| = \text{sign}(\alpha)^T\alpha = 1$ where $\text{sign}(\alpha) = (\omega_1, \omega_2, \dots, \omega_{n-1}, \omega_n, \omega_{n+1}, \dots, \omega_N)^T$ and

$$\omega_n = \begin{cases} \frac{\alpha_n}{|\alpha_n|}; & \text{if } \alpha_n \neq 0 \\ 0; & \text{if } \alpha_n = 0 \end{cases}$$

the MFCLS algorithm is briefly stated as

Step 1: Set $\hat{\alpha}_{\text{MFCLS}} = \hat{\alpha}_{\text{SCLS}}$ from Equation 8.18.
Step 2: Compute λ_1 and λ_2.
Step 3: Compute $\hat{\alpha}_{\text{MFCLS}} = \hat{\alpha}_{\text{SCLS}} - (\mathbf{E}^T\mathbf{E})^{-1}[\lambda_1\mathbf{1} + \lambda_2\text{sign}(\alpha)]$.
Step 4: If $\hat{\alpha}_{\text{MFCLS}}$ has any negative value, go to step 2 else exit.

The MFCLS algorithm utilizes the SCLS solution. Step 4 terminates the algorithm when all the components are nonnegative. Alternatively, a preselected threshold can be used for a fast implementation. For a more detailed derivation, readers are directed to refer [21, p. 184].

The value of abundance in any given pixel range from 0 to 1 (in an abundance map) and there are as many abundance maps as the number of classes. 0 indicates absence of a particular class and 1 indicates presence of only that class in a particular pixel. Intermediate values between 0 and 1 represent a fraction of that class. For example, 0.4 may represent 40% presence of a class in an abundance map and the remaining 60% could be other classes.

8.5 Data

In this section, details about computer-simulated data, Landsat data, WV-2 data, ground data used for validation, and endmembers used in the analysis are discussed.

8.5.1 Computer Simulations

One of the major problems of analyzing the quality of fractional estimation methods is the fact that exact ground truth information about the real abundances at subpixel level for all classes is difficult to obtain [68]. In order to avoid this shortcoming, simulation of imagery was carried out as a simple and intuitive way to perform preliminary evaluation of the techniques. The primary reason for the use of simulated imagery as a complement to real data analysis is that all the details of the simulated images are known. As a result, an algorithm's performance can be examined in a controlled manner.

First, a set of multispectral images was clustered into three groups. From this clustered image where each pixel belongs to any of the three groups, the first test pixel was chosen, (at row 1 and column 1) assuming that group 1 represents endmember 1, group 2 represents endmember 2, and group 3 represents endmember 3. If this chosen test pixel belonged to group 1, then we allocated a maximum random abundance of endmember 1 to this pixel followed by small random abundances to endmembers 2 and 3 by following ANC and ASC. The second test pixel (at location 1, 2) was selected and if this test pixel belonged to group 2 or 3 then endmember 2 or 3 was allocated maximally and the remaining two abundances were given small random proportional values while maintaining the constraints. This was carried out for all the pixels in the image and the final output was three abundance maps, one for each endmember. Now, with the given global spectra of three endmember libraries of substrate, vegetation, and dark objects, (available at http://www.ldeo.columbia.edu/~small/GlobalLandsat/styled-3/index.html) and three abundance maps, the linear mixture Equation 8.1 was inverted to generate computer-simulated synthetic noise-free data of six bands of size 512 × 512. So, artificial band = abundance_map1 × endmember1 + abundance_map2 × endmember2 + abundance_map3 × endmember3.

In a separate set of experiments, noise was added to the data to judge the robustness of the algorithms. We examined the error in the estimate as the noise power increased by 2, 4, 8, 16, 32, 64, 128, and 256. This noise is a random number drawn from a Gaussian distribution where the mean of each endmember is set to 0 and the variability is controlled. Since user-defined means and spread around the mean could be added to the synthetic data in the form of random fluctuations, noise are assumed to be of the form mean + fixed variance, that is, Gaussian noise = mean + random perturbation; random perturbation is a Gaussian random variable of specific variance. A multitude of LMMs were applied with different endmembers (substrate, vegetation, and dark objects) and a series of subpixel class composition estimates were obtained for a pixel of many given spectral response. Table 8.1 lists the statistical properties of the simulated data. In this way, not only the impact of subpixel classification in noise-free data was analyzed, but also the impact of variable noise with respect to the fixed spatial resolution was assessed.

8.5.2 Landsat Data

A spectrally diverse collection of 11 scenes of Level 1 terrain corrected, cloud-free Landsat-5 16 bit (path 43, row 35) for Fresno, California were used in this study. These data were captured on April 4 and 20, May 22, June 7 and 23, July 9 and 25, August 26, September 11 and 27, and October 13 for the year 2008 and were calibrated to exoatmospheric reflectance using the calibration approach and coefficients given by Chander et al. [69]. These scenes were selected because a coincidental set of ground canopy cover was available for a number of surveyed fields within the footprint of Landsat Worldwide Reference System (WRS) path 43, row 35 [70]. The atmospheric reflectance was converted to surface reflectance correcting for atmospheric effects by means of the 6S code implementation in the Landsat Ecosystem Disturbance Adaptive Processing System (LEDAPS) atmospheric correction method [71]. Atmospheric correction reduces the

TABLE 8.1　Statistical Properties of the Computer-Simulated Data

Bands	Minimum	Maximum	Range	Mean	Variance	Variation Coefficient (%)
Computer-Simulated Noise-Free Data						
1	260	1766	1506	753.46	266,644	68.53
2	272	3342	3070	1436.59	931,455	67.18
3	248	4536	4288	1619.09	2,215,320	91.93
4	801	6681	5880	6012.21	236,759	8.09
5	482	6775	6293	3654.63	2,406,200	42.44
6	348	6388	6040	2608.48	3,671,680	73.46
Computer-Simulated Data with Noise ($\mu = 0$ and $\sigma^2 = 2$)						
1	260	1766	1506	753.52	2,66,499	68.51
2	272	3344	3072	1436.57	9,314,33	67.18
3	250	4538	4288	1619	2,215,090	91.93
4	799	6685	5886	6012	2,367,82	8.1
5	481	6776	6295	3655	2,406,350	42.45
6	349	6389	6040	2608	3,672,000	73.47
Computer-Simulated Data with Noise ($\mu = 0$ and $\sigma^2 = 4$)						
1	260	1766	1506	753.6	266,355	68.49
2	273	3346	3073	1436.56	931,397	67.18
3	251	4539	4288	1618.89	2,214,910	91.93
4	797	6689	5892	6011.89	236,797	8.09
5	480	6778	6298	3654.61	2,406,570	42.49
6	349	6391	6042	2608.26	3,672,440	73.47
Computer-Simulated Data with Noise ($\mu = 0$ and $\sigma^2 = 8$)						
1	259	1765	1506	753.69	266,143	68.45
2	273	3350	3077	1436.52	931,384	67.18
3	253	4542	4289	1618.71	2,214,490	91.93
4	792	6697	5905	6011.58	236,885	8.1
5	478	6780	6302	3654.6	2,406,980	42.45
6	351	6396	6045	2608.05	3,673,370	73.49
Computer-Simulated Data with Noise ($\mu = 0$ and $\sigma^2 = 16$)						
1	258	1764	1506	753.91	265,787	68.38
2	274	3357	3083	1436.44	931,386	67.19
3	247	4547	4300	1618.33	2,213,650	91.94
4	784	6713	5929	6010.95	237,117	8.10
5	474	6786	6312	3654.6	2,407,910	42.46
6	353	6405	6052	2607.62	3,675,120	73.52
Computer-Simulated Data with Noise ($\mu = 0$ and $\sigma^2 = 32$)						
1	258	1764	1506	753.91	265,787	68.38
2	274	3357	3083	1436.44	931,386	67.19
3	247	4547	4300	1618.33	2,213,650	91.94
4	784	6713	5929	6010.95	237,117	8.10
5	474	6786	6312	3654.6	2,407,910	42.46
6	353	6405	6052	2607.62	3,675,120	73.52

Continued

TABLE 8.1 (Continued) Statistical Properties of the Computer-Simulated Data

Bands	Minimum	Maximum	Range	Mean	Variance	Variation Coefficient (%)
\multicolumn{7}{c}{Computer-Simulated Data with Noise ($\mu = 0$ and $\sigma^2 = 64$)}						
1	−22	1976	1998	670.589	232,049	71.83
2	−58	3505	3563	1257	837,980	72.83
3	−163	4716	4879	1385.26	1,888,630	99.21
4	−55	6834	6889	5506.74	2,750,490	30.11
5	−72	6951	7023	3234.5	2,680,060	50.61
6	−130	6587	6717	2257.35	3,272,960	80.14
\multicolumn{7}{c}{Computer-Simulated Data with Noise ($\mu = 0$ and $\sigma^2 = 128$)}						
1	−256	2226	2482	670.61	244,276	73.70
2	−295	3689	3984	1256.93	850,354	73.37
3	−472	4880	5352	1385.23	900,980	99.53
4	−228	7045	7273	5507	2,762,550	30.18
5	−223	7159	7382	3234.69	2,692,390	50.73
6	−315	6751	7066	2257.46	3,285,280	80.29
\multicolumn{7}{c}{Computer-Simulated Data with Noise ($\mu = 0$ and $\sigma^2 = 256$)}						
1	−724	2725	3449	670.65	293,295	80.75
2	−771	4138	4909	1256.81	899,574	75.47
3	−1090	5332	6422	1385.18	1,950,160	100.82
4	−574	7672	8246	5506.52	2,811,240	30.45
5	−571	7620	8191	3235.09	2,741,700	51.18
6	−687	7113	7800	2257.69	3,334,520	80.88

perturbations caused by Rayleigh scattering and the absorption of the mixing atmospheric molecules and aerosols [72].

The surveyed field for the ground observations was located within an area of about 25×35 km southwest of the city of Fresno (Figure 8.1). Seventy-four polygons of the fractional vegetation cover were generated from digital photographs taken with a multispectral camera mounted on a frame at nadir view pointed 2.3 m above the ground at the commercial agricultural fields of the San Joaquin Valley (in central California) on the 11 dates mentioned above, except for one date when the Landsat acquisition preceded the ground observation by 1 day. For each date, 2–4 evenly spaced pictures were taken for an area of 100 m × 100 m with center location marked by a global positioning system (GPS) [70]. These fractional measurements belonged to a diverse set of seasonal and perennial crops in various developmental stages, from emergence to full canopy that represented an agricultural scenario/environment in the RS data.

A second set of a pair of coincident clear sky Landsat thematic mapper (TM) data (at 30-m spatial resolution) and WV-2 data (of 2-m spatial resolution) for an area of San Francisco was used to assess the algorithm. San Francisco is chosen for the test site because of its completely different urbanized landscape, having colonial and eclectic mix of building architectures on the steep rolling hills.

WV-2 data were acquired a few minutes after the Landsat TM data acquisition on May 1, 2010 for an area near the Golden Gate Bridge, San Francisco as shown in Figure 8.2. The spectral range of the first four bands of Landsat data has a good correspondence with the WV-2 bands 2, 3, 5, and 6 in terms of wavelength range in the electromagnetic spectrum as given in Table 8.2 which shows that they have a similar mixing space. WV-2 data were converted to top-of-atmosphere (TOA) reflectance values using python program (available at https://github.com/egoddard/i.wv2.toar) in Geographic Resources Analysis Support System

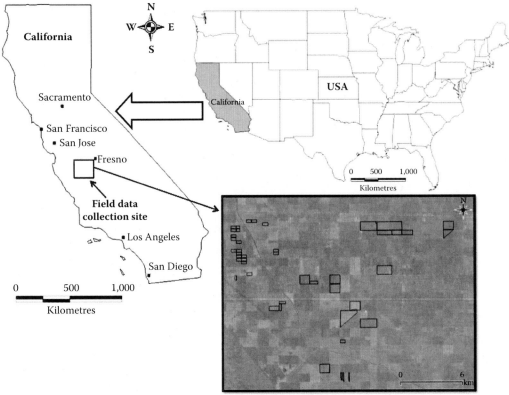

FIGURE 8.1   Study area: San Joaquin Valley in central California. Field data collection site in San Joaquin Valley with surveyed boundaries (marked in black color polygons) overlaid on a false color composite (FCC) of Landsat data from which ground fractional cover were derived for validation.

Geographic Information System (GRASS GIS) 7.1. The Landsat unmixed images were compared with the corresponding WV-2 fraction images for accuracy assessment.

8.5.3 Endmember Generation

Global mixing spaces were sampled by Small and Milesi [73] using a spectrally diverse LC and diversity of biomes with 100 Landsat ETM+ scenes. This defined a standardized set of spectral endmembers of substrate (S), vegetation (V), and dark objects (D). Vegetation refers to green photosynthetic plants, dark objects encompass absorptive substrate materials, clear water, deep shadows, etc., and substrate includes soils, sediments, rocks, and nonphotosynthetic vegetation. For simplicity, we refer substrate, vegetation, and dark objects as "S" (endmember 1 or E1), "V" (endmember 2 or E2), and "D" (endmember 3 or E3) in the rest of this chapter.

The endmembers are found to define the topology of the three-dimensional tetrahedral hull that envelopes ~98% of the spectra found in the global Landsat mixing space. The S-V-D endmember coefficient, in addition to dates and locations of each subscene is available at http://www.LDEO.columbia.edu/~small/GlobalLandsat/. The estimates obtained from the global endmembers have been compared to fractional vegetation cover derived vicariously by linearly unmixing near-coincidental WV-2 acquisitions over a set of diverse coastal environments, using both global endmembers and image-specific endmembers to

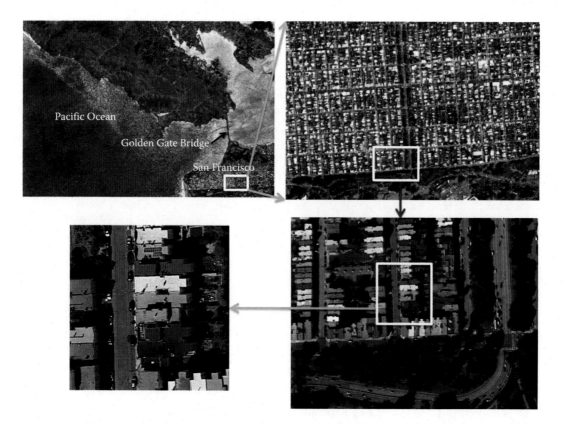

FIGURE 8.2 (See color insert.) Study area: FCC of a part of San Francisco city. Zoomed image of the urban area (marked with rectangles in inset) shows mixing of substrate with vegetation, roads, shadow, and dark objects.

TABLE 8.2   Landsat and WV-2 Band Details Used in Unmixing

	Landsat			WV-2	
Band	Band name	Wavelength (μm)	Band	Band name	Wavelength (μm)
1	Blue	0.45–0.52	2	Blue	0.45–0.51
2	Green	0.52–0.60	3	Green	0.51–0.58
3	Red	0.63–0.69	5	Red	0.63–0.69
4	NIR	0.77–0.90	7	NIR1	0.77–0.895

Abbreviation: NIR, Near-infrared.

unmix the WV-2 images. The strong 1:1 linear correlation between the fractions obtained from the two types of images indicates that the mixture model fractions scale linearly from 2 to 30 m over a wide range of LC types. When endmembers are derived from a large enough sample of radiometric responses to encompass the Landsat spectral mixing space, they can be used to build a standardized spectral mixture model with global applicability [74].

8.6 Methodology

Among the above discussed algorithms in Sections 8.2 through 8.4, UCLS, SCLS, NSCLS, NCLS, NNCLS, FCLS, and MFCLS were implemented in C++ programming language with OpenCV (Open Source

Computer Vision) package (http://opencv.org/) and boost C++ libraries (www.boost.org) on the NEX at the NASA Advanced Supercomputing Facility. MF, MTMF, and CEM were implemented in ENvironment for Visualizing Images (ENVI) image processing software. The SUnSAL programs were obtained from the authors and are also available online for download (http://www.lx.it.pt/~ bioucas/code/sunsal_demo.zip). GRASS—a free and open source package was used for visualization of results and statistical analysis was carried in R statistical package in a Linux system environment. SunSAL and SUnSAL TV were used in the unconstrained form, and CSunSAL and CSUnSAL TV were used with both ANC and ASC enforced. The parameter λ in SUnSAL was set to 10^{-5}, $\delta = 10^{-4}$; λ in SUnSAL TV was set to 5×10^{-4} and λ_{TV} in SUnSAL TV was set to 5×10^{-3} for computer simulated data and 10^{-3} for the Landsat data. The maximum number of iterations was set to 100 and all other parameters were set to default.

For validation of the computer-simulated data, the estimated class proportion maps were compared with the synthetic true abundance maps using visual checks and various other measures such as descriptive statistics (minimum and maximum fractional estimates), Pearson product–moment cc, RMSE, p_s, boxplot, and BDF.

If true abundances are known as in synthetic images, cc and RMSE are calculated between the true abundance and estimated fractional abundance. Considering the true abundance $\alpha = (\alpha_1, \ldots, \alpha_{n-1}, \alpha_n, \alpha_{n+1}, \ldots, \alpha_N)$ and estimated abundance $\hat{\alpha} = (\hat{\alpha}_1, \ldots, \hat{\alpha}_{n-1}, \hat{\alpha}_n, \hat{\alpha}_{n+1}, \ldots, \hat{\alpha}_N)$, cc (r) is defined as

$$r = \frac{\sum_{p=1}^{P}(\alpha_p - \bar{\alpha})(\hat{\alpha}_p - \bar{\hat{\alpha}})}{\sqrt{\sum_{p=1}^{P}(\alpha_p - \bar{\alpha})^2}\sqrt{\sum_{p=1}^{P}(\hat{\alpha}_p - \bar{\hat{\alpha}})^2}} \tag{8.24}$$

r ranges from -1 to 1. 1 implies that a linear equation describes the relationship between α and $\hat{\alpha}$ perfectly, with all the data points lying on a straight line for which $\hat{\alpha}$ increases as α increases. $r = -1$ infers that all data points lie on a line for which $\hat{\alpha}$ decreases as α increases and $r = 0$ means there is no linear correlation between the true and estimated abundances.

RMSE [75] is defined as

$$\text{RMSE} = \frac{1}{N}\sum_{n=1}^{N}\left[\sqrt{\frac{1}{P}\sum_{p=1}^{P}(\hat{\alpha}_{np} - \alpha_{np})^2}\right] \tag{8.25}$$

Smaller the RMSE, better the unmixing result and higher is the accuracy.

Probability of success (p_s) is an estimate of the probability that the relative error power is smaller than a certain threshold [60]. It is defined as $p_s \equiv P\left(\frac{\|\hat{\alpha}-\alpha\|^2}{\|\alpha\|^2} \leq \text{threshold}\right)$. If threshold is 10, and $p_s = 1$, it suggests that the total relative error power of proportional abundances is less than 1/10 with a probability of 1. Here, the estimation result is accepted when $\frac{\|\hat{\alpha}-\alpha\|^2}{\|\alpha\|^2} \leq 0.95$ (5.22 dB). 0.95 is the average of the 99th percentile of all the abundances of the three endmembers for the noise variance 8. At noise variance 8, the SNR which is the logarithm to the base 10 of the ratio of sum of the square of the true abundances $(\alpha_1^2 + \alpha_2^2 + \alpha_3^2)$ to the sum of the square of the difference between the estimated and the true abundances $((\hat{\alpha}_1^2 - \alpha_1^2) + (\hat{\alpha}_2^2 - \alpha_2^2) + (\hat{\alpha}_3^2 - \alpha_3^2))$ turns out to be 5.22 dB. Empirically, we found that when $p_s = 1$, then 1 dB $\leq$ the SNR for the entire pixels in the abundance $\leq$ 8 dB for our data set.

BDF was used to visualize the accuracy of prediction by mixture models. Points along a 1:1 line on the BDF graph indicate predictions that match completely with the real/actual/reference proportions. The smaller the difference between reference and estimated proportions, the closer the point will lie to the diagonal 1:1 line.

8.7 Experimental Results

This section briefs the results from the unmixing algorithms on computer-simulated data and Landsat data of an agricultural landscape near Fresno and an urban agglomeration in San Francisco.

8.7.1 Computer-Simulated Data

Two images from the six-band computersimulated data are shown in Figure 8.3: a (band 3), and b (band 4). The S-V-D endmembers were used to unmix the computer-simulated six bands.

Figure 8.4a through c shows noise-free synthetic abundance maps for endmember 1, 2, and 3. Figure 8.4d through f shows estimated abundance maps obtained for each signature class (for the three endmembers) corresponding directly to the gray scale values for each image from UCLS, Figure 8.4g through i are estimated abundance maps obtained from MTMF, Figure 8.4j through l are estimated abundance maps obtained from SUnSAL, and Figure 8.4m through o are estimated abundance maps from SUnSAL TV with the range of abundance fraction values specified in square bracket [minimum abundance value–maximum abundance value] underneath each figure. Visual inspection of the abundance maps obtained from the six algorithms show that they are similar and the relative fractions of the classes look alike. Also, from the detection point of view, all the six methods performed similarly.

Table 8.3 shows the minimum and maximum abundance values for the three endmembers (E1, E2, and E3) for noise-free data and noise variance 256 for the six unconstrained algorithms. Table 8.4 shows the cc (which were found to be statistically significant at 0.99 confidence level, p-value $<2.2e^{-16}$) between the synthetic true abundance and the estimated abundance for the three endmembers for noise variance 0 and 256. The details of other noise variances are omitted due to space constraints. When minimum and maximum estimated abundance values are compared to the synthetic true abundance, it is observed that for the noise-free data, UCLS, SUnSAL, and SUnSAL TV are closer to true abundance, whereas, at 256 noise level, only UCLS is closer to synthetic true abundances for all the three endmembers, although the minimum abundances are negative and maximum abundances are more than one.

Table 8.4 reveals that for noise-free data, UCLS, SUnSAL, and SUnSAL TV have high cc, low RMSE, and $p_s = 1$. For noise variance 256, only UCLS results had the highest cc and lowest RMSE. It was observed that MF and MTMF had consistent performance overall, although they did not perform very well among the unconstrained algorithms.

Overall, through various measures of accuracy, it was found that all the techniques performed almost equally well at lower noise levels (up to noise variance 32), however, when the noise variance increased, the performance of the algorithms decreased. UCLS performed best among the six unconstrained algorithms,

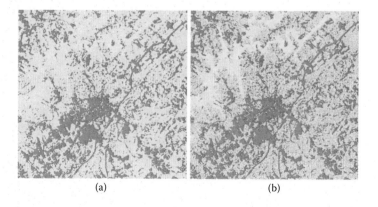

(a) (b)

FIGURE 8.3 A six bands computer-simulated data: (a) band 3, and (b) band 4.

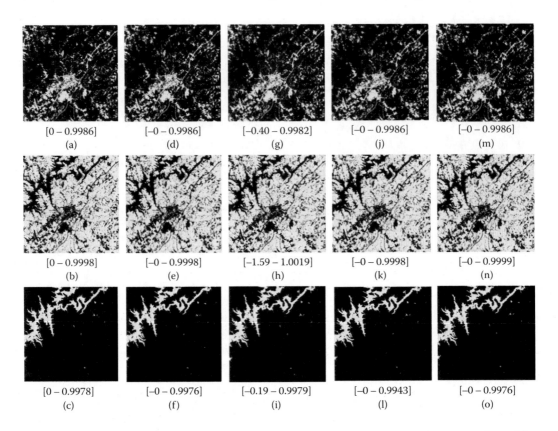

[0 – 0.9986]	[–0 – 0.9986]	[–0.40 – 0.9982]	[–0 – 0.9986]	[–0 – 0.9986]
(a)	(d)	(g)	(j)	(m)
[0 – 0.9998]	[–0 – 0.9998]	[–1.59 – 1.0019]	[–0 – 0.9998]	[–0 – 0.9999]
(b)	(e)	(h)	(k)	(n)
[0 – 0.9978]	[–0 – 0.9976]	[–0.19 – 0.9979]	[–0 – 0.9943]	[–0 – 0.9976]
(c)	(f)	(i)	(l)	(o)

FIGURE 8.4 (a)–(c) synthetic abundance maps for endmember 1, 2, and 3, (d)–(f) abundance maps obtained from UCLS, (g–i) abundance maps obtained from MTMF, (j)–(l) abundance maps obtained from SUnSAL, and (m)–(o) abundance maps obtained from SUnSAL TV from noise-free data. In the figure, black indicates absence of a particular class (the minimum abundance value) and white indicates full presence of that class in a pixel (the maximum abundance value). Intermediate values of the shades of gray represent mixture of more than one class in a pixel. MTMF, mixed-tuned matched filtering; SUnSAL, sparse unmixing via variable splitting and augmented Lagrangian; SUnSAL TV, SUnSAL total variation; UCLS, unconstrained least squares.

followed by MF/MTMF in second place, SUnSAL/SUnSAL TV in third position, and CEM had the worst performance.

Next, the results from constrained algorithms are discussed. Figure 8.5 shows noise-free estimated abundance maps obtained for each signature class corresponding directly to the gray scale values for each image from five selected algorithms: Figure 8.5a through c SCLS, Figure 8.5 d through f NSCLS, Figure 8.5 g through i FCLS, Figure 8.5j through l MFCLS, and Figure 8.5 m through o CSUnSAL with the range of abundance fraction values specified in the square bracket [minimum abundance value–maximum abundance value] underneath each figure. As with the unconstrained algorithms, the various abundance maps obtained from the different constrained algorithms were similar in appearance.

Table 8.5 shows the minimum and maximum values of the abundance maps corresponding to the three endmembers (E1, E2, and E3) for noise-free data and noise variance 256 from the eight constrained algorithms. The details of other noise variances were omitted due to space constraints. It is evident that the constrained algorithms are much better than unconstrained algorithms as they are able to maintain the ANC and ASC to a good extent.

Figure 8.6 shows the cc (statistically significant at 0.99 confidence level, p-value $<2.2e^{-16}$) and RMSE between real abundance and estimated abundances obtained from the eight constrained algorithms for all the three endmembers corresponding to different levels of noise. For endmember 1 and 2, all the methods show a high cc (close to 1) when variance in the noise was increased until 32, beyond which the cc

TABLE 8.3 Minimum and Maximum Values of the Abundance Maps Corresponding to the Three Endmembers (E1, E2, and E3) for Noise Variance 0 and 256 from the Different Unconstrained Algorithms

Algorithm	Minimum			Maximum		
	E1	E2	E3	E1	E2	E3
Synthetic true abundance	0	0	0	0.9986	0.9998	0.9978
Noise $\sigma^2 = 0$						
UCLS	−0.0001	−0.0001	−0.0037	0.9986	0.9999	0.9976
MF	−0.4062	−1.5976	−0.1910	0.9994	1.0018	0.9982
MTMF	−0.4021	−1.5930	−0.1890	0.9982	1.0019	0.9979
CEM	−0.0312	−0.2071	−27.0048	0.9986	8.0605	0.9876
SUnSAL	−0.0001	−0.0001	−0.0036	0.9986	0.9998	0.9943
SUnSAL TV	−0.0001	−0.0001	−0.0037	0.9986	0.9999	0.9976
Noise $\sigma^2 = 256$						
UCLS	−0.1153	−0.1995	−5.7013	1.0644	1.1510	6.3070
MF	−0.5222	−2.0249	−0.3406	1.0942	1.3778	1.1103
MTMF	−0.5216	−2.0228	−0.3408	1.0936	1.3733	1.1107
CEM	−0.2482	−0.1328	−27.6291	1.2665	1.2962	10.1251
SUnSAL	−4.2286	−12.6948	−153.1474	14.0273	11.1966	429.1562
SUnSAL TV	−4.2389	−12.6954	−153.6396	14.0283	11.1965	430.5347

TABLE 8.4 Correlation Coefficient (cc or *r*), RMSE, and Probability of Success (p_s) for Endmember 1, 2, and 3 (E1, E2, and E3) for Noise Variance 0 and 256 for the Unconstrained Algorithms

Algorithm	*r*			RMSE			p_s
	E1	E2	E3	E1	E2	E3	
Noise $\sigma^2 = 0$							
UCLS	1	1	0.99	0	0	0	1
MF	0.99	0.99	0.99	0.28	0.83	0.14	0.68
MTMF	0.99	0.99	0.99	0.28	0.83	0.14	0.68
CEM	0.99	−0.64	0.28	0.06	3.31	11.54	0.01
SUnSAL	1	1	0.99	0	0	0	1
SUnSAL TV	1	1	0.99	0	0	0	1
Noise $\sigma^2 = 256$							
UCLS	0.99	0.87	0.17	0.04	0.20	1.2	0.46
MF	0.99	0.99	0.99	0.29	0.84	0.14	0.68
MTMF	0.99	0.99	0.99	0.29	0.84	0.15	0.68
CEM	0.93	0.40	0.21	0.19	0.46	8.74	0.16
SUnSAL	0.32	0.42	0.12	0.84	0.9	64.39	0.46
SUnSAL TV	0.32	0.42	0.12	0.84	0.9	64.6	0.46

gradually decreases and reaches a minimum of 0.32 for endmember 1, 0.42 for endmember 2, and 0.12 for endmember 3 (for CSUnSAL TV) at maximum noise level. Overall, SCLS, NSCLS, NCLS, and FCLS show a higher cc (~0.99) for endmember 1, SCLS, NSCLS, NCLS, FCLS, MFCLS, and CSUnSAL render a higher cc (~0.87) for endmember 2 and SCLS, NSCLS, FCLS, MFCLS, and CSUnSAL produce a higher cc (~0.81–0.83) for endmember 3 at 256 noise variance (highest noise level).

RMSEs for all the algorithms and endmembers follow a hyperbolic curve. For endmember 1, SCLS, NSCLS, NCLS, and FCLS had least RMSEs (~0.03), whereas CSUnSAL TV exhibited highest error (0.88

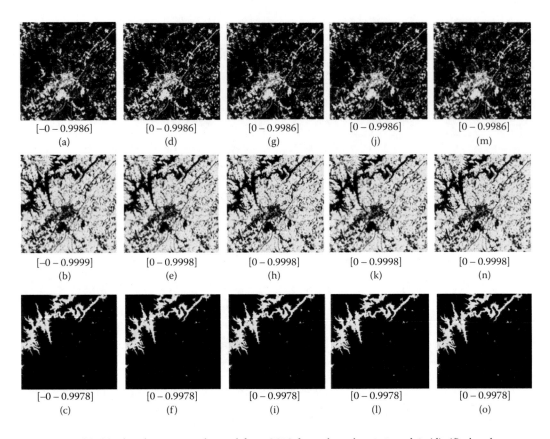

FIGURE 8.5 (a)–(c) Abundance maps obtained from SCLS for endmember 1, 2, and 3, (d)–(f) abundance maps obtained from NSCLS, (g)–(i) abundance maps obtained from FCLS, (j)–(l) abundance maps obtained from MFCLS, and (m)–(o) abundance maps obtained from CSUnSAL from noise-free data. In the figure, black indicates absence of a particular class (the minimum abundance value) and white indicates full presence of that class in a pixel (the maximum abundance value). Intermediate values of the shades of gray represent mixture of more than one class in a pixel. For synthetic abundance of endmember 1, 2, and 3, see Figure 8.4: (a)–(c). CSUnSAL, constrained SUnSAL; FCLS, fully constrained least squares; NSCLS, normalized sum-to-one constrained least squares; MFCLS, modified fully constrained least squares; SCLS, sum-to-one constrained least squares.

for endmember 1, 0.9 for endmember 2, and 64.6 for endmember 3) at maximum noise level. SCLS, NSCLS, NCLS, FCLS, MFCLS, and CSUnSAL gave the lowest RMSE (~0.20) for endmember 2 and SCLS, NSCLS, FCLS, MFCLS, and CSUnSAL had the lowest RMSEs (~0.15) for endmember 3. To a certain noise level (noise variance 32), all the algorithms are robust, however as noise increases in the data, they tend to produce higher RMSEs. Figure 8.7 shows the plot of p_s for the various constrained algorithms against different noise variance levels. Up to noise variance 16, all the algorithms have $p_s = 1$, beyond which it gradually decreases with least for CSUnSAL TV (0.46) for which the quantification results were worse and highest (0.92) for SCLS, NSCLS, FCLS, MFCLS, and CSUnSAL.

Figure 8.8 shows the execution time taken by each of the algorithms for unmixing computer-simulated data for different levels of noise. From the figure, it is obvious that MTMF took maximum time for execution followed by CEM and MF. MTMF performs MNF transformation of the original data followed by unmixing of pixels, background suppression, and then rejection of false positives, so it is more time intensive. SUnSAL and NCLS took comparable time; the complexity of SUnSAL is $O(P^2)$ per iteration [59], and CSUnSAL was found to be slower than SUnSAL, which also has complexity $O(P^2)$ per iteration where P is the total number of pixels. SUnSAL TV and CSUnSAL TV have complexity $O(M^2P) + (MP \log P)$, where M is the number of bands [64] and were slower than SUnSAL. The least squares algorithms (UCLS, SCLS,

TABLE 8.5 Minimum and Maximum Values of the Abundance Maps Corresponding to the Three Endmembers (E1, E2, and E3) for Noise Variance 0 and 256 for the Different Constrained Algorithms

Algorithm	Minimum			Maximum		
	E1	E2	E3	E1	E2	E3
Synthetic Abundance	0	0	0	0.9986	0.9998	0.9978
Noise $\sigma^2 = 0$						
SCLS	−0.0001	−0.0001	−0.0001	0.9986	0.9999	0.9978
NSCLS	0	0	0	0.9986	0.9998	0.9978
NCLS	0	0	0	0.9986	0.9998	0.9976
NNCLS	0	0	0	1	1	0.9978
FCLS	0	0	0	0.9986	0.9998	0.9978
MFCLS	0	0	0	0.9986	0.9998	0.9978
CSUnSAL	0	0	0	0.9986	0.9998	0.9978
CSUnSAL TV	−0.0001	−0	−0.0037	0.9986	0.9998	0.9976
Noise $\sigma^2 = 256$						
SCLS	−0.0939	−0.1992	−0.1593	1.0684	1.151	1.0082
NSCLS	0	0	0	1	1	0.9869
NCLS	0	0	0	1.0506	1.0981	6.2925
NNCLS	0	0	0	1	1	1
FCLS	0	0	0	1.0212	1.0148	0.9904
MFCLS	0	0	0	1	1	0.9904
CSUnSAL	0	0	0	1.0023	1	0.9904
CSUnSAL TV	−4.2388	−12.6954	−153.6288	14.0283	11.1965	430.5271

NSCLS, NCLS, NNCLS, FCLS, and MFCLS) took close to average execution time of all the algorithms for different noise levels. UCLS and SCLS took least time per execution among all.

Considering the various measures of performance discriminators in the above analysis, it was found that overall, FCLS had the highest accuracy among the constrained algorithms followed by SCLS/NSCLS with lower accuracy marginally, and MFCLS/CSUnSAL with slightly lower accuracy in third position in the absence and presence of noise with computer-simulated data. NCLS, NNCLS, and CSUnSAL TV performed worst among all, with poor accuracy measures.

The effect of noise induced in the data on the output of UCLS and FCLS algorithms was analyzed that had superior performance among the unconstrained and constrained algorithms. Figure 8.9 shows box-plot of the distribution of synthetic true abundance (represented as Syn_abn in Figure 8.9) and estimated abundance obtained from UCLS and FCLS for endmember 1, 2, and 3.

It highlights that the abundance maps of endmember 1 has a distribution symmetric about its median, whereas abundance map 2 is right skewed, that is, most of the abundances are toward higher side (~0.9) and abundance map 3 is left skewed. Both abundance map 1 and 3 have outliers. The boxplot also reveals that abundance map 1 and 3 have maximum pixels with lesser abundances (between 0 and 0.2) and fewer pixels with abundance between 0.8 and 1. Spatially, endmember 2 is most dominant in the scene than the other two endmembers. However, the boxplot does not show significant difference between the abundances obtained from UCLS and FCLS. When noise was added to the data, the abundance distribution changed. Figure 8.10 shows the effect of noise on FCLS output for the three endmembers. The whiskers of the boxplots increase monotonically as the variance in the noise increase from 2 toward 256. The outlier for endmembers 1 and 3 increase drastically as maximum noise variance is reached. The pattern and distribution of the abundance maps obtained from the different algorithms with different levels of noise for all the endmembers were similar to Figure 8.4 (abundance maps without noise) and Figure 8.5.

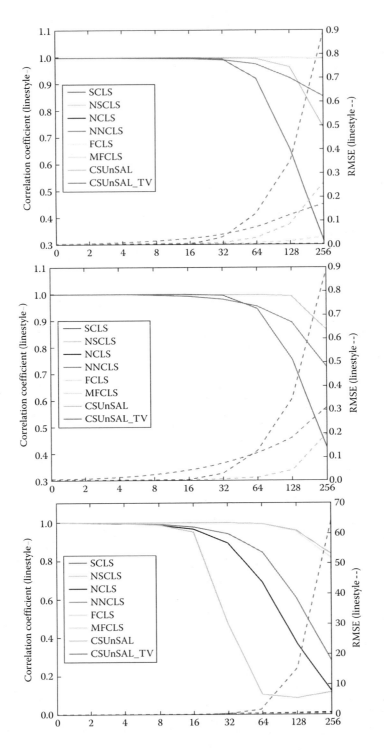

FIGURE 8.6 (See color insert.) Pearson product–moment correlation coefficient (cc on *Y*-axis) and RMSE (secondary *Y*-axis) between abundance values obtained from the eight constrained algorithms and true abundances at different levels of noise. (*X*-axis: Noise variance.) NCLS, nonnegative constrained least squares; NNCLS, normalized nonnegative constrained least squares; RMSE, root mean square error.

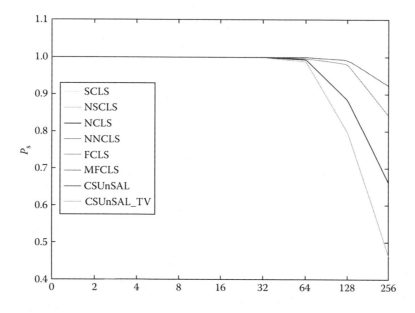

FIGURE 8.7   Plot of probability of success (p_s) for the different constrained algorithms. (X-axis: Noise variance.)

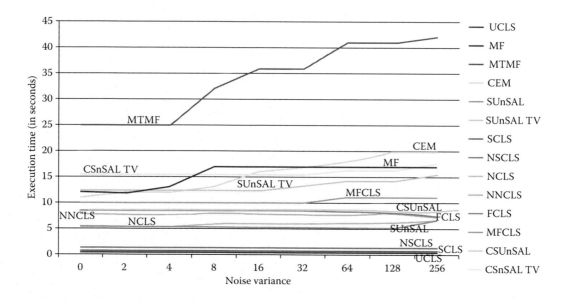

FIGURE 8.8   Time taken by algorithms for unmixing data of different noise levels. CEM, constrained energy minimization; MF, matched filtering; UCLS, unconstrained least squares.

Figure 8.11 shows BDFs of abundance estimated from FCLS against synthetic true abundance for endmember 1 for noise variance 0, 4, 16, 64, 128, and 256.

The points from the true abundance versus estimated abundance from FCLS fell almost in a 1:1 line for all the three endmembers for data without noise (noise variance = 0). It is clear that as the noise increases, the point fall away from the 1:1 line and for noise with variance 128 and 256, most of the points do not lie on the straight diagonal line; they tend to diverge. It indicates that as noise increases, the estimated abundance tends to deviate from the real abundance and the algorithms wrongly predict the class proportions.

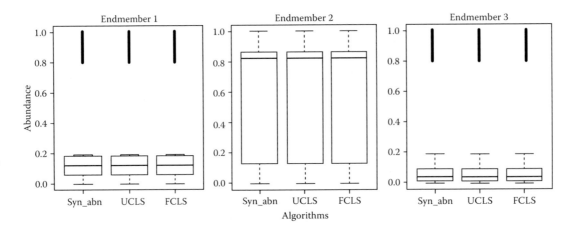

FIGURE 8.9 Boxplots of abundance maps obtained from synthetic abundance (represented as Syn_abn), UCLS, and FCLS for endmember 1, 2, and 3 from noise-free data.

The same trend was observed with all the endmembers and algorithms, for which the BDFs were worse than FCLS. In Section 8.7.2, we discuss implementation of the algorithms on real-world data of different landscapes.

8.7.2 Landsat Data—An Agricultural Setup

Each of the 11 atmospherically corrected Landsat scenes was unmixed using the six unconstrained algorithms and eight constrained algorithms to obtain the abundance estimates within each pixel. Visually, the results obtained for each scene from the different algorithms were identical. Figure 8.12 shows sample estimated abundance maps from one of the Landsat scenes for S, V, and D classes obtained from the FCLS algorithm.

For each scene, the proportions of vegetation fraction in the image were compared with the ground observations. Figure 8.13 shows scatter plots of the real/true vegetation fractions against estimated abundances along with the regression line and R^2 values for UCLS, SUnSAL, FCLS, and MFCLS. Among the unconstrained algorithms, the mean absolute error (MAE) of vegetation fraction for UCLS, SUnSAL, and SUnSAL TV was 0.08, MF was 0.22, MTMF was 0.67, and CEM was 0.11. The cc (statistically significant at 0.99 confidence level, p-value $<2.2e^{-16}$) between fractions of vegetation ground polygons to the fraction abundance estimates for UCLS, MF, SUnSAL, and SUnSAL TV was 0.98, MTMF was 0.57, and CEM was 0.96. UCLS, SUnSAL, and SUNSAL TV gave similar results with MAE close to 8%.

The constrained algorithms, SCLS, NSCLS, NCLS, FCLS, MFCLS, and CSUnSAL had an MAE of 0.08, NNCLS of 0.15, and CSUnSAL TV had an MAE of 0.29. The cc (statistically significant at 0.99 confidence level, p-value $<2.2e^{-16}$) between fractions of vegetation ground polygons and fraction estimates from SCLS, NSCLS, NCLS, FCLS, MFCLS, and CSUnSAL was 0.98, NNCLS was 0.82, and CSUnSAL TV was 0.21. SCLS, NSCLS, NCLS, FCLS, MFCLS, and CSUnSAL gave identical results with an MAE of 8%. Although, MF has a high cc (0.98), its MAE was also high (0.22) when compared to UCLS, SUnSAL and SUnSAL TV. On the other hand, CSUnSAL TV had a low cc (0.21) and a high MAE (0.29), indicating that two different error distributions can have dissimilar range, yet they can have similar means. This is the reason for similar MAEs of MF and CSUnSAL TV with different ccs when compared to the ground vegetation fractions. However, it is to be noted that at a higher decimal order in MAE, MFCLS, FCLS, and CSUnSAL were more accurate than the other methods.

A comparison of the satellite derived vegetation fraction with ground measurements showed that three unconstrained and six constrained mixture models provided a reasonably accurate direct estimation of

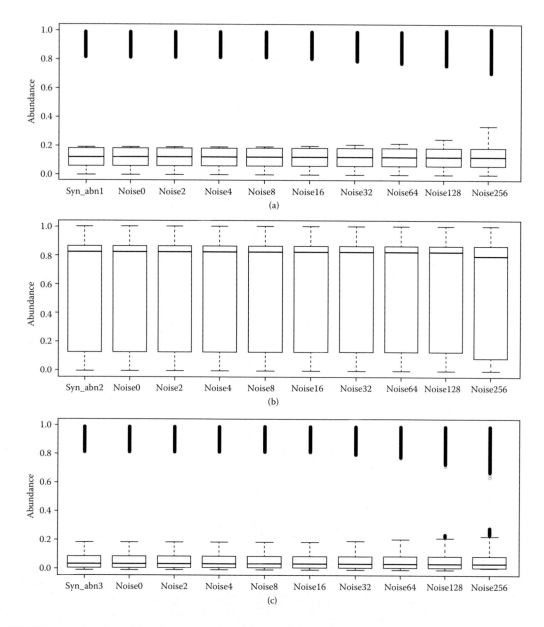

FIGURE 8.10 Boxplots of abundance maps obtained from FCLS with different levels of noise (2, 4, 8, 16, 32, 64, 128, and 256) for endmember 1 (a), endmember 2 (b), and endmember 3 (c).

fractional cover from Landsat data in the context of this study. This indicates that the mixture models were able to accurately reproduce the proportions of S-V-D endmembers in the 11 scenes under investigation.

On an average in ascending order, UCLS took 52 seconds, CEM—166 seconds, MF—190 seconds, MTMF—318 seconds (20 seconds for MNF transformation of the bands and 298 seconds for algorithm execution), SUnSAL—7200 seconds (approximately), SUnSAL TV —118,800 seconds (approximately) for unmixing each scene of Landsat with approximately 7321 rows and 8367 columns. Among the constrained algorithms, SCLS took 66 seconds, MFCLS—147 seconds, NSCLS—201 seconds, NCLS—997 seconds, NNCLS—1053 seconds, FCLS—1215 seconds, CSUnSAL—7200 seconds (approximately), and CSUnSAL TV—118,800 seconds (approximately) to process each Landsat scene. An important point to highlight here is that although SUnSAL, SUnSAL TV, and CSUnSAL gave higher accuracies, they took highest execution

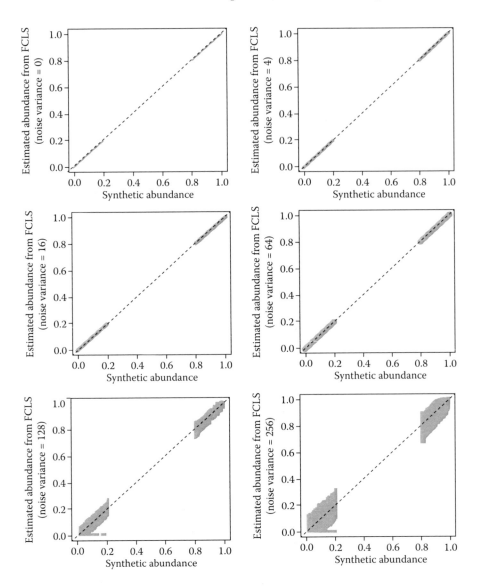

FIGURE 8.11   BDFs of true synthetic abundance against abundance obtained from FCLS for endmember 1 for noise variance 0, 4, 16, 64, 128, and 256. BDF, bivariate distribution function.

time: SUnSAL/CSUnSAL took nearly 2 hours/scene (22 hours for entire Landsat data with 11 scenes), and SUnSAL TV took 33 hours/scene (363 hours for entire Landsat data). Nevertheless, though time exhaustive, CSUnSAL TV did not produce satisfactory results with Landsat data.

8.7.3 Landsat Data—An Urban Setup

Urban areas are currently among the most rapidly changing LC types and the loci of human population activity and are therefore sites of significant natural resource transformation [76]. Usually, urban areas have more heterogeneity with contrasting features and variability compared to other LC classes. Urban environments are particularly difficult to model because of considerable spectral variation and deriving

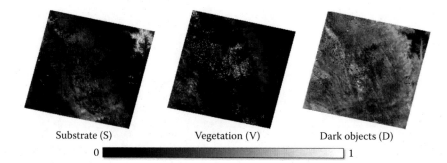

Substrate (S)   Vegetation (V)   Dark objects (D)

0 [gradient bar] 1

FIGURE 8.12 Abundance maps obtained from FCLS for S-V-D classes. In the legend, 0 (black) indicates absence of a particular class and 1 (white) indicates full presence of that class in a pixel. Intermediate values of the shades of gray represent mixture of more than one class in a pixel.

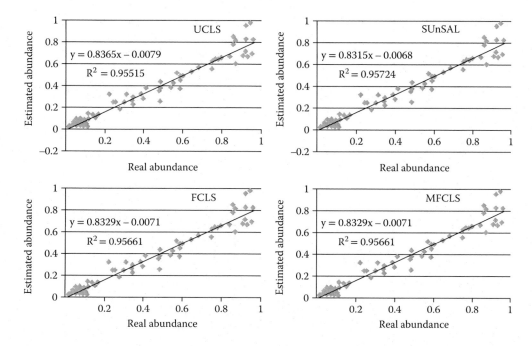

FIGURE 8.13 Scatter plots of vegetation fraction at each of the sampled locations and regression lines of ground estimated (real abundance) versus estimated abundance from UCLS, SUnSAL, FCLS, and MFCLS.

accurate, quantitative measures over urban regions remains a fundamental research challenge due to the great spatial and spectral variability of the materials [77–79].

As discussed earlier, characteristics of the WV-2 mixing space are similar to the Landsat mixing space. Therefore, a pair of coincident Landsat and WV-2 data of San Francisco was unmixed using the same set of global endmembers (S-V-D) to study the performance of the algorithms (Figure 8.14). WV-2 data were unmixed using FCLS, since FCLS was robust against noise with the synthetic data and was among the techniques that gave superior performance in the Landsat data analysis of the agricultural setup. Moreover, its time complexity is reasonable for unmixing, so it emerged as a natural choice.

Figure 8.15 shows the estimated abundance fraction corresponding directly to the gray scale values for each image from Landsat and WV-2 for S-V-D classes. The Landsat fraction estimates are validated by

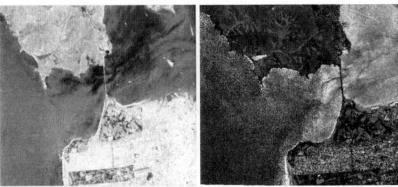

Landsat FCC (R-G-B = 4-3-2)   WV2 FCC (R-G-B = 7-5-3)

FIGURE 8.14   FCC of San Francisco area in Landsat and WV-2 resolution.

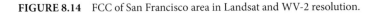

FIGURE 8.15   Endmember fractions of S-V-D from the Landsat data (first row) and WV-2 endmember fractions of S-V-D (second row) from FCLS. In the legend, 0 (black) indicates absence of a particular class and 1 (white) indicates full presence of that class in a pixel. Intermediate values of the shades of gray represent mixture of more than one class in a pixel.

quantifying the degree of correspondence to S, V, and D fraction estimates derived from WV-2 multispectral imagery. A high-resolution WV-2 imagery allows imaging the individual components of the urban area at a significantly higher resolution than the 30 m imagery. As it is evident from Figure 8.16, buildings, sidewalks, streets, and trees can be easily identified in WV-2 FCC. The 2 m resolution is adequate to capture small to medium sized buildings and small streets. This is expected as the higher spatial resolution results in less mixing between individual components of urban scenario such as roofs, streets, trees, pavements, etc. than at 30 m spatial resolution. At 30 m resolution, each 2 m WV-2 pixel is even less than 0.5% of the area within the 30 m full width half maximum of Landsat point spread function. The validity of using

Validation site: Landsat FCC   WV2 FCC

FIGURE 8.16   FCC of validation site near Golden Gate bridge, San Francisco at 30 m (Landsat) and 2 m (WV-2) spatial resolution.

high-resolution fraction abundance maps to assess a moderate resolution fraction depends on whether there is a consistent bias in the mixture model that corrupts both the Landsat and WV-2 estimates in a consistent fashion and consistently erroneous [80]. A subset of the scene having mix of buildings, shadows of buildings, streets, trees, and parks were taken in the two data sets (Landsat and WV-2) for pixel-to-pixel comparison as shown in Figure 8.16.

The Landsat S-V-D fraction estimates were directly compared to WV-2 S-V-D estimates. Here, the sampling operation was duplicated that occurred in the Landsat sensor as per Small and Lu [80]. The 2 m WV-2 fractions were convolved with a Gaussian low-pass filter having 30 m full width half maximum, with the point spread function of the Landsat sensor and resampled to 30 m. Convolution followed by resampling of the high-resolution data to 30 m resolution allows assessing the geo-correction and comparison of the two data sets. Coordinate comparison of the high- and low-resolution data sets at many random pixels corresponding to same spatial location did not reveal any systematic image registration error. The WV-2 S-V-D fractions were then compared to Landsat S-V-D fractions through the MAE and cc for each abundance map.

Among the unconstrained algorithms, MAEs of S, V, and D fractions for UCLS, SUnSAL, and SUnSAL TV was 0.11, 0.07, and 1.99. MF and MTMF had MAE of 0.29 (S), 0.08 (V), and 0.49 (D) and CEM had MAE of 0.29 (S), 0.08 (V), and 0.96 (D). UCLS, SUnSAL, and SUnSAL TV had ccs (statistically significant at 0.99 confidence level, p-value $<2.2e^{-16}$) 0.86 for S, 0.88 for V, and -0.03 for D classes. Note that the third endmember (D) has not been classified properly with the unconstrained algorithms. The minimum and maximum abundance values for the third endmember (D) from UCLS was $(-20.7, 57.2)$, and from SUnSAL and SUnSAL TV were $(-20, 57)$. MF, MTMF, and CEM had a lower cc compared to the other unconstrained algorithms. Figure 8.17 shows scatter plots of the S, V, and D fractions of the true (WV-2) abundances against the estimated abundances along with the regression line and R^2 values for UCLS and SUnSAL. UCLS, SUnSAL, and SUNSAL TV gave similar results with the MAEs close to 11% for S and 7% for V and D classes.

For the constrained algorithms, minimum and maximum abundance values were almost between 0 and 1. SCLS, NSCLS, FCLS, MFCLS, and CSUnSAL had MAE of 0.09 (for S), 0.06 (for V), and 0.06 (for D); other algorithms had higher MAEs. The cc (statistically significant at 0.99 confidence level, p-value $<2.2e^{-16}$) between true fractions (of S, V, and D) and estimated fractions from SCLS, NSCLS, FCLS, MFCLS, and CSUnSAL were 0.87, 0.88, and 0.63; other algorithms had lower cc values. Figure 8.18 shows scatter plots of the S, V, and D fractions of the true (WV-2) abundances against the estimated abundances along with the regression line and R^2 values for FCLS and CSUnSAL. SCLS, NSCLS, FCLS, MFCLS, and CSUnSAL gave similar results with average MAEs of 9% for S and 6% for V and D classes.

Chronologically, UCLS took 65 seconds, CEM and MF—157 seconds, MTMF—586 seconds (30 seconds for MNF transformation of the data and 286 seconds for algorithm execution), SUnSAL—7200 seconds (approximately), SUnSAL TV—118,800 seconds (approximately) for unmixing Landsat scene of 7151 rows and 8241 columns. Among the constrained algorithms, SCLS took 129 seconds, MFCLS—157 seconds,

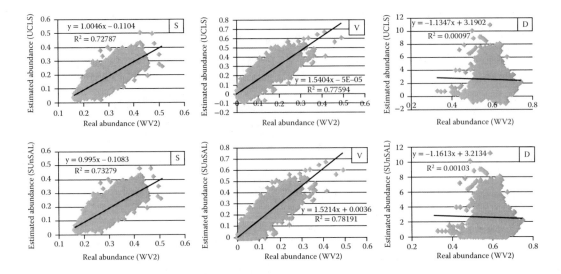

FIGURE 8.17 Scatter plots of S-V-D fractions from WV-2 and Landsat and regression lines of WV-2 estimate (real abundance) versus estimated abundance from UCLS and SUnSAL.

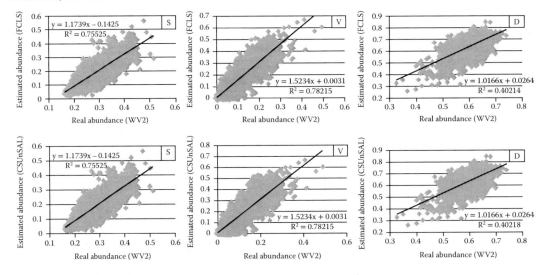

FIGURE 8.18 Scatter plots of S-V-D fractions from WV-2 and Landsat TM and regression lines of WV-2 estimate (real abundance) versus estimated abundance from FCLS and CSUnSAL.

NSCLS—204 seconds, NCLS—977 seconds, NNCLS—1023 seconds, FCLS—1155 seconds, CSUnSAL—7200 seconds (approximately) and CSUnSAL TV—118,800 seconds (approximately) to unmix the Landsat TM data. The SUnSAL family of algorithms was most time exhaustive, although CSUnSAL TV did not produce reasonable results.

8.8 Discussion

Over the past years, many algorithms have been developed for unmixing mixed pixels. In particular, linear spectral unmixing analysis has been widely used for multispectral and hyperspectral image classification as well as for subpixel target detection [81]. The main assumption is that mixed pixels can be expressed in the form of linear combinations of a set of pure pixels (endmembers), which are known in advance, either from an endmember library or extracted from the image itself. In this chapter, we assessed the techniques developed for spectral unmixing—six unconstrained and eight constrained algorithms. The ultimate goal was to analyze the potential of various unmixing algorithms in solving the spectral unmixing problem.

Several unmixing experiments based on computer-simulated data with different levels of noise, 11 Landsat-5 scenes of an agricultural landscape and a Landsat TM data of an urban setup were performed to compare the accuracies and the computational costs of these methods with a set of three global endmembers (S, V, and D). The abundance maps obtained through unmixing temporal 11 Landsat scenes were validated through ground fractional vegetation cover obtained from the field on the same day as that of the satellite pass, and the urban Landsat fraction estimates were assessed by comparing them to high-resolution unmixed abundance maps. Quality of the results were assessed through several indicators—descriptive statistics, cc, RMSE, p_s, boxplot, BDF, MAE between the true abundance and the estimated abundance maps, and execution time. These assessments were an attempt to validate abundance maps produced by spectral unmixing algorithms in an independent and quantitative manner. The results obtained from most of these methods were quite accurate and promising. This chapter reinforces the universal agreement that pixel unmixing enables a better estimation of the LC proportion than per-pixel classification.

Although adequate measures were taken, there could be effect of a few factors contributing to the errors in abundance estimates from the images. For the Fresno area, the acquisitions were taken during clear sky conditions (except for 3 days) and they also coincided with the approximate time of the satellite overpass that took into account the illumination effect. The images were corrected to surface reflectance to curtail the effect of atmospheric noise. To avoid the geolocation errors caused due to misregistration, atmospheric effects, presence of background mixed with substrate, etc. a matrix of 3 × 3 pixels centered over the GPS location was used. Thus, estimates of ground fractional cover from the two to four digital photographs well represented the field conditions within the Landsat IFOV of the 3 × 3 pixels window to which they were compared. Nevertheless, since fraction estimates from digital images were based on image segmentation and thresholding identified from distribution of camera pixels, there could be issues while resolving the ground cover estimates with the mixture model outputs. However, the image derived fraction estimates from the different mixture model algorithms closely matched the ground observations on sparse vegetation conditions, appreciating the fact that vegetation fraction from the image is modeled only for the portion that is illuminated by sunlight and the shaded portions of the canopy are likely to be assigned to the dark fractions.

Given that the dark fraction class represents an endmember, they do not represent a unique set of materials [82], because they may correspond to absorbing surfaces as well as nonilluminated surfaces such as clear and deep or turbid water bodies, which transmit and absorb, respectively, nearly all the radiation falling on it. Some dark component is also present in most substrates in the form of humic materials, tars, dark rocks, soil moisture, etc. In the image, they are present in the form of shadows falling over substrates (soils or rocks) caused by topography, vegetation, canopy roughness with leaves shading other leaves, infrastructure materials, etc., often responsible for variations in the substrate albedo. The algorithms with the available global endmembers accounted for the variance in the soil by substrate and dark object fractions, given the fact that crop conditions were overall very uniform.

For the San Francisco area, most of the urban pixels were either mixed with vegetation (urban forest), roads, shadows, or appear like dark objects because of the different materials used in construction of the terrace as shown in Figure 8.2 (inset). Nevertheless, this study shows that urban reflectance can be accurately modeled with a three-endmember mixture model using Landsat and WV-2 data. Error in San Francisco Landsat data fraction estimation could be either due to error in estimates from different algorithms or geo-registration or both. S-V-D model represented the land surface as independent constituents with different landscape properties such as vegetation and urban. It characterized the fraction of illuminated vegetation, substrate or impervious materials and the shadowed or nonreflective surfaces such as water, roofing tar, etc. High substrate fractions are rational estimates of the impervious surface in developed land in temperate and tropical regions as pervious surfaces are mostly covered by some kind of vegetation and exposed substrate are most likely impervious. Sometimes, if both pervious and impervious surfaces are exposed and illuminated, then this leads to ambiguity; a problematic classification in arid and semiarid regions. Small and Lu [80] argue to use vegetation fraction as a proxy for fractional pervious surface because vegetation cannot thrive on impervious surface, so presence of vegetation implies presence of some amount of pervious surface. Therefore, using detectable vegetation as an indicator of permeable surface can account for the range of natural and built surfaces.

The mixed pixel classification problem is unconstrained without ANC and ASC imposed on the abundance fractions. In such cases, the estimate is used for target detection, discrimination and classification, but not for target quantification [21]. UCLS does not impose any constraints on the abundance vector. So, it may not provide accurate estimates of the abundance fractions and only offers an unconstrained solution. However, it gave reasonably good unmixing results in this study among the unconstrained algorithms. MF uses the covariance matrix of the scene to model the spectral variability of the background. A known target spectrum is projected onto the generalized inverse of the background covariance matrix to derive the MF projection vector. However, false positives are common to MF where materials may be randomly matched if they are rare in a pixel (thus not contributing to the background covariance) because the technique is not subject to the nonnegative and sum-to-one constraints inherent to spectral signals within bounded image pixels [40] and the rare classes do not contribute significantly to the background covariance calculation; thus, they are not nulled properly by the MF method [49]. Systems with low spectral contrast [83] may have increased potential for spectral confusion, which in turn would have introduced irregularity in output MF scores in this study, possibly resulting in MF scores greater than one. These high MF scores are an artifact of class variability and occur as unique mixtures of background materials that falsely represent the class spectra [48].

In MTMF, the units of MNF-space are expressed with respect to noise standard deviations. Therefore, the closeness of a pixel's actual location to its projected location is expressed in term of its spectral variance from the target class. Pixels that have high proportion of target class component have low variance because the target class dominates their spectral characteristics. Therefore, measures of data variance are standardized based on estimates of mixing feasibility [48]. MTMF preserves the ease of calculation of MF and allows for enhanced target class selectivity, excelling at accurate mapping of extremely small subpixel targets with very low false alarm rates [49] and is specifically very useful for detection and discrimination among multiple rare target classes with low spectral contrast between targets and the background. MTMF ranked high in the synthetic data analysis because it had a consistent performance in class detection throughout various noise levels.

Sometimes, when signatures of all the classes are not available, the unidentified sources may include interferes, for which signatures are not easily detectable. In such cases, a well-represented target class signature is not realistic and algorithms like CEM constrain the target of interest while minimizing the energy of unknown signal sources [21]. CEM is considered to be a target detector but may not necessarily imply that it can also classify the target it detected. Here, the detection rate could be quiet high and the classification could be very low. CEM has a disadvantage that it can only detect one target class at a time and is sensitive to the target knowledge that is used in the constraint. It minimizes the interfering effects but does not eliminate undesired targets, so does not achieve its optimal performance. Ideally, a detection model should detect the desired targets, annihilate the undesired targets, and minimize the interfering effects, which is possible through target-constrained interference-minimized filter (TCIMF) [84]. If there is spectral variability and noise, CEM filter may miss these targets and give unsatisfactory output. In this study, CEM has overestimated the abundances. Harsanyi and Chang [42] described CEM and MF to be identical. It has also been referred to as an optimal linear detection method for locating a known signature in the presence of a mixed and unknown background [49].

Unconstrained solutions so far have been used for quantification of classes present in a pixel, but the estimated abundance fractions are not accurate. Subsequently, these abundance estimates do not indicate the corresponding true abundance. It is important to state that for many target detection applications, where high error in the accuracy of the estimated fraction of target class abundance is tolerable, unconstrained algorithms are preferred. To the extent that the estimated abundance from unconstrained algorithms are helpful to discriminate and detect objects from the background, the unconstrained algorithms are useful without ANC and ASC. It is often the case that in target detection applications, fully constrained solutions are not as efficient as unconstrained or partially constrained solutions since the fractional abundance range between 0 and 1 confines the target detection solution.

On the other hand, among the constrained algorithms, SCLS imposes ASC on the abundance fractions with no ANCs. It minimizes the observation error subject to the constraint that sum of abundances add to one at every pixel. Therefore, actually it is possible that abundance may be greater than one for an endmember and negative for other endmembers, in order to equate the sum of the abundances to one. The ASC in SCLS is sometimes a disadvantage and the algorithm performs unsatisfactorily in object detection in a scene with interferers, while trying to constrain the solution. The situation worsens when many target signatures are present in the image with complicated image background. This is avoided by either normalizing the abundance values through NSCLS or by imposing ANC through NCLS (via NNLS algorithm). NCLS ensures the abundance fractions of the target signature to be nonnegative, but discards the ASC. As a result, both methods without either ANC or ASC are not able to estimate target abundance fractions accurately. To this end, it is justifiable to say that the estimated abundance fractions with either ANC or ASC are more accurate than the unconstrained methods and can also be used for target detection. Chang [21] argued that since abundance maps obtained from SCLS sum to one, the magnitude of the abundance fractions are generally small so that ASC is satisfied, and this is especially true when many classes of similar spectra are present in a scene reducing the target detectability. NCLS being unrestricted from ASC, assigns any abundance value, enhancing the target detectability, despite the point that it may not reflect accurate abundance fraction. In such target detection applications, NCLS is preferred over SCLS. In fact, both SCLS and NCLS are considered to render partially constrained solutions.

ANC and ASC can be imposed simultaneously through the FCLS or MFCLS approaches. FCLS makes use of a numerical algorithm based on a least squares approach [85] for optimal solutions. MFCLS modifies the fully constrained linear mixing problem by replacing ANC with absolute ASC (AASC), which replaces ANC in FCLS, while still implementing ASC. It converts a set of inequality constraints to an equality constraint so that a closed form solution can be found through Lagrangian multiplier. Both FCLS and MFCLS use SCLS to obtain a fully constrained solution [21]. The FCLS and MFCLS algorithms used here have higher accuracy among all the algorithms assessed, and MFCLS is also computationally efficient. The FCLS algorithm presented here has slightly higher time complexity since it is computationally iterative in nature.

A more efficient FCLS algorithm has been proposed [86] that used fast NNLS algorithm [87,88] and was computationally efficient than the NNLS algorithm in Lawson and Hanson [66] but its implementation is not discussed here. An application of FCLS to large-scale machine learning with the global endmembers is demonstrated below. FCLS was used for continental unmixing of monthly web-enabled Landsat data (WELD) to study the changes in vegetation and urban areas. Figure 8.19 shows unmixed maps of S-V-D classes for North America (mosaicked with 1673 Landsat scenes of August 2009) obtained from the FCLS algorithm at the NEX, generated in 12 minutes, 41 seconds with 10 nodes and 20 CPUs. These unmixed maps can be used for wetlands delineation, urban growth studies, forest cover mapping, mineral mapping, urban vegetation estimation, etc.

Constrained algorithms are better than unconstrained algorithms in target abundance estimation because they enforce the fractions to be nonnegative and sum-to-one, which is a more practical way of looking and interpreting the LC proportions, as one does not expect to have minus 10% water class in a pixel or 125% of water + park + building classes added together in a pixel, which are obviously meaningless and difficult to interpret.

Sparse unmixing problems are often nonconvex optimization formulations and thus, very difficult to solve. Sparse unmixing techniques convert them to a sequence of simpler ones that are convex optimization problems with sufficient conditions for convergence. Sparse techniques are meant to look fundamentally for the endmembers in a spectral library containing spectra of many materials, with only a few of them present in a pixel, that is, the vector of fractional abundances is sparse and enforce the sparsity of the solution explicitly [60]. However, sparse unmixing techniques demonstrated better performance in this

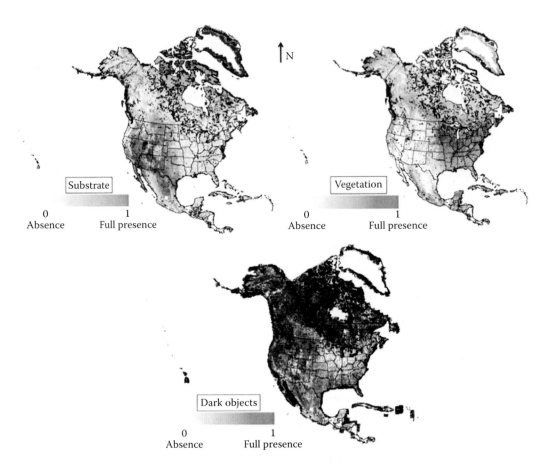

FIGURE 8.19   Abundance maps of substrate, vegetation, and dark objects classes for North America.

study even when the number of endmembers were low, which is often the case and realistic in practice. In the absence of spectral library, these techniques can also be used with image-derived endmembers. SUn-SAL TV accounted for the spatial contextual information while solving the unmixing problem. SUnSAL, SUnSAL TV, CSUnSAL, and CSUnSAL TV programs in MATLAB® were time in-depth. Their equivalent C or C++ code could decrease the execution time. New directions in sparse unmixing such as subspace matching pursuit, structured sparse unmixing, and their applications can be referred in Shi et al. [89,90] and Zhu et al. [91].

The experimental results conducted with both computer-simulated and real data sets are encouraging and the outcomes suggest that among the unconstrained algorithms, UCLS followed by MTMF, MF, and then SUnSAL, SUnSAL TV performed best in decreasing order. UCLS, SUnSAL, and SUnSAL TV performed equally well on Landsat data from both an agricultural and urban landscape. However, SUnSAL and SUnSAL TV have very high time complexity making them slower. Additionally, some more experiments of object/target class detection are required to come to a strong conclusion about unconstrained algorithms, as they are more suitable for applications where target class detection is required rather than target class quantification. For the constrained algorithms, MFCLS, FCLS, CSUnSAL, SCLS, and NSCLS had equally high performance in terms of accuracy for computer-simulated and Landsat data unmixing. In spite of the good performance, it is evident that SCLS can also generate negative abundance values; NSCLS is no longer an optimal solution as it is a normalized value of the SCLS solution, and CSUnSAL has high execution time, so MFCLS and FCLS have advantage over these methods.

Although the mixture models evaluated here differ from one another, a close observation reveals certain specific similarities between them. All the models have a common objective that is accurate estimation of subpixel LC proportion. The purpose of this study was to examine how well the models perform toward achieving this objective and their impact on Landsat data processing. The mixed pixel classification problem was analyzed in the absence and presence of constraints on the abundances. Comparing one relative to another has been very challenging due to lack of standardized data and no defined rules to compare algorithms. Another difficulty arises from the fact that there are no rigorous criteria to substantiate an algorithm [92]. From the results, it is difficult to suggest the best algorithm for unmixing different data sets pertaining to various landscapes. Fortunately, owing to the consistency in the accuracy estimation procedures and standards, this study gives a clear vision of the accuracies of the different algorithms and overall, UCLS, SUnSAL, and SUnSAL TV models proved to fit the data very well, superior than other unconstrained algorithms. FCLS, MFCLS, CSUnSAL, SCLS, and NSCLS were robust with superior performance than other constrained algorithms. The outcome of this study clearly indicates that the large repository of Landsat data can be used to quickly assess state and distribution of S-V-D classes. Mapping of time series S-V-D classes can be used as a periodic investigation tool, which could be valuable information for planning and can be used for physical process models requiring biophysical responses such as S-V-D abundances. Despite our effort to conduct comprehensive and rigorous comparative analysis of various unmixing algorithms, completion is not claimed. A few target detection and unmixing algorithms such as unmixing by simplex projection [93], collaborative sparse unmixing [94], unmixing based on distance geometry model [95], and normal composite model-based unmixing [96] have been proposed in the recent literatures which are the areas of future directions in this domain. Accounting endmember variability [97,98] is beyond the scope of this work, and this study shows that it is reasonable and pragmatic to use a global spectral library based on the most representative endmembers of each category to obtain very accurate unmixing results. Global endmembers have diverse applications in continental to global LC mapping where local endmembers restricted to a particular geographical location may not be easily available. An endmember that is the most representative of its class, in this case of endmember selection, has captured the LC with distinct spectra that occupy smaller areas within the scene. At the same time, it is acknowledged that fractional errors can occur occasionally either when too few endmembers are used [99], resulting in spectral information that cannot be accounted for by the existing endmembers, or too many, in which case minor departures between measured and modeled spectra are often assigned to an endmember that is used in the model, but not actually present.

8.9 Conclusion

Soft classification analyses have attracted considerable attention as a means of solving the mixed pixel problem that is often encountered in remote sensing (RS) applications. The exploitation of LMMs has nurtured a plethora of unmixing algorithms that can be used to estimate the fractions of classes of interest, which may be part of a pixel or smaller than the pixel's spatial resolution in RS data with numerous spectral bands. The exact nature of the unmixing results depends on the classification algorithm and endmember definitions.

In this chapter, survey, definition, and analysis of the state-of-the-art unmixing algorithms were presented for subpixel classification. ANC and ASC were imposed on the abundance fractions. The algorithms were tested on computer-simulated data with different levels of noise, 11 Landsat scenes of an agricultural landscape, and a Landsat data of an urban setup. These approaches were compared for accuracies in both a quantitative and qualitative manner along with the computational costs of these methods with a set of three global endmembers. It was observed that the areas with high fractional abundance of the considered endmember are well delineated while mixed regions with low abundance are more homogeneous. The abundance maps obtained from both unconstrained and constrained algorithms have spatial consistency with good accuracy in the spatial distribution of the classes. The outcomes of this study based on the quantitative and intercomparative assessment of the algorithms reveal a strong conclusion that the constrained algorithms could adequately model the data for unmixing better than the unconstrained unmixing algorithms. The compendium of unmixing algorithms dealt here highlight the increasing complexity of diverse disciplines such as image and signal analysis, algebra, statistics, computer graphics, models, and the need for development of advanced algorithms for large-scale machine learning.

Acknowledgments

We are grateful to NASA Ames Research Center for providing the RS data, computational facilities, and infrastructural support to carry out the analysis. We thank Oak Ridge Associated Universities (ORAU) for funding this research as part of the NASA Postdoctoral Program (NPP). We acknowledge Jose M. Bioucas-Dias, Instituto de Telecomunicações, Instituto Superior Técnico, Portugal for providing the sparse unmixing codes.

References

1. Crósta, A. P. and Moore, J. M., 1989, Geological mapping using Landsat Thematic Mapper imagery in Almeria Province, south-east Spain. *International Journal of Remote Sensing*, vol. 10(3), pp. 505–514.
2. Kenea, N. H., 1997, Improved geological mapping using Landsat TM data, Southern Red Sea Hills, Sudan: PC and IHS decorrelation stretching. *International Journal of Remote Sensing*, vol. 18(6), pp. 1233–1244.
3. Baker, C., Lawrence, R., Montagne, C., and Patten, D., 2006, Mapping wetlands and riparian areas using Landsat ETM+ imagery and decision-tree based models. *Wetlands*, vol. 26(2), pp. 465–474.
4. Lunetta, R. S. and Balogh, M. E., 1999, Application of multi-temporal Landsat 5 TM imagery for wetland identification. *Photogrammetric Engineering & Remote Sensing*, vol. 65(11), pp. 1303–1310.
5. Huang, C., Peng, Y., Lang, M., Yeo, I., and McCarty, G., 2014, Wetland inundation mapping and change monitoring using Landsat and airborne LiDAR data. *Remote Sensing of Environment*, vol. 141, pp. 231–242.
6. Lobell, D. B. and Asner, G. P., 2004, Cropland distributions from temporal unmixing of MODIS data. *Remote Sensing of Environment*, vol. 93, pp. 412–422.
7. Pacheco, A. and McNairn, H., 2010, Evaluating multispectral remote sensing and spectral unmixing analysis for crop residue mapping. *Remote Sensing of Environment*, vol. 114, pp. 2219–2228.
8. Hostert, P., Roder, A., and Hill, J., 2003, Coupling spectral unmixing and trend analysis for monitoring of long-term vegetation dynamics in Mediterranean rangelands. *Remote Sensing of Environment*, vol. 87, pp. 183–197.
9. DeFries, R., Hansen, M., Steiinger, M., Dubayah, R., Sohlberg, R., and Townshend, J., 1997, Subpixel forest cover in central Africa from multisensor, multitemporal data. *Remote Sensing of Environment*, vol. 60, pp. 228–246.
10. Pu, R., Gong, P., Michishita, R., and Sasagawa, T., 2008, Spectral mixture analysis for mapping abundance of urban surface components from the Terra/ASTER data. *Remote Sensing of Environment*, vol. 112, pp. 939–954.
11. Dópido, I., Villa, A., Plaza, A., and Gamba, P., 2011, A quantitative and comparative assessment of unmixing-based feature extraction techniques for hyperspectral image classification. *IEEE Journal of Selected Topics in Applied Earth Observations and Remote Sensing*, vol. 5(2), pp. 421–435.
12. Turner II, B. L., Meyer, W. B., and Skole, D. L., 1994, Global land-use/land-cover change: Towards an integrated study. *Ambio*, vol. 23(1), pp. 91–95.

13. Pielke, R.A. Sr., 2001, Earth system modeling—An integrated assessment tool for environmental studies. In *Present and Future of Modeling Global Environmental Change: Toward Integrated Modeling*, Matsuno, T. and Kida, H., (Eds.). Tokyo: Terra Scientific Publisher, pp. 311–337.

14. Collins, W. J., Bellouin, N., Doutriaux-Boucher, M., Gedney, N., Halloran, P., Hinton, T., Hughes, J., Jones, C. D., Joshi, M., Liddicoat, S., Martin, G., O'Connor, F., Rae, J., Senior, C., Totterdell, I., Wiltshire, A., and Woodward, S., 2011, Development and evaluation of an Earth-system model–HadGEM2. *Geoscientific Model Development Discussion*, vol. 4, pp. 997–1062.

15. Nemani, R., Votava, P., Michaelis, A., Metlon, F., and Milesi, C., 2011, Collaborative supercomputing for global change science. *EOS Transactions*, vol. 92(13), pp. 109–116.

16. Yang, X. and Lo, C. P., 2002, Using a time series of satellite imagery to detect land use and land cover changes in Atlanta, Georgia metropolitan area. *International Journal of Remote Sensing*, vol. 23(9), pp. 1775–1798.

17. Liu, W. and Wu, E. Y., 2005, Comparison of non-linear mixture models: Sub-pixel classification. *Remote Sensing of Environment*, vol. 94, pp. 145–154.

18. Boardman, J. W., 1989, Inversion of imaging spectrometry data using singular value decomposition. In *Proceedings of the 12th Canadian* Symposium on Remote Sensing with Geoscience and Remote Sensing Symposium, *IGARSS*, Vancouver, Canada, IEEE, 10–14 July 1989, vol. 4, pp. 2069–2072. doi:10.1109/IGARSS.1989.577779.

19. Ceamanos, X., Doute, S., Luo, B., Schmidt, F., Jouannic, G., and Chanussot, J., 2011, Intercomparison and validation of techniques for spectral unmixing of hyperspectral images: A planetary case study. *IEEE Transactions on Geoscience and Remote Sensing*, vol. 49(11), pp. 4341–4358.

20. Bioucas-Dias, J. M., Plaza, A., Dobigeon, N., Parente, M., Du, Q., Gader, P., and Chanussot, J., 2012, Hyperspectral unmixing overview: Geometrical, statistical, and sparse regression-based approaches. *IEEE Journal of Selected Topics In Applied Earth Observations and Remote Sensing*, vol. 5(2), pp. 354–379.

21. Chang, C.-I., 2003, *Hyperspectral Imaging Techniques for Spectral Detection and Classification*. New York, NY: Kluwer Academic/Plenum Publishers, pp. 54.

22. Adams, J. B., Sabol, D. E., Kapos, V., Filho, R. A., Roberts, D. A., Smith, M. O., and Gillespie, A. R., 1995, Classification of multispectral images based on fractions of endmembers: Application to land cover change in the Brazilian Amazon. *Remote Sensing of Environment*, vol. 52, pp. 137–154.

23. Kumar, U., Kerle, N., and Ramachandra T. V., 2008, Constrained linear spectral unmixing technique for regional land cover mapping using MODIS data. In: *Innovations and Advanced Techniques in Systems, Computing Sciences and Software Engineering*, Elleithy, K., (Ed.), Berlin: Springer. ISBN: 978-1-4020-8734-9. Paper 87. p. 9.

24. Foody, G. M. and Cox, D. P., 1994, Subpixel land cover composition estimation using a linear mixture model and fuzzy membership functions. *International Journal of Remote Sensing*, vol. 15(3), pp. 619–631.

25. Kanellopoulos, I., Varfis, A., Wilkinson, G. G., and Megier, J., 1992, Land-cover discrimination in SPOT imagery by artificial neural network-a twenty class experiment. *International Journal of Remote Sensing*, vol. 13(5), pp. 917–924.

26. Ju, J., Kolaczyk, E. D., and Gopal, S., 2003, Gaussian mixture discriminant analysis and sub-pixel land cover characterization in remote sensing. *Remote Sensing of Environment*, vol. 84, pp. 550–560.

27. Olthof, I. and Fraser, R. H., 2007, Mapping northern land cover fractions using Landsat ETM+. *Remote Sensing of Environment*, vol. 107, pp. 496–509.

28. Karoui, M. S., Deville, Y., Hosseini, S., and Ouamri, A., 2013, Blind unmixing of remote sensing data with some pure pixels: Extension and comparison of spatial methods exploiting sparsity and non-negativity properties. In *Proceedings of the 8th International Workshop on Systems, Signal Processing and Their Applications* (*WoSSPA*), Algiers, Algeria, 12–15 May, pp. 42–49.

29. Kumar, U., Kumar Raja S., Mukhopadhyay, C., and Ramachandra T. V., 2012, A neural network based hybrid mixture model to extract information from non-linear mixed pixels. *Information*, vol. 3(3), pp. 420–441.

30. Liu, W., and Wu, E. Y., 2005, Comparison of non-linear mixture models: Sub-pixel classification. *Remote Sensing of Environment*, vol. 94, pp. 145–154.

31. Ichoku, C. and Karnieli, A., 1996, A review of mixture modeling techniques for sub-pixel land cover estimation. *Remote Sensing Reviews*, vol. 13, pp. 161–186.

32. Keshava, N., 2003, A survey of spectral unmixing algorithms. *Lincoln Laboratory Journal*, vol. 14(1), pp. 55–78.

33. Parente, M. and Plaza, A., 2010, Survey of geometric and statistical unmixing algorithms for hyperspectral images. In *Proceedings of the 2nd Workshop on Hyperspectral Image and Signal Processing: Evolution in Remote Sensing (WHISPERS)*, Reykjavik, Iceland, IEEE, 14–16 June 2010, pp. 1–4. doi:10.1109/WHISPERS.2010.5594929.

34. Zandifar, A., Babaie-Zadeh, M., and Jutten, C., 2012, Aprojected gradient-based algorithmto unmix hyperspectral data. In *Proceedings of the 20th European Signal Processing Conference (EUSIPCO 2012)*, Bucharest, Romania, IEEE, 27–31 August 2012, pp. 2482–2486. Electronic ISSN: 2076-1465, Print ISSN: 2219-5491, Online ISSN: 2219-5491

35. Shimabukuro, Y. E. and Smith, A. J., 1991, The least-squares mixing models to generate fraction images derived from remote sensing multispectral data. *IEEE Transactions on Geoscience and Remote Sensing*, vol. 29(1), pp. 16–20.

36. Nielsen, A. A., 2001, Spectral mixture analysis: Linear and semi-parametric full and iterated partial unmixing in multi-and hyperspectral image data. *International Journal of Computer Vision*, vol. 42(1), pp. 17–37.

37. Chen, J. Y. and Reed, I. S., 1987, A detection algorithm for optical targets in clutter. *IEEE Transactions on Aerospace Electronic Systems*, vol. AES-23(1), pp. 46–59.

38. Clark, R. N., Swayze, G. A., Livo, K. E., Kokaly, R. F., Sutley, S. J., Dalton, J. B., McDougal, R. R., and Gent, C. A., 2003, Imaging spectroscopy: Earth and planetary remote sensing with the USGS Tetracorder and expert systems. *Journal of Geophysical Research*, vol. 108(E12): 5131, pp. 5–1 to 5–44. doi:10.1029/2002JE001847.

39. Kruse, F. A., 2008, Expert system analysis of hyperspectral data. In *Proceedings of the SPIE Defense Security, Algorithms Technologies for Multispectral, Hyperspectral, Ultraspectral Imagery XIV, Conference DS43*, 16–20 March, Orlando, FL: Orlando World Center Marriott Resort and Convention Center, Paper Number: 6966–25.

40. Boardman, J. W., 1998, Leveraging the high dimensionality of AVIRIS data for improved subpixel target unmixing and rejection of false positives: Mixture tuned matched filtering. In *Proceedings of the 5th JPL Geoscience Workshop*, R.O. Green (Ed.), Pasadena, CA: NASA Jet Propulsion Laboratory, pp. 55–56.

41. ENVI Tutorial: Advanced Hyperspectral Analysis, 2014, pp. 27–30. URL: http://www.exelisvis.com/portals/0/pdfs/envi/Adv_Hyperspectral_Analysis.pdf [Last accessed: February 20, 2015, 04:15:00 p.m.].

42. Harsanyi, J. C. and Chang, C.-I., 1994, Hyperspectral image classification and dimensionality reduction: An orthogonal subspace projection. *IEEE Transaction on Geoscience and Remote Sensing*, vol. 32(4), pp. 779–785.

43. Stocker, A. D., Reed, I. S., and Yu, X., 1990, Multidimensional signal processing for electro-optical target detection. In *Proceedings of the SPIE 1305, Signal and Data Processing of Small Targets 1990*, 1 October, vol. 1305, p. 218.

44. Yu, X., Reed, I. S., and Stocker, A. D., 1993, Comparative performance analysis of adaptive multispectral detectors. *IEEE Transactions on Signal Processing*, vol. 41(8), pp. 2639–2656.

45. Mitchell, J. J. and Glenn, N. F., 2009. Subpixel abundance estimates in mixture-tuned matched filtering classifications of leafy spurge (Euphorbia esula L.). *International Journal of Remote Sensing*, vol. 30(23), pp. 6099–6119.

46. Bowles, J., Palmadesso, P., Antoniades, J., Baumback, M., and Rickard, L. J., 1995, Uses of filter vectors in hyperspectral data analysis. In *Proceedings of SPIE Infrared Spaceborne Remote Sensing III*, Sandiego, CA, pp. 148–157.

47. Tseng, Y.-H., 2000, Spectral unmixing for the classification of hyperspectral images. *International Archives of Photogrammetry and Remote Sensing*, vol. 33, Part B7, Amsterdam 2000, pp. 1532–1538.

48. Mundt, J. T., Streutker, D. R., and Glenn, N. F., 2007, Partial unmixing of hyperspectral imagery: Theory and methods. In *Proceedings of the ASPRS Annual Conference*, 7–11 May 2007, Tampa, FL, Bethesda, MD: American Society of Photogrammetry and Remote Sensing, pp. 46–57.

49. Boardman, J. W. and Kruse, F. A., 2011, Analysis of image spectrometer data using–dimensional geometry and a mixture-tuned matched filtering approach. *IEEE Transactions on Geoscience and Remote Sensing*, vol. 49(11), pp. 4138–4152.

50. Heinz, D. C., Chang, C.-I., and Althouse, M. L. G., 1999, Fully constrained least squares based linear unmixing. In *Proceedings of the IEEE International Geoscience and Remote Sensing Symposium*, 28 June–2 July, Hamburg, Germany, pp. 1401–1403.

51. Du, Q., Ren, H., and Chang, C.-I., 2003, A comparative study for orthogonal subspace projection and constrained energy minimization. *IEEE Transactions on Geoscience and Remote Sensing*, vol. 41(6), pp. 1525–1529.

52. Harsanyi, J. C., Farrand, W., and Chang., C.-I., 1994a, Detection of subpixel spectral signatures in hyperspectral image sequences. In *Proceedings of the Annual Meeting, American Society of Photogrammetry & Remote Sensing*, Reno, NV, pp. 236–247.

53. Harsanyi, J. C., Farrand, W., Hejl, J., and Chang, C.-I., 1994b. Automatic identification of spectral endmembers in hyperspectral image sequences. In *Proceedings of the International Symposium on Spectral Sensing Research (ISSSR)*, San Diego, CA, 10–15 July, pp. 267–277.

54. Ren, H., Du, Q., Chang, C.-I., and Jensen, J. O., 2004, Comparison between constrained energy minimization based approaches for hyperspectral imagery. In *Proceedings of the IEEE Workshop on Advances in Techniques for Analysis of Remotely Sensed Data, 2003*, vol. 244(248), pp. 27–28.

55. Settle, J., 2002, On constrained energy minimization and the partial unmixing of multispectral images. *IEEE transactions on Geoscience and Remote Sensing*, vol. 40(3), pp. 720–721.

56. Chang, C.-I., Du, Q., Sun, T. S., and Althouse, M. L. G., 1999, A joint band prioritization and band correlation approach to band selection for hyperspectral image classification. *IEEE Transactions on Geoscience and Remote Sensing*, vol. 37(6), pp. 2631–2641.

57. Chang, C.-I., Du, Q., Chiang, S.-S., Heinz, D., and Ginsberg, I. W., 2001, Unsupervised sub-pixel target detection in hyperspectral imagery. In *Proceedings of the SPIE Conference on Algorithms for Multispectral, Hyperspectral and Ultraspectral Imagery VII*, 20–24 April, Orlando, FL, pp. 370–379.

58. Harsanyi, J. C., 1993, *Detection and Classification of Subpixel Spectral Signatures in Hyperspectral Image Sequences, Department of Electrical Engineering*. Baltimore County, MD: University of Maryland.

59. Bioucas-Dias, J. M. and Figueiredo, M. A. T., 2010, Alternating direction algorithms for constrained sparse regression: Application to hyperspectral unmixing. In *Proceedings of the 2nd Workshop on Hyperspectral Image and Signal Processing: Evolution in Remote Sensing (WHISPERS)*, 14–16 June 2010, vol. 1(4), pp. 14–16.

60. Iordache, M.-D., Bioucas-Dias, J. M., and Plaza, A., 2011, Sparse unmixing of hyperspectral data. *IEEE Transactions on Geoscience and Remote Sensing*, vol. 49(6), pp. 2014–2039.

61. Eckstein, J. and Bertsekas, D., 1992, On the Douglas-Rachford splitting method and the proximal point algorithm for maximal monotone operators. *Mathematical Programming*, vol. 5, pp. 293–318.

62. Gabay, D. and Mercier, B., 1976, A dual algorithm for the solution of non-linear variational problems via finite-element approximations. *Computer and Mathematics with Applications*, vol. 2, pp. 17–40.

63. Chen, S., Donoho, D., and Saunders, M., 1995, Atomic decomposition by basis pursuit. *SIAM Review*, vol. 43(1), pp. 129–159.

64. Iordache, M. D., Bioucas-Dias, J. M., and Plaza, A., 2012, Total variation spatial regularization for sparse hyperspectral unmixing. *IEEE Transactions on Geoscience and Remote Sensing*, vol. 50(11), pp. 4484–4502.

65. Chambolle, A., 2004, An algorithm for total variation minimization and applications. *Journal of Mathematical Imaging and Vision*, vol. 20(1/2), pp. 89–97.

66. Lawson, C. L. and Hanson, R. J., 1995, *Solving Least Squares Problems*, Philadelphia, PA: SIAM.

67. Haskell, K. H. and Hanson, R. J., 1981, An algorithm for linear least squares problems with equality and nonnegativity constraints generalized. *Mathematical Programming*, vol. 21, pp. 98–118.

68. Rumelhart, D. E., Hinton, G. E., and Williams, R. J., 1986, Learning representations by back-propagating errors. *Nature*, vol. 323, pp. 533–535.

69. Chander, G., Markham, B. L., and Helder, D. L., 2009, Summary of current radiometric calibration coefficients for Landsat MSS, TM, ETM+, and EO-1 ALI sensors. *Remote Sensing of Environment*, vol. 113, pp. 893–903.

70. Johnson, L. and Trout, T., 2012, Satellite-assisted monitoring of vegetable crop evapotranspiration in California's San Joaquin Valley. *Remote Sensing*, vol. 4, pp. 439–455.

71. Masek, J. G., Vermote, E. F., Saleous, N., Wolfe, R., Hall, F. G., Huemmrich, F., Gao, F., Kutler, J., and Lim, T. K., 2006, A Landsat surface reflectance data set for North America, 1990–2000. *IEEE Geoscience and Remote Sensing Letters*, vol. 3(1), pp. 68–72.

72. Vermote, E. F., Tanré, D., Deuzé, J. L., Herman, M., and Morcrette, J. J., 1997, Second simulation of the satellite signal in the solar spectrum: An overview. *IEEE Transactions on Geoscience and Remote Sensing*, vol. 35, pp. 675–686.

73. Small, C. and Milesi, C., 2013, Multi-scale standardized spectral mixture models. *Remote Sensing of Environment*, vol. 136, pp. 442–454.

74. Small, C., 2004, The Landsat ETM plus spectral mixing space. *Remote Sensing of Environment*, vol. 93, pp. 1–17.

75. Nascimento, J. M. P. and Dias, J. M. B., 2005, Vertex component analysis: A fast algorithm to unmix hyperspectral data. *IEEE Transaction on Geoscience and Remote Sensing*, vol. 43(4), pp. 898–910.

76. Lambin, E. F., Turner, B. L., Geist, H. J., Agbola, S. B., Angelsen, A., Bruce, J. W. et al., 2001, The causes of land-use and land-cover change: Moving beyond the myths. *Global Environmental Change*, vol. 11, pp. 261–269.

77. Forster, B. C., 1985, An examination of some problems and solutions in monitoring urban areas from satellite platforms. *International Journal of Remote Sensing*, vol. 6, pp. 139–151.

78. Lu, D. and Weng, Q., 2004, Spectral mixture analysis of the urban landscape in Indianapolis with Landsat ETM+ imagery. *Photogrammetric Engineering and Remote Sensing*, vol. 70, pp. 1053–1062.

79. Xian, G. and Crane, M., 2005, Assessments of urban growth in the Tampa Bay watershed using remote sensing data. *Remote Sensing of Environment*, vol. 97, pp. 203–215.

80. Small, C. and Lu, J. W. T., 2006, Estimation and vicarious validation of urban vegetation abundance by spectral mixture analysis. *Remote Sensing of Environment*, vol. 100, pp. 441–456.

81. Sabol, D. E., Adams, J. B., and Smith, M. O., 1992, Quantitative subpixel spectral detection of targets in multispectral images. *Journal of Geophysical Research*, vol. 97(E2), pp. 2659–2672.

82. Tompkins, S., Mustard, J. F., Pieters, C. M., and Forsyth, D. W., 1997, Optimization of endmembers for spectral mixture analysis. *Remote Sensing of Environment*, vol. 59, pp. 472–489.

83. Okin, G. S., Roberts, D. A., Burray, B., and Okin, W. J., 2001, Practical limits on hyperspectral vegetation discrimination in arid and semiarid environments. *Remote Sensing of Environment*, vol. 77, pp. 212–225.

84. Ren, H. and Chang, C.-I., 2000, Target-constrained interference-minimized approach to subpixel target detection for hyperspectral images. *Optical Engineering*, vol. 39(12), pp. 3138–3145.

85. Scharf, L. L., 1991, *Statistical Signal Processing*. Reading, MA: Addison-Wesley.

86. Heinz, D. C. and Chang, C.-I., 2001, Fully constrained least squares linear spectral mixture analysis method for material quantification in hyperspectral imagery. *IEEE Transactions on Geoscience and Remote Sensing*, vol. 39(3), pp. 529–545.

87. Bro, R. and Jong, S. D., 1997, A fast nonnegativity-constrained least squares algorithm. *Journal of Chemometrics*, vol. 11(5), pp. 393–401.

88. Chang, C.-I. and Heinz, D. C., 2000. Constrained subpixel target detection for remotely sensed imagery. *IEEE Transactions on Geoscience and Remote Sensing*, vol. 38(3), pp. 1144–1159.

89. Shi, Z., Zhai, X., Borjigen, D., and Jiang, Z., 2011, Sparse unmixing analysis for hyperspectral imagery of space objects. In *Proceedings of the International Symposium on Photoelectronic Detection and Imaging 2011: Space Exploration Technologies and Applications, Proceedings of SPIE*, John C., Zarnecki, Carl A. Nardell, Rong Shu, Jianfeng Yang, Yunhua Zhang (Eds.), vol. 8196, p. 81960Y-1.

90. Shi, Z., Tang, W., Duren, Z., and Jiang, Z., 2014, Subspace matching pursuit for sparse unmixing of hyperspectral data. *IEEE Transactions on Geoscience and Remote Sensing*, vol. 52(6), pp. 3256–3274.

91. Zhu, F., Wang, Y., Xiang, S., Fan, B., and Pan, C, 2014, Structured sparse method for hyperspectral unmixing. *ISPRS Journal of Photogrammetry and Remote Sensing*, vol. 88(2014), pp. 101–118.

92. Chang, C.-I. and Ren, H., 2000, An experiment-based quantitative and comparative analysis of target detection and image classification algorithms for hyperspectral imagery. *IEEE Transactions on Geoscience and Remote Sensing*, vol. 38(2), pp. 1044–1063.

93. Heylen, R., Burazerovic, D., and Scheunders, P., 2011, Fully constrained least squares spectral unmixing by simplex projection. *IEEE Transactions on Geoscience and Remote Sensing*, vol. 49(11), pp. 4112–4122.

94. Iordache, M. D., Bioucas-Dias, J. M., and Plaza, A., 2014, Collaborative sparse regression for hyperspectral unmixing. *IEEE Transactions on Geoscience and Remote Sensing*, vol. 52 (1), pp. 341–354.

95. Pu, H., Xia, W., Wang, B., and Jiang, G-M., 2014, A fully constrained linear spectral unmixing algorithm based on distance geometry. *IEEE Transactions on Geoscience and Remote Sensing*, vol. 52(2), pp. 1157–1176.

96. Zhang, B., Zhuang, L., Gao, L., Luo, W., Ran, Q., and Du, Q., 2014, PSO-EM: A hyperspectral unmixing algorithm based on normal compositional model. *IEEE Transactions on Geoscience and Remote Sensing*, vol. 52(12), pp. 7782–7792.

97. Song, C., 2005, Spectral mixture analysis for subpixel vegetation fractions in the urban environment: How to incorporated endmember variability? *Remote Sensing of Environment*, vol. 95, pp. 248–263.

98. Powell, R. L., Roberts, D. A., Dennison, P. E., and Hess, L. L., 2007, Sub-pixel mapping of urban land cover using multiple endmember spectral mixture analysis: Manaus, Brazil. *Remote Sensing of Environment*, vol. 106, pp. 253–267.

99. Roberts, D. A., Gardner, M., Church, R., Ustin, S., Scheer, G., and Green, R. O., 1998, Mapping chaparral in the Santa Monica Mountains using multiple endmember spectral mixture models. *Remote Sensing of Environment*, vol. 65, pp. 267–279.

9

Semantic Interoperability of Long-Tail Geoscience Resources over the Web

9.1 Introduction.. 175
9.2 Scientific Long-Tail Web Resources 176
   Characteristics of the Scientific Data and Models • Considerations for Web Resources
9.3 Interoperability of Resources: Semantic Web 180
   Typologies of Interoperability • Semantic Web: Linked Data and Web Services
9.4 Semantic Interoperability in Geoscience 183
   Semantic Heterogeneity Constrain • Semantic Interoperability Initiatives in Geosciences • Semantic Annotation • Finding and Connecting Scientific Web Resources • Semantic Knowledge Discovery
9.5 Geosemantic Framework 191
   Overview of Functionalities • Architecture

Mostafa M. Elag

Praveen Kumar

Luigi Marini

Scott D. Peckham

Rui Liu

9.1 Introduction

Scientists today have to manipulate resources such as data and models, which originate and reside in multiple autonomous and heterogeneous repositories over the Web. Several resource management systems have emerged within geoscience communities for sharing long-tail data, which are collected by individual or small research groups, and long-tail models, which are developed by scientists or small modeling communities. While these systems have increased the availability of resources within geoscience domains, deficiencies remain due to heterogeneity in the methods that are used to describe, encode, and publish information about resources over the Web. This heterogeneity limits our ability to access the right information in the right context so that it can be efficiently retrieved and understood by humans and machines—without the scientist's mediation. A primary challenge of the Web today is the lack of semantic interoperability among the massive number of resources that already exist and are continually being generated at rapid rates. The Semantic Web (SW) holds the promise to build a Web-scale topology of linked and interoperable resources, which will allow users to search simultaneously across many different and distributed information structures. This chapter focuses on defining the long-tail concept in the context of scientific resources, identifying the role of the SW in increasing the interoperability of long-tail resources, analyzing the reasons for their semantic heterogeneity, and introducing the design and architecture of a Geosemantic framework for addressing the semantic heterogeneity challenges associated with the interoperability of these long-tail resources.

9.2 Scientific Long-Tail Web Resources

Data availability has seen a dramatic increase in the last two decades across different domains such as business, art, science, and engineering. Early websites have been used to publish data for particular scientific communities or projects with no established metadata standards that describe, locate, and manage data resources [1], especially in distributed network environments such as the Web. Therefore, most of the data on the Internet were not easily discoverable or reusable and lacked the information required to facilitate their synthesis or integration with models for scientific analysis.

The Internet's rapid growth has resulted in a huge increase in information generated and shared by scientific communities. This information explosion has created an equally huge, but unmet, need for tools and approaches to make geoscience resources fully interoperable and reusable over the Web. Indeed, recent transdisciplinary research initiatives within many scientific communities depend on the synthesis of data from multiple resources, at multiple scales, and across scientific disciplines. We are now at a point where our ability to collect data far outstrips our capabilities to effectively analyze it using existing technologies, and where the inadequacy of tools available for describing and sharing data leads to heterogeneity in the way data are organized, described, and encoded.

Scientists are now encouraged to share data and models (henceforth referred to as resources) over the Web, including an unstructured and uncurated collection of spreadsheets, documents, images, numerical models, etc. Unstructured collections is one of the major challenges in information technology because it is not straightforward to extract information out of these collections and transform it into actionable knowledge [2]. The file system model is considered an unstructured data model, which provides a chance to preserve resources that cannot be constrained by a schema such as curation of models in Web repositories. Large collections of distributed and heterogeneous Web resources are often referred to in scientific communities as long-tail resources [3]. Discoverability of these resources is defined as the ability to navigate within the unstructured content of resources and track their relationships.

Due to the ubiquity of the World Wide Web, searching for data that fit a particular model seems much easier today than it has ever been. A user can search the Web for suitable data, which on many occasions results in multiple hits for data or the repositories that curate these data. Unfortunately, despite a large number of potential hits, we are often frustrated by the lack of quality in the search outcomes, which are not often well-aligned with our search goals. Searching is one of the most popular applications on the Web that is based on the occurrence of a specific object (e.g., a word in a document) or availability of a specific attribute (e.g., collected by a specific organization). In the context of the Web, advanced search engines like Google augment the information retrieval mechanism by storing information about the Web page structure and content, which allows retrieval of Web pages that are better related to search criteria. The availability of large amounts of structured information about a wide range of unstructured Web pages provides an emerging solution for treating unstructured data over the Web [4].

Creating a well-structured and descriptive information profile that follows a standard schema for each scientific resource is the key for indexing and linking resources over the Web. An information profile defines the metadata associated with a Web resource based on a metadata schema. The quality of information profiles varies and ranges from standard information that is used by large data centers to ad hoc labeling that is used by small research groups. Publishing information profiles for resources over the Web has become one of the focus areas for many research communities (e.g., [5–7]). However, only 5% of the scientific resources are stored using standard information schemas [8]. It is rare to find a scientific resource uploaded to the Web with descriptive information. Incomplete information about resources results in inconsistency in vocabularies pertaining to names of variables, units, spatial characterization, query formats, retrieval methods, and Web transfer protocols [9]. Lack of information makes automating the integration of scientific resources over the Web particularly challenging.

A primary challenge that hinders the seamless integration of resources over the Web is the semantic heterogeneity among resources that span across various geoscience disciplines. Semantics represents the agreement among resources and tools on the meaning of terms and concepts that are used to describe the characteristics of a resource. For example, use of the Dublin Core identifier "dc: creator" to define the relationship between a creator *X* and a document *Y* allows the machine to automatically understand this relationship. SW technologies rely on identifying the relationships among resources using standards such as Dublin Core to promote the accessibility and reusability of resources by Web applications. SW allows a machine or a Web application to understand the information profile of a resource and do sophisticated inference processing [10]. Application of the SW technologies requires structured information about each resource that describes its content, behavior, function, and relationships with other resources [1]. Therefore, there is an immediate need for techniques and methods to create this standard information profile for each resource and allow the synthesis of the existing information profiles.

9.2.1 Characteristics of the Scientific Data and Models

Currently, there are tremendous amounts of scientific data available online [8]. As illustrated in Figure 9.1, the relationship between the volume of resources and their dispersion represents a proxy for the variability and heterogeneity of information flows across geosciences disciplines. The left side of the curve represents the "large" or "big" data, which are usually characterized by being relatively more homogeneous, well-defined, continuously maintained, and easy to reuse, such as remote sensing data produced by NASA. On the right side of the curve, individual researchers and small groups provide a large variety of scientific data. Two data reuse patterns can be identified in this graphic: (1) large scientific agencies produce standardized data that are self-descriptive and easy to reuse and (2) small research groups individually produce a small volume of data, which is often complex and harder to (re)use. Scientists have used the term "long tail" [8] to indicate the lower quantity but higher complexity of available data, which is usually unstructured, uncurated, harder to find, and less frequently reused. "Dark," "Gray," and "Wide" are synonyms for long-tail, which reflects that these data cover a broad range of the scientific data production and are currently underused [8].

Often, individual scientists and small groups collect long-tail data to address specific scientific issues that usually have limited geographic or temporal range [11,12]. However, a large number of such collections together constitute a large database that is of immense value to the scientific community. Such data are

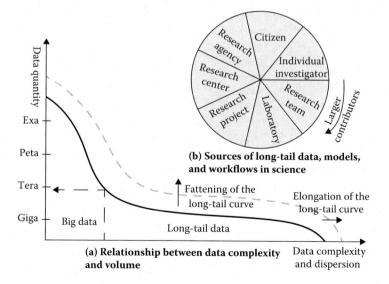

(b) Sources of long-tail data, models, and workflows in science

(a) Relationship between data complexity and volume

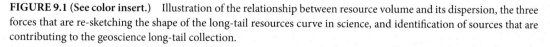

FIGURE 9.1 (See color insert.) Illustration of the relationship between resource volume and its dispersion, the three forces that are re-sketching the shape of the long-tail resources curve in science, and identification of sources that are contributing to the geoscience long-tail collection.

complex in that they encompass a heterogeneous collection with many dimensions, coordinate systems, scales, variables, providers, users and scientific contexts [11,13,14] (Figure 9.2). Long-tail data are often relegated to hard drives or servers, which are generally only available on local area networks. Sometimes they are published on the Internet in less formal ways and are contained in Web servers as files in directories or curation Web services. In a similar vein, we coin the term "long-tail models" to characterize a heterogeneous collection of models and/or modules developed for targeted problems by individuals and small groups, which together provide a large valuable collection [9,15–18] (Figure 9.2). Such models are also complex in that they incorporate differing variable names and units for the same concept, run at different time steps, use differing naming and reference conventions (e.g., angles with respect to east or north), etc. The ability to integrate long-tail models with long-tail data across the geoscience field will provide a transformative opportunity for the community where not only can models be coupled, but it will also be possible to discover and use data in model application contexts of space, time, and scientific questions. Harnessing the complex, heterogeneous, and extensive set of distributed resources is essential to represent, understand, predict, and manage heterogeneous but interconnected Earth system processes.

Three drivers are reshaping the distribution of long-tail resources curve in science (Figure 9.1): (1) advancements in resource production tools, (2) availability of resource distribution technologies over the Internet, and (3) multidisciplinary research collaborations. Advances in computational, sensing, information, and modeling technologies allow more scientists to contribute more resources. This high production rate elongates the curve of the long-tail scientific resources (horizontal expansion in Figure 9.1). The availability of easy and accessible digital storage, publication, and curation repositories provides scientists with the tools and environments to share, distribute, and reproduce more resources. For example, Data Observation Network for Earth (DataOne, https://www.dataone.org/what-dataone) enhances the discoverability of well-described Earth observation data over the Web, and Sustainable Environment Actionable Data (SEAD, http://sead-data.net/) provides data services to manage and preserve long-tail data. In addition, multidisciplinary collaboration advances the sharing and reuse of available data, models, and tools, which leads to resource proliferation. These two forces fatten the long-tail curve of scientific resources (vertical expansion in Figure 9.1).

Driving motivations for advancing the interoperability of resources over the Web are (1) increasing the productivity of scientists and research groups, (2) repurposing qualified resources for new research objectives, (3) providing flexibility for interdisciplinary research and collaboration, and (4) enhancing the

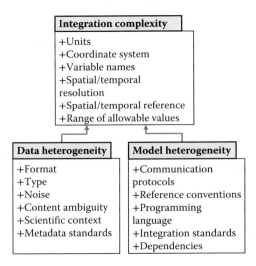

FIGURE 9.2 Classification of heterogeneity elements that are common in data and modeling communities. These types of heterogeneity hinder the development of a unified information system for describing resources. The complexity of using data by models is characterized in how to preprocess data to be suitable for model requirements.

quality of available resources. Advances in the interoperability of scientific resources will maximize the use of our extensive but widely dispersed data and information.

9.2.2 Considerations for Web Resources

One of the key considerations for long-tail resources is their immense value for the advancement of science. These data are irreplaceable and expensive to reproduce [3,13]; they are generated from various sources such as scientific experiments, field observations, model simulations, or derived from other data sets. In many cases, scientists will not be able to reproduce long-tail data because they record a unique event. On the other hand, a long-tail model is typically used to simulate specific physical, chemical, biological, or other processes. Such models are valuable because scientists spend time and effort as well as collaborate to develop, test, validate, and integrate models into different integration workflows. The content of the same model may vary from one project to another by adding or removing a piece of a subroutine, or changing the assumptions and conditions to fit the new project needs. This continuous change in the models' content minimizes their reusability. Therefore, to ensure the maximal use of a model, it is necessary to describe its content and operational requirements in a way that a computer can understand.

Reproducibility of results coming from the analysis of data or models is another consideration for long-tail resources. Independent scientists, that is, scientists who are not involved in the production of resources but use them, should be able to reproduce the same results if they retrace the production workflow steps [19]. Currently, most of the resources' curation effort focuses on the physical preservation of resources more than linking existing and new resources. An advanced preservation of the products should aim to preserve the resources in a Web-accessible format, allow their connectivity to related scientific activities, and store the log of changes that are associated with the resource. Reproducibility requires a machine-understandable description of the relationships between resources and conditions that defined their production workflow [19].

Confidence in the quality of information used to describe a resource impacts its reusability. This reason leads to an important consideration for resource curation: How do we identify the reliability of the information that is used to describe a resource? The reliability of information used for resource description depends on using standard information protocol to define resources and preserving the changes in their content, especially if many users are contributing to the resource definition [20]. Long-tail resources are poorly curated and follow different information models, which decreases their reliability [3,13]. Additionally, if we assume that users are willing to use a standard information schema to describe their resources, the heterogeneity between information standards minimizes reusability of resources. Heterogeneity between schemas minimizes confidence in the quality of hits returned by search engines, which to a lesser extent decreases reusability across geoscience disciplines. Many initiatives in the geoscience community have proposed a wide range of solutions, including crosswalks, translation algorithms, information registries, and specialized data dictionaries to achieve interoperability between equivalent standards. Yet despite some genuine advances, it is still not easy to identify the common elements in different information standards to allow mapping between information standards [21].

Building a flexible information system that is capable of increasing the interoperability of scientific resources requires (1) creation of tools for associating descriptive information with preserved research outcomes, (2) allowance of crosswalks between related information standards, (3) leveraging the SW technologies to organize resources over the Web and publish their contextual relationships, and (4) provision of a low-barrier technology for scientists who are not experts in information systems to contribute their information or update the existing information. Long-tail resources are important in science, but they are not meaningful until they became interoperable among related scientific disciplines. Interoperable preservation of resources over the Web will allow scientists to mine data or incorporate models in their studies. The SW technologies will allow scientists to connect the current silos of distributed resources, which have nonuniform curation systems. This will allow users to easily find resources related to their search context with higher precision compared to the current status.

9.3 Interoperability of Resources: Semantic Web

ISO/IEC 2382-01, Information Technology Vocabulary, defines interoperability as "the capability to communicate, execute programs, or transfer data among various functional units in a manner that requires the user to have little or no knowledge of the unique characteristics of those units." Interoperability allows scientists and scientific communities to share and reuse their resources and connect distributed systems. It minimizes the need for building and maintaining large and costly warehouses to hold data; that is, it will minimize the need for a centralized approach to managing distributed resources. Interoperability is a core concept for enabling the decentralization of resources over the Web, where each resource has the information required to make it self-descriptive and machine processable. With an appropriate level of interoperability, data can be harvested in real time from distributed information systems. Interoperability represents perhaps the most significant paradigm shift in how resources and information are managed and utilized since the emergence of the Internet [22].

The Web provides a seamless environment where users can navigate through published resources. There is a strong contrast between the reality of the seamless Web and the prevailing culture of information dissemination over the Web. Usually, scientific communities manifest their resources individually on the Web using information models, which only satisfy their domain requirements and do not integrate with the universal standards. Therefore, there is a need for a new perspective that facilitates communication among resources that are curated in different servers and fit the needs of the Web environment. In the following sections, we will elaborate on the different interoperability levels and explain the role of the SW to achieve interoperability among geoscience resources.

9.3.1 Typologies of Interoperability

The typologies of interoperability between information systems are classified based on the interpretation level of data and information exchanged between the information systems [23]. The levels of conceptual interoperability model (LCIM) defines six levels of interoperability between systems [23,24]. This model can be used to characterize the interoperability problem among scientific resources over the Web. The interoperability levels vary from no connection to conceptual interoperability (Table 9.1). We suggest adding the automatic level of interoperability on top of these existing characterizations. In this level, the systems are conceptually interoperable, and the contents of the data being exchanged are error-free (e.g., no noise,

TABLE 9.1 Level of Conceptual Interoperability Model (LCIM)

Level	Interoperability definition	Description
L7	Automatically interoperating	Systems are conceptually interoperable, and they are exchanging objects' content that is error-free.
L6	Conceptually interoperating	Systems are completely aware of each other's information, processes, contexts, and modeling assumptions.
L5	Dynamically interoperating	Systems can reorient information consumption based on understood changes to meaning, due to changing context.
L4	Pragmatically interoperating	Systems are aware of the context and meaning of exchanged information.
L3	Semantically interoperating	Systems are exchanging a set of terms that they can semantically parse.
L2	Syntactically interoperability	Systems have a protocol to exchange the right forms of data in the right order, but the meaning of data elements is not established.
L1	Technically interoperating	Systems can exchange data.
L0	No	NA

Source: Wenguang Wang et al., *Proceedings of the 2009 Spring Simulation Multiconference*, Society for Computer Simulation International, p. 168, 2009.

no missing values, no anomalous values, no values out of expected bounds, etc.). Promotion of the interoperability between two systems from one category to another requires fulfilling the requirements of the previous category. Currently, we can argue that most of the geoscience systems have achieved the syntactic interoperability level (L2) over the Internet, where they can exchange information. Systems that are syntactically interoperable use a predefined data format and/or schema to communicate with each other. The lower bound of L2 is using a predefined data format, which is known as structural communication between systems. Interaction at this bound relies on the data format, and any change in the format breaks the systems' interoperability. The upper bound of L2 depends on storing data in commonly accepted and agreed-upon information systems (e.g., relational database and/or markup languages). An agreement on the data syntax is defined between the syntactically interoperable systems. Syntactic interoperability is a prerequisite for semantic interoperability.

Semantic interoperability refers to the ability of a resources management system to understand the information models of related systems. Semantics is at the center of Web intelligence, as it intricately defines the way in which information is exchanged, classified, and conceptualized. Semantics studies the meaning of words and represents their interpretation based on the context of use. It focuses on the conception of meaning—how the meaning of an element is constructed, interpreted, and illustrated. Since we express our resources and information with human language, we need to understand the sources of semantic heterogeneity. The focus here is on the interpretation of vocabularies and consolidating their definitions across related scientific domains over the Web. Advancing the semantic interoperability among resources is the gateway to promoting their connectivity, reusability, and discoverability over the Web.

9.3.2 Semantic Web: Linked Data and Web Services

Two standards are the core of the World Wide Web: the Hypertext Transfer Protocol (HTTP), which is used as the communication protocol [25,26], and the Hypertext Markup Language (HTML), which is the format for content representation used by Web browsers. HTML is used to semantically mark up the structure of a Web page to provide cues for the visual presentation of the content included in the page. It was designed for human consumption through mediation by user agents such as the Web browser, but it is not optimal for machine consumption [27].

Web services were introduced to characterize machine-to-machine communication built over HTTP [28]. A set of standards has been developed by the World Wide Web Consortium (W3C) working groups over time to describe machine processable formats and communications protocols such as the Web Services Description Language (WSDL) and the Simple Object Access Protocol (SOAP). Over time, some of these standards have fallen out of popularity because of their complexity and have been replaced by simpler software architecture styles such as Representational State Transfer (REST); conceptually it is built upon the same principles of HTTP and provides methods to facilitate the development of Web services. Most RESTFul web services are written as plain HTTP services using eXtensible Markup Language (XML) or JavaScript Object Notation (JSON) for content representation. XML was developed by the W3C for data exchange over the Web [29]. It is a flexible, self-describing markup language format and provides a hierarchical syntax structure for encoding data and their relationships. However, using XML can produce syntactic heterogeneity because data can be organized in many different ways. JSON is a lightweight interchange format native to the Web, easy for both humans and machines to read and write. Although Web application programming interface (API) can refer to both types, over time it is becoming synonymous with the RESTFul approach. Both RESTFul and API use simple protocols, but they have shortcomings in self-description methods, and built-in ways to support interoperation among services. Today many resources are available through Web API, but the format in which these resources are made available and described varies greatly.

The SW movement led by W3C attempts to provide methods for associating semantics with data, to overcome Web APIs' heterogeneity and the lack of standards for embedding structured data in HTML. SW technologies provide standards and languages to describe data, rules for interpretation of information associated with data, and methods to include external rules from related knowledge representation systems [10]. One family of specifications is the Resource Description Framework (RDF) data model, which is designed to provide conceptual descriptions and an information model for Web resources. Similar to the entity-relationship model popular in relational databases, RDF is based on the idea of making statements about Web resources in the form of *subject–predicate–object* triples. RDF represents resources as graph nodes and their associated properties as labeled arcs. RDF subjects and predicates have to be defined as Uniform Resource Identifier (URIs), while objects can be URIs or literal strings. This property makes every RDF statement globally identified and, as a result, more easily portable between databases. It helps to avoid collisions and mismatches when trying to aggregate and normalize data from different sources. Furthermore, RDF does not require the definition of a schema before creating new statements. This enables a flexibility that is not typical of relational databases.

RDF in Attributes (RDFa) [30] is another attempt at solving the problem of embedding machine-readable data within HTML using RDF. It provides markup attributes to augment existing markup with machine-readable hints. RDFa follows the lightweight approach of *microformats* to the semantic markup of HTML pages in such a way that the content of a Web page is readable by humans and machines alike. Although many *microformats* exist, RDFa includes the advantages of RDF and the wider SW echo system. RDFa is also the basis on which other *microformats* are built, such as the Open Graph protocol [31] created by Facebook to make it easier for Facebook to include external pages in its social graph. Efforts by search companies such as *schema.org* exist to standardize *microformats* for search engine retrieval that include both new standards, *microformats* in this case, and the adaption of existing standards (RDFa).

The Linked Data paradigm emerged in the context of SW technologies for publishing and sharing data over the Web. It connects related individual Web resources in a graph database, where resources represent the graph nodes, and an edge connects a pair of nodes. Generally, a graph has (1) global metrics, which describe the characteristics of the graph as a whole such as graph diameter and number of fully connected subgraphs, and (2) individual analysis, which seeks to establish relationships between individuals based on their information profiles, such as creating a coauthor relationship between scientist (data nodes) if they publish a journal article. The Linked Data approach allows a machine to navigate among resources by tracing the edges that correspond to search criteria. It aims to link Web resources together seamlessly to enable structured and well-described information to transfer between users and allow the machine to understand the content and context of a Web resource. Implementation of the Linked Data paradigm to organize scientific resources has a great potential to increase their interoperability.

Publishing and linking scientific resources using SW technologies require that the user community follows the three Linked Data principles (Figure 9.3). First, each resource needs to be represented using a unique Uniform Resource Identifier (URI), which consists of (1) a Uniform Resource Locator (URL) to

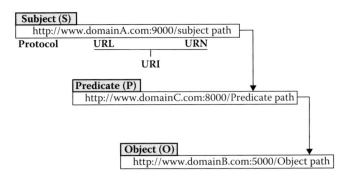

FIGURE 9.3 Representation of the Linked Data principles: a unique *URI* for each Web resource, relationships between resources [represented in triple format (S,P,O)], and *HyperText Transfer Protocol (HTTP)* (as universal access mechanism for resources). An example demonstrates the relationship between a subject that is stored in server A, an object that is stored in server B, and a predicate that is stored in server C.

define the server path over the Web, and (2) a Uniform Resource Name (URN) to describe the exact name of the resource. Second, the relationships between resources are described using the triple format, where a subject (S) has a predicate (P) with an object (O). A predicate is either an undirected relationship (bidirectional), where it connects two entities in both ways or a directed relationship (unidirectional), where the presence of a relationship between two entities in one direction does not imply the presence of a reverse relationship. The triple format is the structure unit for the Linked Data system. Finally, the HTTP is used as a universal access mechanism for resources on the Web [32].

For machine-to-machine communication, patterns used by SW technologies have evolved over time to be organized under the Linked Data approach (Figure 9.3). This approach provides best practices for building Web services using RDF, HTTP, and URIs. Relationships between URIs allow services to interpret information that is defined by distributed services. Accessing a semantic object allows the Web agents to navigate between services using HTTP as the communication protocol. This is similar to a user navigation between links in any Web browser. Machines can now follow a link in the data and rely on a global knowledge graph, instead of using the semantics of specific service.

The graph-oriented nature of RDF also helps in building framework-based knowledge representation and ontologies. The Web Ontology Language (OWL) [32,33] and RDF Schema (RDFS) are two knowledge-base representations built on RDF that provide basic ways to define ontologies used to structure RDF graphs. They provide a mechanism for defining internal relationships between properties and concepts [34] and allow the creation of engines for reasoning for RDF data. They are machine-readable languages that were designed to process information on the Web instead of just representing information for human use [35]. RDFS extends the RDF approach for representing data by introducing means to model classes, properties, and hierarchies of classes and properties [35]. RDFS is considered to be a lightweight ontology language that can be used for creating taxonomies [35]. RDFS provides vocabularies to describe relationships between classes, relationships between properties, and relationships between classes and properties. However, it does not provide a mechanism for defining internal relationships between properties and concepts (e.g., explicit cardinalities in properties relationships and union in classes) [34]. OWL is the standard language for representing knowledge on the Web. It is a machine-readable language that was designed to process information on the Web instead of just representing information to human users [35]. OWL can be used to express more sophisticated relationships between terms, compared to RDFS. OWL can explicitly represent the meaning of terms in vocabularies and relationships. OWL is recommended by the W3C as an ontology language. It is XML-based, applies RDF syntax, and is compatible with most querying languages [22,32,34].

9.4 Semantic Interoperability in Geoscience

Migration from the Web to SW encourages scientists to adapt SW technologies in their research, such as running software through a Web browser and combining services with data. For example, a new paradigm emerged to use models as a scaffold to integrate data sources on different spatiotemporal resolutions (e.g., [12,36]). Meanwhile, this migration has provided a new level of well-structured information, which requires the advancement of the current information retrieval mechanisms or the development of new mechanisms [37]. This section analyzes the sources of semantic heterogeneity over the Web, provides examples of the main semantic interoperability initiatives in geoscience, and briefly summarizes our understanding of the role of semantic enrichment in advancing resource discovery approaches.

9.4.1 Semantic Heterogeneity Constrain

Semantic heterogeneity occurs when there is a disagreement about the meaning of the label used to describe a resource or one of its parts, ambiguity about the interpretation of the content of a resource, or its usage pattern among related parties. Among the scientific communities, four sources of semantic heterogeneity can be identified: (1) *Structural*—that is, the syntax of the language used to describe and code the resources, including the natural language, programming languages, and data encoding languages; (2) *Content*—that is, the method and vocabularies used to describe the elements within each resource, including the naming process and missing information inside a file; (3) *Conceptual*—that is, the heterogeneity between related domains in the conceptualization of their information system, including heterogeneity in the structure of concepts, their level of granularity and relationships, and the terminologies that are used to describe these concepts; and (4) *Contextual*—that is, the mismatch between schemas used to map resources between information systems [38].

Figure 9.4 demonstrates the sources of semantic heterogeneity among resources and provides illustrative examples from the water resources domain. *Structural* heterogeneity, for instance, prevents integration between models that are coded in different programming languages. It begs the question—how to make a model usable by another programming language without the need to recode the model's content? *Conceptual* heterogeneity comes from the inappropriate use of the terms to describe a resource over the Web. A considerable percentage of semantic heterogeneity arises due to the inconsistency of concepts among related scientific domains. For example, adding a *water flow* annotation to a data set is not enough to make it self-descriptive for a machine. But, annotating the same data set with a standard name (SN), which inherits its definition from a domain ontology, can guide the machine to a proper consumption pattern of this data. *Content* heterogeneity refers to how the information is organized and described within each data file,

Structure

Encoding
Ingest encoding mismatch: ANSI vs. ASCII
Ingest encoding lacking
Query encoding mismatch
Query encoding lacking
Languages
Script mismatch:
Parsing/morphological: Arabic vs. Latin
Syntactical and semantics errors

Content

Naming
Case sensitivity: Uppercase vs. Camel case
Synonyms: Geo and earth
Acronyms: Watershed name vs. location
Homonyms: Watershed name and county
Misspellings: As stated
Missing or mismatch ID: *Truncated URIs*
Missing data: *As stated*
Element ordering: *As stated*

Contextual

Schematic discrepancy:
Element-value to element-label mapping
Attribute-value to element-label mapping
Attribute-value to attribute-label mapping
Scale or units
Measurement type: meter vs. feet
Units: m vs. cm
Precision: *5.01 vs. 5.00001*
Data representation
Primitive data type: literals vs. URIs
Data format: Type of delimitation

Conceptual

Naming
Case sensitivity: Uppercase vs.
Camel case
Synonyms: Runoff vs. surface runoff
Acronyms: Temperature vs. T
Homonyms: County and city name
Misspellings: As stated
Generalization/specialization:
"Velocity" is it air or water
Aggregation
Intra-aggregation: Clustering data based on attributes
Inter-aggregation: Exchanging data between systems
Internal path discrepancy
missing item
Content discrepancy: Velocity column in an excel file
Missing content: Velocity in different location within a watershed
Attribute list discrepancy: Difference in attribute completeness between data sets
Missing attribute: Multiple use of same attribute
Type mismatch: Absence of indices (tags) that refer to the related domains domain
Constraint mismatch: When attributes referring to the same thing have different cardinalities

FIGURE 9.4 Sources and types of semantic heterogeneity among scientific resources over the Web. (Adapted from AI3, adaptive information, adaptive innovation, and adaptive infrastructure, 2013. http://www.mkbergman.com/1825/why-clojure/.)

such as arrangement of a data column within a spreadsheet and the mismatch in symbols used to describe each column header.

9.4.2 Semantic Interoperability Initiatives in Geosciences

The problem of determining the semantic similarity or equivalence of data items across heterogeneous resources or collections has long been recognized within the geoscience community. In 1967, the U.S. Office of Naval Research and Department of Defense had already a project called TEST (Thesaurus of Engineering and Scientific Terms) [39], the results of which were published by the Engineers Joint Council (1969) in a book that can still be purchased online. Since then, there have been many efforts, across many different science domains, to introduce some level of semantic standardization. These efforts take many different approaches and are usually restricted to a fairly narrow scientific domain. Some are best described as Controlled Vocabularies (CVs), while others may be viewed as ontologies or data models.

CVs are essentially lists of standardized labels for object names, quantity or attribute names, and other terms that are important for a particular scientific domain. It provides unique, descriptive, and unambiguous labels for the entities that are contained within a given resource, and is simultaneously both human- and machine-readable. Most of these CVs were designed to be human-readable and have no intention of being machine-understandable or parsable. Usually, CVs use domain-specific jargon and concepts without further explanation. Many CV initiatives take a fairly unstructured approach, making no effort to categorize terms or break concepts into parts (e.g., objects, quantities, operations, adjectives, etc.). For example, there may be no distinction made between object names and quantity names, with both appearing in the same list. Sometimes an object name, like a chemical substance name, is meant to be understood as a particular quantity of that substance (e.g., mass concentration in water) without this being made explicit, that is, by relying on a human understanding of the scientific context.

9.4.2.1 CUAHSI Variable Names

In the domain of hydrology, the Consortium of Universities for the Advancement of Hydrologic Science, Inc. (CUAHSI, https://www.cuahsi.org/) *Controlled Vocabulary*—a set of 13 separate CV [40]—has been developed for describing quantities measured at a point over time, that is, time series data. One of these sets is the CUAHSI *VariableName*; it provides 614 standardized quantity names that support hydrologic observations. All but 136 of these names are for water chemistry and biota. Like many CVs, these are very much a work in progress and were designed to be extended by the hydrologic community. However, CUASHI names in no way span the breadth of quantities used in the domain of hydrology, including very few names for the hydrology-related disciplines such as snow hydrology, open channel flow, hydraulics, infiltration, evaporation, soil physics, glaciology, and hydrometeorology.

9.4.2.2 NASA Global Change Master Directory

In the geosciences, one of the first large efforts to create a CV began at NASA in 1995 and is called the NASA Global Change Master Directory (GCMD). The latest version (8.0) of the GCMD Science Keywords component consists of 2542 entries, organized hierarchically into earth science domains and subdomains and sometimes ending in a quantity name [41]. The GCMD continues to be influential and was part of the inspiration for the Semantic Web for Earth and Environmental Terminology (SWEET) ontology [42]. A goal of the SWEET ontology was to further organize earth science terminology by breaking the terminology used to describe a complex concept or quantity into distinct parts. For example, using SWEET, the single-string Climate and Forecasting Standard Name (CFSN) "tendency_of_mole_concentration_of_dissolved_inorganic_phosphorus_in_sea_water_ due_to_biological_processes" is stored as a so-called *multi-attribute* parameter name:

- *Property*: mole_concentration
- *Science_Processing*: tendency (i.e., time derivative)

- *State*: dissolved, inorganic
- *Substance*: phosphorous
- *Medium*: sea_water
- *Process*: biological_processes

However, the parts of a specific quantity name, which may refer to one or more objects, often does not map neatly to the attribute types available in SWEET. When one considers the wide variety of objects that can occur in a model or data set and the quantities associated with them, it seems unlikely that a general set of attribute types can be identified for classifying them beyond the more fundamental attributes of quantities such as objects, parts of objects, adjectives, base quantities, and mathematical operations.

9.4.2.3 Climate and Forecasting Standard Names

The CFSNs [43] are used in the domain of ocean and atmosphere modeling for labeling output variables from models that have been saved into NetCDF files [44]. This set of naming conventions extends and generalizes the Cooperative Ocean Atmosphere Research Data Service (COARDS) conventions that were introduced in 1995 [45]. Each name consists of a single string, with words separated by underscores, as in the previous SWEET example. While there are guidelines, there are no hard and fast rules for constructing CFSNs. Some names represent a quantity name while others contain object names, operations, assumptions, and other semantic constructs. The CFSNs contain terms for quantities that describe the ocean and atmosphere, along with a limited set of hydrologic terms. There is no intention of adding terms for other geoscience domains because these are considered out of scope. Although created for the purpose of uniquely labeling quantities stored in NetCDF files, the Earth System Modeling Framework (ESMF, http://www.earthsystemmodeling.org/) uses a subset of these names to support model coupling [46].

While these CFSNs are intended to be unambiguous and support mathematical operations (called transformations in CF, and applied to the whole name), they are not suitable for use as a domain-independent "*lingua franca*" for the geosciences. To be specific, CFSNs: (1) use domain-specific terminology (e.g., "tendency" instead of "time derivative"); (2) have complicated construction guidelines (vs. rules) that are not applied consistently; (3) do not include provisions for dimensionless numbers, mathematical and physical constants, empirical parameters, or reference quantities; (4) do not have a natural grouping (e.g., alphabetical) of related names; (5) include assumptions in the name using "assuming"; and (6) have a scope that is limited to atmosphere and ocean modeling.

There are currently 2514 CFSNs, with 183 new proposed names under review (some since 2010). The names have been grouped into categories with Atmospheric Chemistry as the largest category with 968 names. However, a large fraction of these result from selectively applying 8–26 quantity name patterns (e.g., *atmosphere mass content of X*) to each of about 90 chemical species (e.g., ammonia, benzene, carbon dioxide, toluene, xylene.) There are 661 names that include the *due to* construct to identify one possible source contribution to a given quantity. Other constructs that are meant to capture an assumption are *expressed as*, *assuming*, and *excluding* and are used in 197 names. Mathematical operations are supported but are used in very few names, except for *tendency of*, used in 690 names. Other name categories are Atmospheric Dynamics (146), Cloud (128), Hydrology (176), Ocean Dynamics (32), Radiation (179), Sea Ice (64), and Surface (229). The purpose of this summary is not to be overly critical of but rather to clarify the relatively narrow scope of these names, despite their number and widespread use.

9.4.2.4 Community Service Dynamic Modeling System Standard Names

The Community Service Dynamic Modeling System (CSDMS) Standard Names (CSNs) [47] were introduced in 2013 as an alternative to the CFSN with a somewhat similar purpose but were designed to overcome many of the shortcomings of the CFSN and of other CVs that are less structured and domain-specific. They were created to solve the semantic mediation problem that arises in the context of component-based, "plug-and-play" geoscience modeling. Different models—each with its own "internal vocabulary" of input and output variable names—map these internal variable names to the CSN in a "hub-and-spoke" approach where CSN serves as a domain-independent lingua franca. Since this mapping is made by the model developers as a key part of implementing a CSDMS basic model interface (BMI) for their model, it allows the CSDMS modeling framework to automatically and reliably connect a model that provides a given variable to other models that wish to use it in their computations.

CSNs are designed to (1) obey specific construction rules, (2) be domain independent, (3) be unambiguous, (4) distinguish between many different parts of the name, and (5) produce names that are both human- and machine-readable. Since CSN has structure and is constrained by rules, it provides a concise definition for a scientific quantity that appears in many CVs such as "heat capacity." However, "heat capacity" by itself is ambiguous because there are many different types of heat capacity. Some vocabularies include the term "specific heat capacity," but this is still ambiguous because it could be either a mass-specific (per unit mass) or volume-specific (per unit volume) heat capacity. It also matters whether the heat capacity is *isobaric* (associated with a constant pressure) or *isochoric* (associated with a constant volume). Only quantity names that clearly specify each of these attributes can be unambiguously mapped to a specific quantity name such as "volume_specific_isobaric_heat_capacity." This example is typical in that most familiar quantity names require one or more adjectives before they become unambiguous. Other common examples are compressibility, concentration, density, hardness, latitude, and viscosity.

The CSN conventions are based on a relatively small number of robust rules for constructing unambiguous quantity names that are both human- and machine-readable. Each name is a single string that contains two main parts: a chain of object names and a specific quantity name. The chain of object names begins with a general or container object, followed by a nested set of "subobjects" or parts of objects and ends with the specific object or object part to which the quantity name part applies (e.g., "automobile_engine_cylinder"). This leads to a natural alphabetical ordering that puts the different quantities that are used to describe a particular object close to one another. Each quantity name ends with a base quantity name, which may be thought of as a quantity type and often implies particular units. Both the quantity name and the base quantity name are drawn from two separated and carefully constructed CVs. Quantity names can also be prefixed with one or more mathematical operations, also drawn from a relatively small but robust *Controlled Vocabulary*, to generate a new quantity name with a narrower scope. Many quantity names consist of a process name (another CV) paired with a base quantity name, like "melt_volume_flux."

CSNs are constructed using five special delimiters that make it easy to decompose them into their parts. The object and quantity parts of names are separated by a *double underscore*. Separate words in either the object or quantity part of a name are separated by *single underscores*, unless the words need to be treated as a single object, in which case a *hyphen* is used to bind them, as in "carbon-dioxide." A *tilde* character is used to attach a chain of one or more modifiers or adjectives to each object or subobject in the object part of the name. These adjectives are attached to the right of the object name, which contributes to a natural alphabetical ordering of the names. The fifth delimiter is the word "of," which appears at the end of every operation name; this reflects a human speech pattern and is both human- and machine-parsable. Note that the object part of a name could include chemical compounds and these sometimes use commas and apostrophes.

The naming conventions of the CSN have been tested for broad applicability across science domains and result in names that are easily understood by scientists, but that are also easily machine-parsed with a simple Python script. Considerable research has gone into building these conventions and making sure that constructed names are unambiguous and consistent. There are currently over 2500 distinct variable names. Crosswalks between the CSN and other CVs, including the CFSNs, are being developed as part of the National Science Foundation (NSF) EarthCube project called Earth System Bridge. In addition to standardized *variable* names, the CSN also includes a large collection of standardized *assumption* names that can be used to describe and classify models.

Another interesting and ongoing project is Commonwealth Scientific and Industrial Research Organization (CSIRO) Spatial Information Services Stack Vocabulary Service (SISSVoc) [48], which is developing a Linked Data API for Simple Knowledge Organization System (SKOS) vocabularies. Like the CSNs, there is a focus on being able to accurately encode multiple names in a standard way as a composition of parts.

9.4.3 Semantic Annotation

We classified the semantic interoperability between resources over the Web into five classes based on the ability of one resource to programmatically reuse and understand the information model associated with another resource (Table 9.2). The Interoperable class includes resources that follow global metadata standards (e.g., Dublin Core). Reusing this type of resources is straightforward and usually can be done programmatically. The Semi-Interoperable class defines the interoperability between a resource that follows a global standard and another one that compiles to domain-level standards, henceforth described as partially-standardized resources. Semantic mediation between the two standards is necessary to make resources interpretable for each other. The Potential-Interoperable class describes the interoperability between two partially-standardized resources, where each resource is defined using its domain concepts and vocabularies. The One-Sided Interoperable class identifies the interoperability between a non-standardized resource and another resource that is not supported with metadata, henceforth defined as non-standardized resources. In this class, frequent scientist intervention is required to allow the non-standardized resource to programmatically interpret and process the partially-standardized resources. Finally, the Non-Interoperable class groups resources that are not supported with information. While it is difficult to quantify the cost that results from the lack of semantic interoperability, we believe it is necessary to leverage the partially-standardized and *non-standardized* resources to the standardized class (Table 9.2).

Annotation of Web resources can leverage partially-standardized and non-standardized resources to the *standardized* level. Raw hypertext information can be converted into semantic-based information that is understandable and accessible by remote Web agents. Building this type of information will empower Web Mining agents to make intelligent decisions such as creating relationships between URIs of resources, discovering of association rules between resources, and auto-categorization of resources. Thus, knowledge discovery tools will manipulate semantic annotation to extract useful knowledge, instead of manipulating the data itself. For instance, using vocabularies from the GeoNames ontology for tagging of images and videos triggers the development of Geo-miners, which group resources in new geo-collections and recognize the relationships between users that share the same locations.

Semantic Annotation

A semantic annotation T is a formal annotation that machines can understand based on a defined ontology. It consists of a triple (T_s, T_p, T_o), where T_s is the annotated subject (a Web resource or one of its sections), T_o is the annotating object (ontological term, which is an SN defined by reference ontology), and T_p is the predicate (the annotation relation) that defines the type of the relationship between T_s and T_o (e.g., Model input).

TABLE 9.2 Classes of Semantic Interoperability among Resources over the Web and the Description of Each Class

Class	Description	Examples
Interoperable	Two scientific resources are using global standard vocabularies to describe their attributes.	Use of SWEET unit ontology to describe data or model units.
Semi-interoperable	One resource is using global standard vocabularies, and other is using domain-level vocabularies.	Water velocity versus the Community Service Dynamic Modeling System Standard Names (CSN)[a] definition for stream water velocity.
Potential-Interoperable	Both resources are using their domain-level vocabularies.	Use of temperature to describe water or soil temperature.
One-sided Interoperable	One resource is using domain-level vocabularies while the other has no information.	
Non-Interoperable	Both resources have no information.	

[a] http://csdms.colorado.edu/wiki/CSDMS_Standard_Names.

Semantic Annotation of a Web resource is defined as an association of one or more SNs with a Web resource. It can describe the content of a resource using an ontology-based annotation, or more precisely, by inserting pointers to appropriate ontology. Semantic Annotation increases the discoverability and connectivity of resources by allowing cross-domain annotation; that is, a resource can be annotated for different purposes using different ontologies and should be retrieved using a common framework. It prepares resources for reuse and allows Web clients in identifying the contextual relationship among Web resources. For example, semantic annotation supports Web clients in identifying the contextual spatial relationship between resources that share the same watershed. Semantic annotation provides a platform for inference engines and algorithms to process information and discover new facts about resources. Because many information systems do not provide information about their resources, we have to build tools that can analyze the data and infer an information profile for each resource based on the available ontologies. Semantic annotation tools are important to generate such information by matching of concepts and attributes of an ontology to information included in semi-structured data files such as identification of the spatial location of a data file. Generally, automatic analysis of files is error-prone, but it is the primary key for harnessing distributed and heterogeneous resources.

9.4.4 Finding and Connecting Scientific Web Resources

A truly vast number of resources are available to scientists on the Web. Unfortunately, it is currently not easy or straightforward to find resources that are relevant to a given problem; even if they are available, it is hard to connect them together (or "integrate them") into a functional unit that can be used to solve the problem of interest. When connecting them is possible, it requires a significant amount of human time and effort by someone who has extensive technical knowledge (e.g., formats, interfaces, etc.) as well as knowledge of one or more science domains. Much of this work can, in principle, be automated, but first resources must be equipped with standardized information that describes their contents and interfaces to make them machine-understandable. Depending on the use case, it may also be necessary for this information to capture relationships between entities to allow machine reasoning.

It is helpful to think in terms of three major "use cases" that occur when a scientist wants to find and connect resources (whether doing so over the Web or on a local machine). The first use case is *discovery*, where the scientist is searching for resources that are relevant to a given problem. In this case, semantic technology is required to assess *semantic similarity*. At a minimum, this requires some thesaurus for mapping the search terms provided by the user to a set of synonyms and other closely related terms. Word-Net http://wordnet.princeton.edu/ uses a similar approach, including the conceptual-semantics and lexical relationships between terms (nouns, verbs, etc.) that are required to create a network of meaningful words and concepts. One difficulty here is the lack of uniqueness because different people tend to create different associations and hierarchies in their minds between terms. These may be characterized as "world views" and their non-uniqueness and personal preference is a major challenge; it can make the underlying database a source of contention. This contention can be particularly increased, if there is a requirement for concepts and terms to be organized into a hierarchical, that is, tree-like data structure. One way around this issue is to use *annotation* instead of trees, that is, to annotate terms with any number of relevant annotations from a CV list.

A second major use case is *semantic mediation* or matching. This arises in the context of connecting resources, where one resource (e.g., a data set or model) provides some sort of data item and another resource (e.g., model) needs to use it to perform a calculation. Scientists aim to connect the various heterogeneous resources needed to solve a given problem in a simple, flexible, *"plug-and-play"* manner. The connectivity between resources may be arbitrarily complex, both graph-like and iterated in time. Moreover, the passing of data between the resources may occur in machine memory (often referred to as "tight coupling"), through a file I/O or across the Web ("loose coupling"). In this use case the focus is on *semantic equivalence* instead of *semantic similarity*. The resource that wants to ingest data from another resource has to be assured that it is getting the expected and required data—there is no room for error or mismatches. This is a very different type of semantic problem than the discovery use case.

The underlying data structure is required to be in the form of a lookup table that contains pairs of equivalent terms, drawn from different vocabularies. This data structure might be implemented as a Python dictionary or a SKOS [49] file, which can express equivalence and different degrees of similarity. The primary difficulty for this use case is that there are a large number of CVs, as well as "resource-specific" or "internal" vocabularies that may use domain jargon, standard symbols, or abbreviations for the data items. In addition, a term in one controlled or internal vocabulary may have no counterpart in another vocabulary so that a mapping between them is not even possible. Creating mappings (or lookup tables) between all possible pairs of controlled (or internal) vocabularies is not scalable or even possible—a fool's errand. While RDF allows a resource to expose its internal vocabulary with others as "preferred labels," there are no ways to solve the problem of creating accurate semantic mappings between different sets of preferred labels. The most reasonable, sustainable, and scalable solution to this problem is to identify the semantic relationships (crosswalks) among CVs using a highly expressive information model, which serves as a *lingua franca*. This is the approach that is being pursued by the CSNs [47], and significant progress has been made in the context of model coupling.

A third major use case can be described as *knowledge representation*, which is used to enable *machine reasoning*. This is the most challenging of the three cases, because it depends on capturing many different types of relationships (functional, hierarchical, semantic, etc.) between terms. Children are first taught the names of things in the world—a set of unique labels or identifiers in their native language. They are later taught how those things relate to one another, for example, "dogs chase cats," "dogs chase cars," "cats eat birds," and "men may have beards." Note that objects are nouns and the things that objects can *do*, as well as the relationships that can exist between objects, are expressed as *verbs* (or more technically, *predicates*). The number of relationships between things is far larger than the number of individual things. Given enough pairwise relationships between things, it becomes possible to perform chains of reasoning on their relationships. Reasoning computes the relationships between things such as "part of," "synonymous," and "specialization of." The SW initiative uses Description Logic, which captures the terminological relationships between things as *assertions*. This is similar to how mathematicians use deductive reasoning, chaining together lemmas and theorems to create proofs for new assertions.

9.4.5 Semantic Knowledge Discovery

Knowledge discovery seeks to extract implicit and uncover potentially useful information from a standardized collection of data using Machine Learning and Data Mining techniques [50,51]. Knowledge discovery is traditionally used for analysis of large amounts of data or the information associated with these data to recognize unknown patterns and identify new relationships among data sets. Mining resources over the SW may start off in one collection of resources, and navigate using the connections between resources to end up in a different collection. Building a semantic knowledge discovery approach is one layer above the information retrieval mechanisms that is used by search engines [31]. In semantic knowledge discovery, search engines have to be connected with databases and knowledge bases, which include domain ontologies and the logic rules required to discover patterns, trends, and dependencies between resources. In addition, the query statement has to be analyzed both syntactically and semantically to retrieve the Web results that are more relevant to the user's query. Knowledge discovery techniques have been successfully applied not only to structured data, that is, databases, but also to semi-structured and unstructured data including text, graphs, images, and video [52]. For example, in the digital library community, knowledge discovery methods have been used to identify the coauthorship network by parsing the social graph of the scientists to identify the common publication between authors and their scientific interests. Therefore, using the same analogy of combining the knowledge discovery techniques with predefined logic rules can improve the interoperability among geoscience resources, that is, build the SW.

Semantic heterogeneity represents one of the primary challenges in Federated Information Retrieval (FIR) systems [53]. Currently, there are billions of triples (relationships) of data publicly available in the Web [54]. Although centralized databases can be used to collect and query large volumes of Web data, this approach requires significant disk space, dynamic synchronization between distributed data prepositions and a central repository, as well as optimization for the query statements [55]. FIR is considered a primary alternative for centralized databases [56]. It can exploit multiple disparate information systems with one query. An FIR system provides a single endpoint to multiple information resources and returns the related data in a standard and homogeneous format. It consists of a three-tier approach with independent participants constituting the back end. An information catalog or schema is used to act as a mediator between the user's query and the participant information systems. Users interact with the FIR system through an interface, which could be an HTML form or a Web service, that is, front end.

Leveraging the crosswalks or semantic relationships between SNs in the FIR mechanism can enhance the search capabilities of the FIR systems [57]. Precisely, using the crosswalks between SNs can help in semantic mediation between the SNs that a user provides in a query argument and the SN used to annotate a resource in its information system. Development of this approach requires the annotation of domain ontologies with their semantic relationships with related ontologies across domain boundaries. SKOS ontology provides the information model required to link and manage the SN through the SW [58]. In this context, ontologies can be considered another resource that should be searched during the FIR process. The federated search engine has to query the related ontologies to retrieve a list of semantically linked SNs. A fundamental limit of this approach is the absence of precise and robust crosswalks between domain ontologies due to the need for collaboration among scientists of related domains to build these semantic relationships.

9.5 Geosemantic Framework

In this section, we introduce the Geosemantic framework, which provides a set of Web services to support semantic interoperability among geoscience resources including data, models, and SNs. The Geosemantic framework is intended to enable structured information to flow between resources without violation of its meaning in order to achieve better semantic interoperability. It provides methods to enrich, share, link, and evaluate the information associated with individual resources. The Geosemantic framework adopts the Linked Data and RESTFul microservices approaches to advance the interoperability of distributed geoscience resources. The Linked Data approach is used for linking resources at Web scale that are easier to parse by independent data and model providers. It allows different providers (servers) to continue to use local definitions while still providing a way for consumers (clients) to ingest information from various sources. RESTFul Web services provide a resource-oriented architecture using standard and common interfaces that are highly compatible with Linked Data. Combining both approaches simplifies the task of contributing new functionality to the scientific community with the goal that development cycles can be shortened, and the number of people contributing to it can easily increase beyond individual teams. Instead of one monolithic solution, a distributed and nimble solution has the potential to result in steady growth created by the community, not unlike that of the Internet over the past 30 years.

9.5.1 Overview of Functionalities

The framework considers semantic annotation with standards and CVs as the primary approach to enrich the semantics of partially-standardized and non-standardized resources (Figure 9.5). In the Geosemantic framework, semantic annotation considers two categories: (1) user-based annotations, where a user in any data management environment chooses the annotating object (T_o) and annotation predicate (T_p) from the framework's knowledge base to annotate a subject (T_s), and (2) automatic annotation, which is suitable for files that have a known structure and information about its content in the human-based format (e.g., ASCII and tabular data files). In automatic annotation, the information is extracted from the document, mapped to a suitable standard, then shared using the Linked Data format.

The Geosemantics framework provides semantic alignment services between the information profile associated with resources for data synthesis, model and data integration, as well as models coupling. This

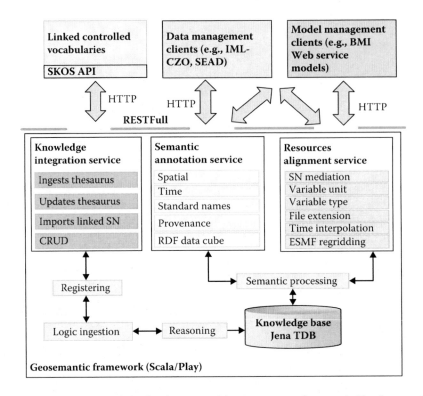

FIGURE 9.5 (See color insert.) High-level architecture of the Geosemantic framework. The framework uses micro services architecture and provides each service as an endpoint through a RESTFul API. KIS ingests standards and can annotate them back. A unique endpoint is provided for each registered standard. The Logic Ingestion component infers the relationships among registered standards. Reasoning among ingested standards is based on the Jena Reasoners and Rules Engine. The Semantic Processing component handles the requests to the knowledge base and the communication among micro services.

alignment is used to programmatically perform the semantic mediation and matching between resources that are supposed to interact, such as models, as well as model and data integration. The semantic alignment checks the consistency between the information profile associated with two resources and provides the conversion tools required to map the source profile to the target profile. Semantic alignment maps the fields and values that are related to space, time, file format, variable names, and unit attributes of resources.

Another key functionality of the framework is to allow semantic linkage between SNs by annotating their relationship based on SKOS Ontology. Cross-domain interoperability requires the presence of links, which map the relationship between terms of SNs, such as the relationship between CUASHI: rainfall and CSN: precipitation. The Geosemantic framework developed an information system to represent each SN as a single unique Web entity and describe its attributes including the SNs' relationships with other SNs. This information system supports semantic mediation across domain-specific vocabularies. Each SN is indexed by an HTTP URL (Linked Data essential rule), has its HTML page that displays its attributes, and facilitates the visual navigation among related SNs. The framework provides a RESTFul API (http://hcgs.ncsa.illinois.edu/skosmos), which relies on the Skosmos controlled vocabulary browser [59] to enable a programmatic search across and retrieval of the information profiles associated with the SN. This information system allows navigation among domain-specific vocabularies, advances the programmatic search capabilities for SNs, and provides a structured display of concept hierarchies. In addition, it provides the technology required for building a multilingual user interface.

Scientists' intervention is essential to evaluate and ensure the quality of the SNs and their links. The Geosemantic framework does not provide a new user interface to define and edit standard vocabularies and their connections, but it relies on existing tools (e.g., [60,61]) to help scientists to build the crosswalks between SNs. For users with experience in information systems, the Protégé (http://protege.stanford.edu/) or NeOn (http://www.neon-project.org) ontology editors could be used to define ontologies on the desktop and then manually upload them to the Geosemantics knowledge base. But, these tools are suitable for information experts and do not keep track of the changes in the knowledge base. In addition, to provide a semantic collaborative environment for users to link standard vocabularies and create new terms, the framework includes a Geosemantics Wiki, which is an extension of the *Semantic MediaWiki* (https://semantic-mediawiki.org/), to incorporate ontologies that are registered in the knowledge base of the Geosemantic framework. The Geosemantics Wiki is an easy-to-use, low-learning curve interface for users to produce, share, and track semantic annotations.

Technically, the design of the framework is constrained by the objectives of maximizing the framework flexibility to adopt new technologies and its scalability to serve a broad range of geoscience communities. Provision of well-defined interfaces (e.g., APIs and standardized Web service interfaces) is required to allow third parties to embed Geosemantics Web services in their applications. The framework can use and integrate ontologies and thesauruses of standard vocabularies regardless of their serialization language to decrease the dependency on domain-level solutions and ensure the capability of the framework to ingest new standards. The framework components are loosely coupled; that is, components can be interconnected without the need to change their contents. Scientists will be able to use existing services or plug in a new service with additional functionality. The framework considers the scalability issues from the following aspects: type and amount of resources required for knowledge discovery, and the size and granularity level of their meta-information.

9.5.2 Architecture

The Geosemantic framework has been designed to address the previously mentioned functionalities, particularly annotating, integrating, and reasoning about the integration of geoscience resources. It aims to close the semantic loop among data, models, and SNs. The framework serves three types of Web clients: data management, model coupling framework, and SN clients. A basic assumption for the framework is the availability of resources over the Web. Although this might not yet be the case for all data and models, many initiatives in the geoscience communities are allowing publication of their resources or moving toward using the Web as an environment for publishing and integrating disparate resources (e.g., [17,62–65]).

The Geosemantic framework is a set of RESTFul Web services using JSON-LD data format for the body of the service calls. The current version has been implemented using the Play Web application framework [66] and the Apache Jena middleware [67]. Play provides a lightweight server for applications written in Java and Scala, a complete RESTFul API supporting all standard HTTP actions, and Web session management. Jython [68], an implementation of Python over the Java Virtual Machine, is used to provide the extensibility required to include scripts written in Python into a Java environment. Jena provides libraries to manipulate RDF triples, serialize and desterilize JSON-LD, and reason on the relationships among RDF triples using logic rules.

Figure 9.5 shows the general architecture of the Geosemantic framework and illustrates the information flow from the distributed knowledge sources, through the framework, and to its clients. The architecture is made of three building blocks (each block may include one or more component): a knowledge base, three sets of services, and a pipeline for simple reasoning and manipulation of the RDF triples going into the knowledge base. The knowledge base stores the URIs of registered elements as graph nodes and can create URIs for the elements that are not serviced. Entities from the distributed ontologies are loaded in the knowledge base. The pipeline ensures the consistency of information flow in the framework. The service layer provides interfaces for humans and machines to the underneath components of the framework. Each service represents a standalone set of functions implemented as RESTFul Web service endpoints that can be used individually or combined in a workflow.

9.5.2.1 Knowledge Base

The Geosemantic framework uses a corpus of ontologies that are defined in its knowledge base. The Apache Jena Tuple DataBase (Jena TDB) is used as a persistent RDF graph storage for the framework. Jena TDB is a high-performance RDF store and is compatible with the Jena SPARQL query engine. The knowledge base level provides a triple store for storing meta-information of ontologies and list of the registered SNs. The backbone ontology of the knowledge base, that is, upper level, consists of the three main classes *Space, Time,* and *Variable.* The *Space* and *Time* classes use vocabularies from W3C Geospatial ontologies (http://www.w3.org/2005/Incubator/geo/XGR-geo-ont-20071023/) and W3C Time ontology, (http://www.w3.org/TR/owl-time/) respectively. The *Variable* class defines the SNs and their semantic relationships according to SKOS ontology.

9.5.2.2 Knowledge Management Pipeline

The knowledge management pipeline consists of four components that are responsible for handling the ontologies used by the framework. First, the ontologies are stored in the register, a catalog for storing a URI of an ontology and the related query statement that is required to retrieve information from the ontology. For example, the CSNs' ontology URI is stored in the register with its related query statement, which is required to retrieve the SN entities and not the whole ontology. The Register allows mutual information exchange between the Ontology namespace and the knowledge base of the framework. Second, the Logic ingestion component ingests the semantic links that are established between SNs and checks their consistency based on the SKOS ontology. It updates the Knowledge Integration Services (KIS) with the semantic links associated with each SN to annotate its information profile. Third, a reasoner is built on top of the Jena reasoner and rule engines to allow the semantic reasoner in the discovery of resources and the alignment of their information profiles. The reasoner validates and asserts facts about the ontologies stored in the knowledge base using first-order logic reasoning. In addition, the reasoner is responsible for the semantic knowledge discovery by issuing sophisticated queries that inject the knowledge graph associated with an SN in the query statement. Finally, the semantic processor harvests the knowledge base to provide functions for semantic mediation, SPARQL query, semantic similarity, and composition of RDF tags.

9.5.2.3 Services Layer

A full documentation of each service with example applications is available at (http://geosemantics. hydrocomplexity.net). Figure 9.6 illustrates an example of using the microservice responsible for retrieving SNs that match a user input such as wind speed by searching across register ontologies. The framework's services are designed to use a shared knowledge base. This approach minimizes the number of endpoints required for communication and provides a simpler API for each service. Each service is composed of a set of microservices that can be executed alone or in a workflow. The Semantic Annotation Services (SAS) annotates resources with their spatiotemporal context, variable, and provenance relationships, either by running automatic extractors based on the data files' MIME type (e.g., GeoTIFF and CSV types) or by providing an interactive interface for manual annotation. SAS is currently deployed in the Clowder data management system (https://clowder.ncsa.illinois.edu), which a is scalable data repository for sharing, organizing, and analyzing long-tail data. Clowder serves as a backbone for many NSF-funded projects such as Sustainable Environmental Actionable Data http://sead-data.net/), Intensively Managed Landscape Critical Zone Observatory data management system http://data.imlczo.org/geodashboard/), and the Great Lakes Monitoring system (https://greatlakesmonitoring.org/geodashboard/).

The Resource Alignment Service (RAS) is a set of microservices used to align the data with the requirements of a model using the semantic annotations available on both resources. When the KIS returns a partial match between model requirements and data, the RAS attempts to find a path between the current data file and the required data file. For example, in the case of temporal realignment, the RAS could

GET /hcgs/sas/vars/list

Synopsis: Searches for variable names across all the stored graphs containing the given keywords and returns the results in JSON-LD format.

Query Parameters:
- **term** *(string)* – keywords separated by the character "+."

Example request

```
GET /hcgs/sas/vars/list?term=wind+speed     HTTP/1.1
Host: hcgs.ncsa.illinois.edu
Accept: */*
```

Example response

```
HTTP/1.1 200 OK
Vary: Accept
Content-Type: application/json
```

```
{
   "@context" : [
"odm2" : "http://vocabulary.odm2.org/variablename",
"csn" : "http://csdm.colorado.edu/wiki/CSN_Searchable_List
   ],
"variableName"  :{
        "csn:earth_surface_wind__range_of_speed",
        "csn:land_surface_wind__reference_height_speed",
        "csn:land_surface_wind__speed_reference_height",
        "csn:projectile_origin_wind__speed"
        "odm2:windGustSpeed",
        "odm2:windSpeed"
            }
}
```

FIGURE 9.6 Excerpts from the Semantic Annotation Services (SAS) illustrating the request and response for the microservice *vars*, which is responsible for retrieving Standard Names that matches the request's argument received from a user.

interpolate data at minutes interval to data at hourly interval using temporal interpolation algorithms such as linear. RAS minimizes heterogeneity between data storage format and the format required by a model (e.g., converting from NetCDF to geoTiFF) by leveraging the Brown Dog services [69] for the file format conversions. Building an extensive list of alignment services that covers most common cases is not a trivial task that requires the collaboration of the community to turn their tools into standard services and make them available for the community. We have begun prototyping specific algorithms for the most common cases.

The KIS is used to ingest and register the SN with the framework's knowledge base. It collects meta-information about ontologies and feeds this information to the Reasoner and Logic Ingestion components. This service is supported by providing an API for external resources to register information about how to access such services. This information is stored in the knowledge base. The knowledge discovery API responds to four query types: (1) given a data source what models are compatible with it, (2) given a model resource what data sources are compatible with it, (3) given a model resource what other models can be coupled with it, and (4) given a data source what other data sources are related to it. Using the Jena generic rule reasoner and a set of rules built around the SKOS *exactMatch* predicate, an inference model of the original semantic annotations on both models and data is created at query time. These inference models include derived links between variables that can lead to matches between models and data that would not have been present in the original data model. Similar to the federated query approach, when the data is not present in the knowledge base, the KIS queries external sources, such as the Clowder data management system, for data that matches a set of annotations derived from the inference model. In the case of multiple results for the derived annotations, the KIS merges the lists of results by taking the union of the results based on the SKOS exact match rules and the intersection of the query parameters originally specified.

All services are written to be stateless to make it easier to deploy multiple instances of each next to each other and allow easy horizontal scalability. The NGINX Web server (http://nginx.org/) is deployed as a proxy in front of each service to load balance across multiple instances of the services. As client traffic increases, more instances can be added to the pool to service a greater load. All resources are universally identified using URLs so that different instances of each service can be maintained and crosswalks between different instances maintained by different communities are still valid. This places the burden of scaling on distributed instances instead of a centralized one. The RESTFul and Linked Data approaches support a distributed approach similar to that of the World Wide Web. Although we are currently the only maintainers of a set of these services, the long-term goal is to encourage specific communities to deploy specific instances maintaining resources relevant to each community and links to related instances.

Acknowledgments

Support from NSF grants "ICER-1440315," "ACI-0940824," "ACI-1261582," "EAR-1331906," and "EAR 1417444" are gratefully acknowledged. The authors also wish to acknowledge Peishi Jiang, a research assistant in the Department of Civil and Environmental Engineering at University of Illinois, and Leslie Hsu, an associate research scientist in Lamont-Doherty Earth Observatory at Columbia University, for their helpful discussions and comments.

References

1. Praveen Kumar. *Hydroinformatics: Data Integrative Approaches in Computation, Analysis, and Modeling*. Boca Raton, FL: Taylor & Francis, 2006.
2. Robert Blumberg and Shaku Atre. The problem with unstructured data. *DM Review*, 13(42–49):62, 2003.
3. Research Information Network. *To Share or Not to Share: Publication and Quality Assurance of Research Data Outputs: Main Report*. Research Information Network, 2008.
4. Dragan Gasevic, Jelena Jovanovic, and Vladan Devedzic. Enhancing learning object content on the semantic web. In *Proceedings IEEE International Conference on Advanced Learning Technologies*, 2004, pages 714–716, IEEE. http://ieeexplore.ieee.org/document/1357633/.
5. James Myers, Margaret Hedstrom, Dharma Akmon et al. Towards sustainable curation and preservation: The SEAD project's data services approach. In *e-Science (e-Science), 2015 IEEE 11th International Conference on*, pp. 485–494. IEEE, 2015.
6. DG Tarboton, R Idaszak, JS Horsburgh et al. Hydroshare: An online, collaborative environment for the sharing of hydrologic data and models. In *AGU Fall Meeting Abstracts*, volume 1, p. 1510, 2013.
7. Leslie Hsu, Raleigh L Martin, Brandon McElroy, Kimberly Litwin-Miller, and Wonsuck Kim. Data management, sharing, and reuse in experimental geomorphology: Challenges, strategies, and scientific opportunities. *Geomorphology*, 244:180–189, 2015.
8. MJ Earl and DF Feeney. Opinion: How to be a CEO for the information age. *Sloan Management Review*, 41(2): 11, 2000.
9. S Peckham, E Hutton, and B.Norris. A component-based approach to integrated modeling in the geosciences: The design of CSDMS. *Computers & Geosciences*, 53: 3–12, 2013.

10. Tim Berners-Lee, James Hendler, Ora Lassila et al. The semantic web. *Scientific American*, 284(5):28–37, 2001.

11. Ian Foster, Daniel S Katz, Tanu Malik et al. Wagging the long tail of earth science: Why we need an earth science data web, and how to build it, 2012.

12. Mostafa M Elag, Jonathan L Goodall, and Anthony M. Castronova. Feedback loops and temporal misalignment in component-based hydrologic modeling. *Water Resources Research*, 47(12), 2011.

13. P Bryan Heidorn. Shedding light on the dark data in the long tail of science. *Library Trends*, 57(2):280–299, 2008.

14. Benjamin L Ruddell, Ilya Zaslavsky, David Valentine et al. Sustainable long term scientific data publication: Lessons learned from a prototype observatory information system for the Illinois River basin. *Environmental Modelling & Software*, 54:73–87, 2014.

15. G H Leavesley, S L Markstrom, M S Brewer et al. The modular modeling system (MMS)—The physical process modeling component of a database-centered decision support system for water and power management. *Water, Air, & Soil Pollution*, 90(1–2):303–311, 1996.

16. RM Argent, A Voinov, T Maxwell et al. Comparing modelling frameworks–a workshop approach. *Environmental Modelling & Software*, 21(7):895–910, 2006.

17. AM Castronvoa, JL Goodall, and MM Elag. Models as web services using the Open Geospatial Consortium (OGC) Web Processing Service (WPS) standard. *Environmental Modelling & Software*, 41: 72–83, 2013.

18. Mostafa Elag and Jonathan L Goodall. An ontology for component-based models of water resource systems. *Water Resources Research*, 49(8):5077–5091, 2013.

19. Idafen Santana-Perez, Rafael Ferreira da Silva, Mats Rynge et al. Leveraging semantics to improve reproducibility in scientific workflows. In *The Reproducibility at XSEDE Workshop*. 2014.

20. Qi Li, Yaliang Li, Jing Gao et al. A confidence-aware approach for truth discovery on long-tail data. *Proceedings of the VLDB Endowment*, 8(4), 425–436, 2014.

21. Carol Jean Godby. A crosswalk from ONIX version 3.0 for books to MARC 21, 2012. http://www.oclc.org/content/dam/research/publications/library/2012/2012-04.pdf

22. P. Hitzler, M. Krötzsch, and S. Rudolph. *Foundations of Semantic Web Technologies*. Boca Raton: Chapman & Hall/CRC, 2009.

23. Andreas Tolk and James A Muguira. The levels of conceptual interoperability model. In *Proceedings of the 2003 Fall Simulation Interoperability Workshop*, volume 7. Citeseer, 2003.

24. Wenguang Wang, Andreas Tolk, and Weiping Wang. The levels of conceptual interoperability model: Applying systems engineering principles to M&S. In *Proceedings of the 2009 Spring Simulation Multiconference*, p. 168. Society for Computer Simulation International, 2009.

25. Tim Berners-Lee, Mark Fischetti, and Michael L Foreword By-Dertouzos. *Weaving the Web: The Original Design and Ultimate Destiny of the World Wide Web by its Inventor*. San Francisco: HarperInformation, 2000.

26. Ivan Herman, Ben Adida, Manu Sporny et al. RDFa1.1 primer–second edition. Working group note, W3C, Aug 2013. http://www.w3.org/TR/rdfa-primer/.

27. Jens Rasmussen. *Information Processing and Human-Machine Interaction. An Approach to Cognitive Engineering*. North-Holland, 1986.

28. Francisco Curbera, Matthew Duftler, Rania Khalaf et al. Unraveling the web services web: An introduction to SOAP, WSDL, and UDDI. *IEEE Internet Computing*, 6(2):86, 2002.

29. T Bray, J Paoli, CM Sperberg-McQueen et al. Extensible markup language (XML). *World Wide Web Journal*, 2(4):27–66, 1997.

30. Ora Lassila and Ralph R Swick. Resource description framework (RDF) model and syntax specification. 1999. https://www.w3.org/TR/1999/REC-rdf-syntax-19990222/.

31. The open graph protocol, 2014. http://ogp.me/.

32. G Antoniou and F Harmelen. Web ontology language: OWL. In *Handbook on Ontologies*, Berlin, Heidelberg: Springer Berlin Heidelberg, pp. 91–110, 2009.

33. Ian Horrocks, Peter F Patel-Schneider, and Frank Van Harmelen. From SHIQ and RDF to OWL: The making of a web ontology language. *Web Semantics: Science, Services and Agents on the World Wide Web*, 1(1):7–26, 2003.

34. J Garrido and I Requena. Proposal of ontology for environmental impact assessment: An application with knowledge mobilization. *Expert Systems with Applications*, 38(3):2462–2472, 2011.

35. D Fensel, FM Facca, E Simperl et al. *Semantic Web Services*. Heidelberg: Springer, 2011.

36. Michael C Dietze, David S Lebauer, and Rob Kooper. On improving the communication between models and data. *Plant, Cell & Environment*, 36(9):1575–1585, 2013.

37. Bettina Berendt, Andreas Hotho, and Gerd Stumme. Towards Semantic Web mining. In *The Semantic Web? ISWC 2002*, pp. 264–278. Springer, 2002.

38. AI3, adaptive information, adaptive innovation, and adaptive infrastructure, 2013. http://www.mkbergman.com/1825/why-clojure/.

39. Engineers Joint Council. *Thesaurus of Engineering and Scientific Terms*. Engineers Joint Council, 1969.

40. I Zaslavsky, D Valentine, R Hooper et al. Community practices for naming and managing hydrologic variables. In *Proceedings of the AWRA Spring Specialty Conference on GIS and Water Resources*, New Orleans, LA, March, pp. 26–28, 2012.

41. LM Olsen, G Major, K Shein et al. NASA/global change master directory (GCMD) earth science keywords. *Version*, 6(0.0):0, 2007.

42. Robert G Raskin and Michael J Pan. Knowledge representation in the Semantic Web for earth and environmental terminology (SWEET). *Computers & Geosciences*, 31(9):1119–1125, November 2005.

43. Brian Eaton, Jonathan Gregory, Bob Drach et al. NetCDF climate and forecast (CF) metadata conventions, 2003.

44. Jonathan Gregory. The CF metadata standard. *CLIVAR Exchanges*, 8(4):4, 2003.

45. SD Woodruff, HF Diaz, JD Elms et al. COADS release 2 data and metadata enhancements for improvements of marine surface flux fields. *Physics and Chemistry of the Earth*, 23(5):517–526, 1998.

46. Rocky Dunlap, Leo Mark, Spencer Rugaber et al. Earth system curator: Metadata infrastructure for climate modeling. *Earth Science Informatics*, 1:131–149, November 2008.

47. Scott Dale Peckham. The CSDMS standard names: Cross-domain naming conventions for describing process models, data sets and their associated variables. In *7th International Conference on Environmental Modeling and Software*. International Environmental Modelling and Software Society, 2014. http://cfconventions.org/cf-conventions/v1.6.0/cf-conventions.html.

48. Simon JD Cox, Jonathan Yu, and Terry Rankine. SISSVOC: A linked data API for access to SKOS vocabularies. *Semantic Web*, 7(1):9–24, 2014. http://www.semantic-web-journal.net/system/files/swj658.pdf.

49. Simon JD Coxa, Jonathan Yua, and Terry Rankineb. SISSVoc: *A linked data API for SKOS vocabularies*. Cambridge, Mass: MIT Press

50. David Hand, Heikki Mannila, and Padhraic Smyth. *Principles of Data Mining*. Adaptive computation and machine learning series, MIT Press, 2001.

51. Usama Fayyad, Gregory Piatetsky-Shapiro, and Padhraic Smyth. The KDD process for extracting useful knowledge from volumes of data. *Communications of the ACM*, 39(11):27–34, 1996.

52. Dunja Mladenić, Marko Grobelnik, Blaž Fortuna et al. *Knowledge Discovery for Semantic Web*. Berlin: Springer, 2009.

53. David George. Understanding structural and semantic heterogeneity in the context of database schema integration. *Journal of the Department of Computing, UCLAN*, 4:29–44, 2005.

54. Yingjie Li. A federated query answering system for semantic web data, Lehigh University Bethlehem, PA, 2013.

55. Ricardo Baeza-Yates, Berthier Ribeiro-Neto. *Modern Information Retrieval*, volume 463. New York, NY: ACM press, 1999.

56. Jens Graupmann, Michael Biwer, and Patrick Zimmer. Towards federated search based on web services. In *BTW Conference*. Citeseer, 2003.

57. Diego Collarana, Christoph Lange, and Sören Auer. Fuhsen: A platform for federated, RDF-based hybrid search. In *Proceedings of the 25th International Conference Companion on World Wide Web*, WWW '16 Companion, pp. 171–174, Republic and Canton of Geneva, Switzerland, 2016. International World Wide Web Conferences Steering Committee.

58. Alistair Miles and Sean Bechhofer. SKOS simple knowledge organization system reference. W3C recommendation, 18:W3C, 2009.

59. SKOSMOS. https://github.com/NatLibFi/Skosmos. Accessed: 2016-04-21.

60. V Presutti, A Gangemi, S David et al. Neon deliverable d2. 5.1. A library of ontology design patterns: Reusable solutions for collaborative design of networked ontologies. *NeOn Project. http://www.neon-project.org*, 2008.

61. Natalya F Noy, Michael Sintek, Stefan Decker et al. Creating semantic web contents with Protégé-2000. *IEEE Intelligent Systems*, 16(2):60–71, 2001.

62. Carlos Granell, Laura Díaz, and Michael Gould. Service-oriented applications for environmental models: Reusable geospatial services. *Environmental Modelling & Software*, 25(2):182–198, 2010.

63. JS Horsburgh, DG Tarboton, M Piasecki et al. An integrated system for publishing environmental observations data. *Environmental Modelling & Software*, 24(8):879–888, 2009.

64. Jonathan L Goodall, Bella F Robinson, and Anthony M Castronova. Modeling water resource systems using a service-oriented computing paradigm. *Environmental Modelling & Software*, 26(5):573–582, 2011.

65. JL Goodall, JS Horsburgh, TL Whiteaker et al. A first approach to web services for the National Water Information System. *Environmental Modelling & Software*, 23(4):404–411, 2008.

66. Play Framework-Build Modern. Scalable web apps with Java and Scala. http://www.playframework.com/

67. Apache Jena. https://jena.apache.org/. Accessed: 22 April 2016.

68. Samuele Pedroni and Noel Rappin. *Jython Essentials*. " O'Reilly Media, Inc.", 2002.

69. Smruti Padhy, Greg Jansen, Jay Alameda et al. Brown dog: Leveraging everything towards autocuration. In *IEEE Big Data*, 2015.

Index

A

Abundance nonnegativity constraint (ANC), 134, 141, 164–165
Abundance sum-to-one constraint (ASC), 134, 141, 142, 165
Adaptive mixture of experts, 65
ADMM, *see* Alternating direction method of multipliers
Alternating direction method of multipliers (ADMM), 137
Amazon Machine Image (AMI), 123–124
Amazon Web Services (AWS), 114
    computing infrastructure, 122–124
AMI, *see* Amazon Machine Image
Analogue Techniques, 62
ANC, *see* Abundance nonnegativity constraint
ANNs, *see* Artificial neural networks
API, *see* Application programming interface
Application programming interface (API), 181
Artificial neural networks (ANNs), 58, 67–68, 82
Art scalable machine learning, state-of, *see also* Statistical downscaling (SD), in climate
ARVI, *see* Atmospherically resistant vegetation index
ASC, *see* Abundance sum-to-one constraint
ASR, *see* Automatic speech recognition
Atmospherically resistant vegetation index (ARVI), 116
"Autoencode," 66
Autoencoders, 67
Automated regression-based statistical downscaling, 58
Automatic speech recognition (ASR), 114
Average model, 26
AWS, *see* Amazon Web Services

B

Backpropagation neural network (BPNN), 114
Bayesian methods, 61, 62
BCSD, *see* Bias correction spatial disaggregation
BDF, *see* Bivariate distribution function
Bias correction spatial disaggregation (BCSD), 58
BIOCLIM, 84
Bivariate distribution function (BDF), 148, 155, 158

Black-box method, 109
Boosted regression trees (BRTs), 82
BPNN, *see* Backpropagation neural network
BRTs, *see* Boosted regression trees

C

Canadian Institute for Advanced Research (CIFAR), 114
CEM, *see* Constrained energy minimization
CFSNs, *see* Climate and Forecasting Standard Names
CIFAR, *see* Canadian Institute for Advanced Research
Classical estimation approaches, 22
Classification trees (CTs), 82
Class imbalance problem, 79
Climate and Forecasting Standard Names (CFSNs), 186, 187
Climate applications
    multitask sparse structure learning and, 21–26
    sparse group lasso and, 16–20
Climate data, challenges in
    grid distortion, 38
    handling missing observations, 39
Climate extremes, downscaling of, 65–66
Climate hazards, 1
Climate model data, 33, 45–46
    hindcast performance, 49–50
    preprocessing, 46
    range of β values, 47–48
    regional analysis, 50–53
    spatial influence parameters, 47
    temperature anomalies, 46–47
Climate models, 55
Climate networks, for covariate selection, 63–64
Climate Research Unit (CRU), 26
Climate risk management, network science in, 3–6
Climate science, 13
Cloud microphysics, 97
Cloud Model 1 (CM1), 97
Clowder data management system, 194
CM1, *see* Cloud Model 1
CMIP5 models, *see* Coupled Model Intercomparison Project Phase 5 models

CNN, *see* convolutional neural network

COARDS, *see* Cooperative Ocean Atmosphere Research Data Service

Cold climates, 29

Collaboratory for Adaptation to Climate Change, 85

Community Service Dynamic Modeling System (CSDMS) Standard Names (CSNs), 186–187

Computational cost, 66

Computer-simulated data

abundance maps, representation of, 150, 152

bivariate distribution function, 155, 158

boxplots of abundance maps, 153, 156, 157

correlation coefficient, RMSE and probability of success, 149–152, 154, 155

execution time plot, 152, 155

minimum/maximum values of abundance maps, 149–151, 153

six bands, 149

statistical properties of, 143–145

Conceptual heterogeneity, 184

Conditional random field (CRF), 114, 119–121

Consortium of Universities for the Advancement of Hydrologic Science Inc. (CUAHSI), 185

Constrained energy minimization (CEM), 136–137, 164

Constrained sparse regression (CSR), 137

Constrained SUnSAL (CSUnSAL), 133, 152, 162

Constrained SUnSAL total variation (CSUnSAL TV), 133, 152, 156

Content heterogeneity, 184

Contextual heterogeneity, 184

Continental United States (CONUS), 113–114

Controlled Vocabularies (CVs), 185, 187

CONUS, *see* Continental United States

Convolutional neural networks (CNNs), 68–69, 114

Cooperative Ocean Atmosphere Research Data Service (COARDS), 186

Correlative niche models, 73–76

deployment of, 84–85

evaluation, 80–81

performance of, 79

presence/absence methods, 81–83

presence/background methods, 83

presence-only methods, 79–80, 83–84

spatial extent, 77–78

spatial sampling, 79

spatial scale, 78–79

Coupled Model Intercomparison Project Phase 5 (CMIP5) models, 57

Covariate selection

climate networks for, 63–64

techniques, 57

CRF, *see* Conditional random field

CRU, *see* Climate Research Unit

CSR, *see* Constrained sparse regression

CTs, *see* Classification trees

CUAHSI, *see* Consortium of Universities for the Advancement of Hydrologic Science Inc.

CVs, *see* Controlled Vocabularies

D

Data handling techniques, for large-scale data, 109

Data mining techniques, 97

Data Observation Network for Earth (DataOne), 178

Data science developments, 59–62

DBN, *see* Deep belief network

Deep belief network (DBN), 67–69, 113–114

classification using, 117–119

Deep learning

Amazon Web Services computing infrastructure, 122–124

NASA Earth Exchange high-performance computing architecture, 114, 121–122

overview, 113–114

results and comparative studies, 124–127

for statistical downscaling, 66–67

Dependency measures, 64

Dimensionality reduction, 59

Distance-based methods, 84

DOMAIN, 84

Downscaling

of climate extremes, 65–66

techniques, categories of, 56

Dublin Core, 177

Dynamical downscaling, 56

E

Earth System Bridge, 187

Earth System Modeling Framework (ESMF), 186

Ecological niche factor analysis (ENFA), 83

Ecological niche models (ENMs), 74

Elastic Cloud Compute (EC2), 114, 124

El Nino Southern Oscillation (ENSO) precipitation index, 57

ENFA, *see* Ecological niche factor analysis

Enhanced vegetation index (EVI), 116

ENMs, *see* Ecological niche models

ENSO precipitation index, *see* El Nino Southern Oscillation precipitation index

ESMF, *see* Earth System Modeling Framework

EUBrazilOpenBio initiative, 85

EVI, *see* Enhanced vegetation index

Expanded downscaling, 58

Extensible markup language (XML), 181

F

FCLS, *see* Fully constrained least squares

Feature extraction, 116–117

Federated Information Retrieval (FIR) system, 190–191
Finite impulse response (FIR) filter, 136
FIR system, *see* Federated Information Retrieval
Fixed kriging, 61
Fixed-share
    algorithm, 34
    represented as Markov random field, 35–36
Fully constrained algorithms, 140–142
Fully constrained least squares (FCLS), 141–142, 156–159

G

GAMs, *see* Generalized additive models
GARP, *see* Genetic algorithm for rule-set production
GCMD, *see* Global Change Master Directory
GCMs, *see* General circulation models
GCMs combination, *see* Global climate models combination
GDMs, *see* Generalized dissimilarity models
General circulation models (GCMs), 33, 35, 38, 44, 55–58
Generalized additive models (GAMs), 83
Generalized dissimilarity models (GDMs), 83
Generalized linear models (GLMs), 83
Genetic algorithm for rule-set production (GARP), 83
Geosemantic framework, 191
    architecture, 192–196
    functionalities of, 191–193
Geosemantics Wiki, 193
Gibbs energy function, 120
Gibbs sampling, 39
GLMs, *see* Generalized linear models
Global Change Master Directory (GCMD), 185–186
Global circulation models (GCMs), 9
Global climate models (GCMs) combination, 13–14, 20–21
    experiments on, 26–29
Greedy layer-wise training algorithm, for deep belief networks, 118

H

HABITAT, 84
Hidden Markov model (HMM), 34
High-dimensional scenarios, regression in, 14–15
High-performance computing (HPC)
    architecture, 122
    infrastructure, 114
Hindcast, performance on, 49–50
HMM, *see* Hidden Markov model
HTML, *see* Hypertext Markup Language
HTTP, *see* Hypertext Transfer Protocol
"Hub-and-spoke" approach, 187

Hypertext Markup Language (HTML), 181, 182
Hypertext Transfer Protocol (HTTP), 181, 183

I

Impacted systems, 4
Intergovernmental Panel on Climate Change (IPCC), 13, 33, 46
Interoperable class, 188

J

JavaScript Object Notation (JSON), 181
Jena Tuple DataBase (Jena TDB), 194

K

KIS, *see* Knowledge Integration Services
Knowledge discovery, 190–191
Knowledge Integration Services (KIS), 192, 194–196
Knowledge representation, in Semantic Web, 190
Kriging, 61
    multivariate version of, 69

L

Land cover (LC) mapping, 132–133
Landsat data, *see also* Computer-simulated data
    agricultural setup, 156–159
    San Joaquin Valley, California, 145–147
    thematic mapper and WV-2 data, 145, 147
    urban setup, 158–162
Landsat Ecosystem Disturbance Adaptive Processing System (LEDAPS), 143
Large data sets
    algorithms for, 109
    for machine learning, 108–109
Large-scale machine learning
    preparing simulations for, 103–106
    severe weather simulations, 97–98
    spatiotemporal relational random forests, 98–103
Lasso, 67
LCIM, *see* Levels of conceptual interoperability model
Learn-α hierarchy, 37, 38
Least squares regression, 23–25
LEDAPS, *see* Landsat Ecosystem Disturbance Adaptive Processing System
Levels of conceptual interoperability model (LCIM), 180
Linear mixture model (LMM), 132
Linear models, 58
Linear regression, 26
Linked Data approach, 182–183, 191, 196
LIVES, 84
LLMs, *see* Low-level mesocyclones
LMM, *see* Linear mixture model

Long-tail web resources, 176–179
Low-level mesocyclones (LLMs), 96

M

Machine learning community, 79
Markov chain Monte Carlo (MCMC), 39, 44
Markov random field (MRF), 33, 34
    effect of α_{space} in, 47
    energy functions in, 37–38
    fixed-share represented as, 35–36
    inference in, 39
    simulated data, 44–45
    tracking climate models, 39–40
MARS, *see* Multivariate adaptive regression splines
Matched filtering (MF), 135
MATLAB® compiler runtime (MCR), 124
Maxent, 83
MCMC, *see* Markov chain Monte Carlo
MCR, *see* MATLAB® compiler runtime
Mechanistic simulation models, 74
Mesocyclone objects, 104–105
MF, *see* Matched filtering
MFCLS, *see* Modified fully constrained least squares
Mixed National Institute of Standards and Technology
    database (MNIST), 114
Mixed-tuned matched filtering (MTMF), 135–136, 164
Mixture of experts, 65
MNIST, *see* Mixed National Institute of Standards and
    Technology database
Mobile radars, 97
Modeling approaches, development of, 73
Modified fully constrained least squares (MFCLS), 142
MRF, *see* Markov random field
MSSL, *see* Multitask sparse structure learning
MTL, *see* Multitask learning
MTMF, *see* Mixed-tuned matched filtering
Multimodel regression, with spatial smoothing, 26–27
Multiple linear regression, 58
Multitask learning (MTL), 15, 63, 64–65
Multitask sparse structure learning (MSSL), 64
    general, formulation, 22
    least squares regression, 23–25
    notation and preliminaries, 21–26
    residual precision structure, 25–26
    structure estimation, 21–22
Multivariate adaptive regression splines (MARS), 83
Multivariate linear correlations, 69

N

NAIP, *see* National Agriculture Imagery Program
NASA, *see* National Aeronautics and Space
    Administration
    Global Change Master Directory, 185–186

NASA Earth Exchange high-performance computing
    (NEX HPC) architecture, 114, 121–122
National Aeronautics and Space Administration (NASA),
    114
National Agriculture Imagery Program (NAIP), 113, 121
NCEP Reanalysis I data set, 67
NCLS, *see* Nonnegative constrained least squares
NDVI, *see* Normalized difference vegetation index
Neighborhood-augmented Tracking Climate Models
    (NTCM), 40–42
    effect of β values in, 47–48
    time complexity of, 42–43
"Neighboring" pixels, 119
Network science
    in climate risk management, 3–6
    network-based technology stack, 6–8
    overview, 1–3
    telescoping systems of systems, 8–9
Network technology stack (Net tech Stack), 6–8
NEX HPC architecture, *see* NASA Earth Exchange
    high-performance computing architecture
NGINX Web server, 196
NNCLS, *see* Normalized nonnegative constrained least
    squares
NNLS, *see* Nonnegative least squares
Non-interoperable class, 188
Nonnegative constrained least squares (NCLS), 140–141
Nonnegative least squares (NNLS), 140
Normalized difference vegetation index (NDVI), 116
Normalized nonnegative constrained least squares
    (NNCLS), 141
Normalized sum-to-one constrained least squares
    (NSCLS), 140
NTCM, *see* Neighborhood-augmented Tracking Climate
    Models
Numerical weather prediction (NWP) models, 96

O

Occurrence data/presence data, 73
OLS, *see* Ordinary least squares
One-sided interoperable class, 188
Online update, of training database, 121
Ordinary least squares (OLS), 18–19
OWL, *see* Web Ontology Language

P

Parameter precision structure, 22
Partially constrained algorithms, 140–142
Partially labeled data, 79
PCA, *see* Principal component analysis
PCDMI, *see* Program for ClimateModel Diagnosis and
    Intercomparison
Pearson correlation, 67

PGDM, *see* Physics-guided data mining
Physics-guided data mining (PGDM), 63
"Plug-and-play" geoscience modeling, 186, 189
Potential-interoperable class, 188
Presence/absence methods, 81–83
Presence/background methods, 83
Presence-only methods, 79–80, 83–84
Principal component analysis (PCA), 59, 66
Probability of success (p_s), 148, 151, 155
Program for ClimateModel Diagnosis and
    Intercomparison (PCDMI), 57

Q

Quantile regression neural networks, 58

R

Random forest (RF), 82
RAS, *see* Resource Alignment Service
RBMs, *see* Restricted Boltzmann machines
RCMs, *see* Regional climate models
RDF, *see* RDF in Attributes; Resource Description
    Framework
RDF in Attributes (RDFa), 182
RDF Schema (RDFS), 183
Real-world infrastructure systems, 7
Rear-flank downdraft (RFD), 96
Regional climate models (RCMs), 56
Regression methods, 58
Relational probability tree (RPT), 100
Representational State Transfer (REST), 181, 191, 196
Residual precision structure, 25–26
Resilience management framework, 4–6
Resource Alignment Service (RAS), 194–195
Resource Description Framework (RDF), 182
REST, *see* Representational State Transfer
Restricted Boltzmann machines (RBMs), 67, 114
    contrastive divergence algorithm for training, 119
RFD, *see* Rear-flank downdraft
Rootmean square errors (RMSEs), 18–19, 27–28, 67
RPT, *see* Relational probability tree

S

SAS, *see* Semantic Annotation Services
SAT, *see* Surface air temperature
Scale-free networks, 7
Scientific long-tail web resources, 176–179
SCLS, *see* Sum-to-one constrained least squares
SD, *see* Statistical downscaling
SDMs, *see* Species distribution models
Semantic annotation, 188–189
    automatic/user-based annotations, 191
Semantic Annotation Services (SAS), 194, 195

Semantic heterogeneity, 183–185
Semantic interoperability
    classes of, 188
    definition of, 180, 181
    in geosciences, 185–187
    typologies of, 180–181
Semantic knowledge discovery, 190–191
Semantic mediation/matching, 189–190
Semantic similarity, 189
Semantic Web (SW), 175, 177, 183
Semantic Web for Earth and Environmental
    Terminology (SWEET), 185–186
Semi-interoperable class, 188
Semivariogram analysis, 79
Severe weather simulations, 97–98
SGL and hierarchical norms
    ocean variable selection for brazil, 19–20
    prediction accuracy of climate, 18–19
Simple Knowledge Organization System (SKOS), 187
Simple Queuing Service (SQS), 114
SKOS, *see* Simple Knowledge Organization System
SPACES, *see* Spatial Portal for Analysis of Climatic
    Effects on Species
Sparse group lasso, and climate applications, 16–20
Sparse regression, 60–61
Sparse structured estimation
    multitask learning (MTL), 15
    structured regression, 14–15
Sparse unmixing via variable splitting and augmented
    Lagrangian (SUnSAL), 137–138
Spatial evaluation of climate impact on the envelope of
    species (SPECIES), 82
Spatial extent, 77–78
Spatial influence parameters, 47
Spatial Information Services Stack Vocabulary Service
    (SISSVoc), 187
Spatial lattice, extension to, 36–37
Spatial Portal for Analysis of Climatic Effects on Species
    (SPACES), 85
Spatial resolution, observational and environmental data,
    78–79
Spatial sampling, 79
Spatiotemporal data mining, advancements in
    climate networks for covariate selection, 63–64
    computational cost, 66
    downscaling extremes, 65–66
    mixture of experts, 65
    multitask learning, 64–65
Spatiotemporal global climate model tracking, 33
    climate data, challenges in, 38–39
    climate model data *see* Climate model data
    experiments, 43–44
    fixed-share algorithm, 34
    fixed-share as Markov random field, 35–36

inference in Markov random field, 39–40
learning parameters, 37–38
Neighborhood-augmented Tracking Climate Models,
    40–43
simulated data, 44–45
spatial lattice, 36–37
Spatiotemporally relational attributed data, 99–100
Spatiotemporal relational probability tree (SRPT), 98
    training, 100–102
Spatiotemporal relational random forests (SRRF),
    98–103
Spearman correlation, 67–68
SPECIES, *see* Spatial evaluation of climate impact on the
    envelope of species
Species distribution models (SDMs), 74
    presence-only data, 79–80
    theory and concept, 74–76
    using correlative niche models *see* Correlative niche
        models
SQS, *see* Simple Queuing Service
SRL, *see* Statistical relational learning
SRM, *see* Statistical region merging
SRPT, *see* Spatiotemporal relational probability tree
SRRF, *see* Spatiotemporal relational random forests
Statistical downscaling (SD), in climate
    current state, 57–59
    deep belief networks, 67–69
    deep learning for, 66–67
    downscaling techniques, categories of, 56
    physics-guided data mining, 63
    recent developments in, 59–62
    spatiotemporal data mining, advancements in,
        63–66
    uncertainty within data, 57
Statistical models, 74
Statistical region merging (SRM), 114
    algorithm, 115–116
Statistical relational learning (SRL), 98
Storm objects, 103–104
Stressors, 4
Strong vortices, identifying, 104–106
Structural heterogeneity, 184
Structured estimation, in high dimensions
    global climate models combination, experiments on,
        26–29
    multitask sparse structure learning and climate
        applications, 21–26
    sparse and structure learning, 14–15
    sparse group lasso and climate applications,
        16–20
Structured regression, 14–15
Structure learning, 14–15
Sum-to-one constrained least squares (SCLS), 140, 152,
    165

SUnSAL, *see* Sparse unmixing via variable splitting and
    augmented Lagrangian
SUnSAL total variation (TV), 138–139
Supercells, 95
Support vector machines (SVMs), 82
Surface air temperature (SAT), 14
Sustainable Environment Actionable Data (SEAD), 178
SVMs, *see* Support vector machines
SWEET, *see* Semantic Web for Earth and Environmental
    Terminology
System-based resilience approach, 1

T

TCMs, *see* Tracking climate models
Teleconnections, 64
Telescoping systems of systems, 8–9
Temperature anomalies, 46–47
Thesaurus of Engineering and Scientific Terms (TEST),
    185
Three-dimensional shapes, 102
Tornado objects, 105
Tracking climate models (TCMs), 34
Transmission lines, 63
TreeGen, 58
True skill statistic (TSS) value, 107
Truncation, 77
TSS, *see* True skill statistic value

U

Unconstrained least squares (UCLS), 134, 156, 164
Unconstrained unmixing algorithms
    constrained energy minimization, 136–137
    matched filtering, 135
    mixed-tuned matched filtering, 135–136
    sparse unmixing via variable splitting and augmented
        Lagrangian, 137–138
    SUnSAL total variation, 138–139
    unconstrained least squares, 134
Uniform resource identifier (URI), 182, 183, 194
Unmixing algorithms
    computer-simulated data *see* Computer-simulated
        data
    endmember generation, 146–147
    Landsat data *see* Landsat data
    merits and demerits, 163–167
    methodology, 147–148
    partially and fully constrained, 140–142
    unconstrained *see* Unconstrained unmixing
        algorithms
Unsupervised segmentation, 115–116
URI, *see* Uniform resource identifier
U.S. National Academy of Engineering (NAE), 9

V

VDAC technique, *see* Vortex detection and
    characterization technique
Very high-resolution (VHR) land cover, 113
Virtual private cloud (VPC), 123–124
Vortex detection and characterization (VDAC)
    technique, 104–105
VPC, *see* Virtual private cloud

W

"Warn-on-Forecast," 96
Web Ontology Language (OWL), 183
World Economic Forum (WEF), 9
World Wide Web Consortium (W3C), 181

X

XML, *see* Extensible markup language